GW01326528

MAMMALIAN ENERGETICS

MAMMALIAN

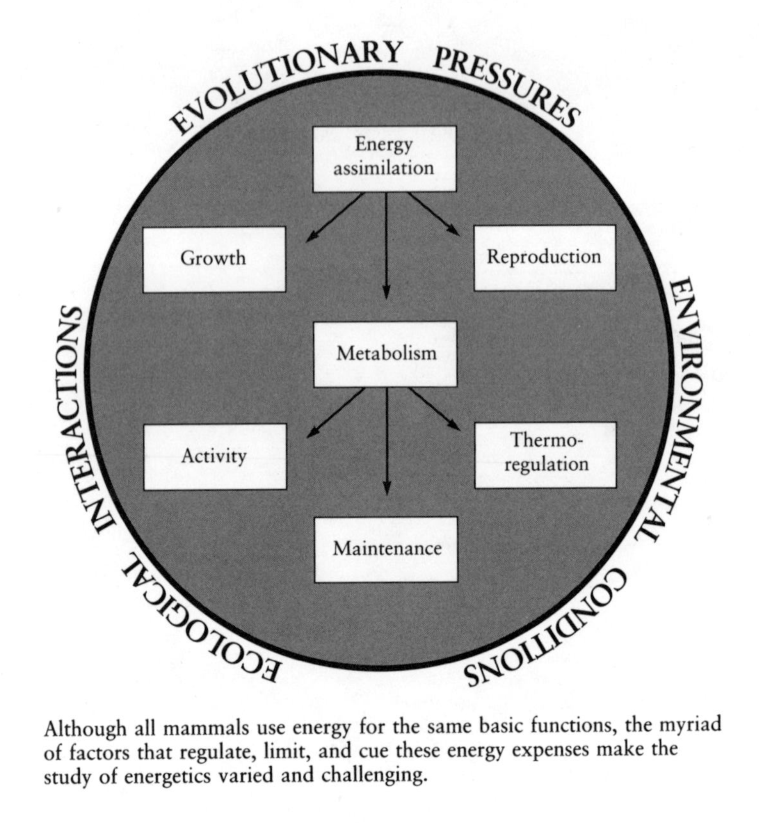

Although all mammals use energy for the same basic functions, the myriad of factors that regulate, limit, and cue these energy expenses make the study of energetics varied and challenging.

ENERGETICS

INTERDISCIPLINARY VIEWS OF METABOLISM AND REPRODUCTION

EDITED BY

THOMAS E. TOMASI
Department of Biology
Southwest Missouri State University

TERESA H. HORTON
Department of Biological Sciences
Kent State University

COMSTOCK PUBLISHING ASSOCIATES
a division of Cornell University Press
ITHACA AND LONDON

Copyright © 1992 by Cornell University

All rights reserved. Except for brief quotations in a review, this book, or parts thereof, must not be reproduced in any form without permission in writing from the publisher. For information, address Cornell University Press, 124 Roberts Place, Ithaca, New York 14850.

First published 1992 by Cornell University Press.

Printed in the United States of America

♾ The paper in this book meets the minimum requirements of the American National Standard for Information Sciences—Permanence of Paper for Printed Library Materials, ANSI Z39.48-1984.

Library of Congress Cataloging-in-Publication Data

Mammalian energetics : interdisciplinary views of metabolism and reproduction / edited by Thomas E. Tomasi and Teresa H. Horton.
p. cm.
Includes bibliographical references and index.
ISBN 0-8014-2659-6 (alk. paper)
1. Bioenergetics. 2. Mammals—Metabolism. 3. Mammals—Reproduction. 4. Physiology, Comparative. I. Tomasi, Thomas E. (Thomas Edward), 1955– . II. Horton, Teresa H. (Teresa Helen), 1956– .
QH510.M36 1992
599′ .016—dc20 91-55568

WE DEDICATE THIS BOOK TO PATRICIA J. BERGER AND
NORMAN C. NEGUS, WHO ENCOURAGED US AS THEIR ACADEMIC
OFFSPRING AND CARED ENOUGH TO BE OUR FRIENDS.

Contents

Preface

It has been said that if studies of physiology are to be thoroughly incorporated into evolutionary biology, biologists will need to make direct statements about the costs and benefits of physiological processes to the survival of individuals and the production of young. The diversity of pathways available to mammals for allocating energy and scheduling reproduction, however, presents a complex maze rather than a clear path connecting physiology and evolution. Because of the fundamental importance of reproduction and energetics to the survival of species and individuals, this volume focuses on these areas of physiology. A common theme throughout the chapters, stated or implied, is the costs and benefits of physiological processes within and among mammalian species.

In his analysis of areas for future research in physiological ecology, M. E. Feder (1987) stated that "the central question of organismal biology . . . is, How can we explain the evolution of diverse complex organisms? . . . In the past, ecological physiology usually emphasized understanding the function and significance of individual adaptations . . . as opposed to understanding the evolution of organismal complexity." One goal of the present volume is to advance understanding of complexity and variability at the organismal level and emphasize their significance. At our request, the contributors all deal with variation, either among or within species, and attempt to ascertain whether this variation is adaptive. Only by analyzing this diversity will biologists be able to answer the broader questions addressed herein.

While presenting data on specific physiological processes in an evolutionary context, the chapters also demonstrate how assumptions, explicit or implicit, about the use of energy by mammals can influence the interpretation of data and the fundamental design of experiments. Assumptions play a critical role in studies of evolutionary biology. By definition, assumptions about a system are the principles that either are generally accepted as being true about the system or are proposed as the ground rules of a hypothesis about how the system works. The primary assumptions about energetics in evolutionary theory are that the process of living is expensive and that there are trade-offs in the allocation of energy among maintenance, growth, and reproduction. A related assumption is that physiological characteristics have persisted because they are beneficial to the animal in an evolutionary sense. Implicit in this assumption is the belief that the observed characteristic results in the enhancement of reproductive output through more efficient use of energy or improved survival or both. All these assumptions have provided a valuable and productive point of departure for the study of animal physiological ecology (Bennett, 1987).

Our volume starts with a historical perspective of the field of energetics followed by a discussion of minimum and (acute) maximum metabolic requirements in the order Rodentia and an in-depth treatment of swimming as an example of locomotor energetics. Next, Chapter 4 addresses modifications of the utilization rates of thyroxine, which might accompany the evolution of different metabolic rates, and Chapter 5 considers how morphological and physiological measurements can be used to infer the energetic status of wild-caught mammals. The next two chapters examine the use of heterothermy by mammals: Chapter 6 addresses the use of torpor to circumvent seasonal energetic deficiencies, and Chapter 7 discusses whether altricial mammalian young are capable of maintaining a constant body temperature. The remaining chapters focus on reproductive strategies of mammals. Chapter 8 provides an overview of the role that natural selection may play in maintaining variability in responses to environmental cues, and Chapter 9 presents evidence for genetic variability in responses to specific environmental cues. Chapter 10 reviews the data available on measurements of the costs of reproduction and discusses whether there are sufficient data to support current assumptions concerning the costs of reproduction.

Some chapters are overviews with relatively wide scopes, while others address more specific questions. We make no claim to have covered all areas of comparative energetics; instead we have endeavored to illustrate the wide variety of questions, approaches, and interrelationships within mammalian biology. Several books published in the 1980s should be consulted for a broader view of the field of physiological ecology and reproductive energetics (see Bronson, 1989; Feder et al., 1987; Loudon and Racey, 1987).

Comparative energetics in particular—and physiological ecology in general—is by nature interdisciplinary. It unites studies of biophysics, biochemistry, physiology, morphology, behavior, and evolutionary theory. We hope that this volume will be read and appreciated by biologists from all those disciplines and that it will encourage them to ponder how the diversity of energetic strategies may pertain to their own areas of study. We further hope that the concepts and viewpoints presented in the following chapters will serve as a foundation for the next decade of research into the diversity and utility of energetic strategies.

This volume resulted from two symposia held at the 1988 meeting of the American Society of Mammalogists at Clemson University, Clemson, South Carolina. Several of the chapters are revised versions of presentations made at the symposia; others were added to extend the scope of the book. We thank the American Society of Mammalogists for their support of the symposia.

All the chapters in this book were critically reviewed, both by anonymous reviewers and by the reviewers named in the chapter acknowledgments. Their comments have greatly contributed to the quality of this volume.

We are grateful to the authors for their hard work in preparing their manuscripts and to Cornell University Press for publishing the results.

THOMAS E. TOMASI

Springfield, Missouri

TERESA H. HORTON

Kent, Ohio

Literature Cited

Bennett, A. F. 1987. The accomplishments of ecological physiology. Pp. 1–8 *in* New Directions in Ecological Physiology (M. E. Feder, A. F. Bennett, W. W. Burggren, and R. B. Huey, eds.). Cambridge University Press, Cambridge. 364 pp.

Bronson, F. H. 1989. Mammalian Reproductive Biology. University of Chicago Press, Chicago. 325 pp.

Feder, M. E. 1987. New directions in ecological physiology: conclusions. Pp. 347–351 in New Directions in Ecological Physiology (M. E. Feder, A. F. Bennett, W. W. Burggren, and R. B. Huey, eds.). Cambridge University Press, Cambridge. 364 pp.

Feder, M. E., A. F. Bennett, W. W. Burggren, and R. B. Huey, eds. 1987. New Directions in Ecological Physiology. Cambridge University Press, Cambridge. 364 pp.

Loudon, A. S. I., and P. A. Racey, eds. 1987. Reproductive Energetics in Mammals. Symposia of the Zoological Society of London 57. Clarendon Press, Oxford. 371 pp.

Brian K. McNab

1. Energy Expenditure: A Short History

The study of mammalian energy expenditure is an important component of contemporary physiology and ecology, and has played a pivotal role in the history of biology. Originally mammalian energetics was examined to determine whether the metabolism of animals was similar to combustion in the physical world or was a process unique to organisms, that is, a "vital" property of organisms. With that question resolved in favor of the physical basis of "animal heat," animal energetics, with its consideration of heat production and loss, became a prime candidate for the application of physics to organisms. Such work led in many directions. One was a biophysical approach to temperature control, which in recent years has addressed the endocrinological and neurological bases of temperature regulation and energy expenditure at both the systemic and cellular levels of organization.

In contrast, comparative mammalian energetics has increasingly had an ecological and evolutionary orientation. Much ecological theory in the last few years has concerned the acquisition and allocation of resources, including energy. Associated with this development has been the shift from measuring the cost of temperature regulation in mammals under standard laboratory conditions to measuring the cost of locomotion (Fish, MacMillen and Hinds, this volume) and reproduction (Hill, Thompson, this volume) and to estimating field energy expenditures from field time budgets. Most recently the measurement of energy expenditures in the field has become possible.

Department of Zoology, University of Florida, Gainesville, Florida 32611.

In this chapter I give a short history of the intellectual developments in mammalian energetics, adopting a holistic, ecological approach. Such a history is equivalent to starting at the roots of a complex intellectual tree and following selected branches, arbitrarily discarding many branches that lead in other directions. The analogy of intellectual development with the structure of a tree emphasizes the interconnectiveness found among concepts and fields that at first glance might appear unrelated. Of necessity, this chapter must skim the highlights.

A history of a field is simultaneously a history of ideas and a history of people. These two histories are obviously connected, and yet to some extent independent. I am not convinced that history is made principally by a few prominent individuals, but suggest that the development of a field (or of an idea) is the product of many subtle, complex interactions among individuals, technology, and the intellectual (and political) times in which people live. Individuals that are discussed below should be taken to represent an abbreviation for these complex interactions. (I mean no disrespect to those whose contributions have been omitted, but here emphasize contributors who have suggested generalizations; those individuals who have principally contributed data are best treated in chapters dedicated to the facts of mammalian energetics.) Although not comprehensive, this historical sketch will demonstrate the international nature of the study of mammalian energetics, and by implication, physiological ecology, a conclusion diametrically opposed to Bennett's (1987, p. 2) view that "the center of activity of the field has always been the United States."

The Eighteenth Century

Before the eighteenth century, those interested in animal heat were primarily concerned with whether it represented a property unique to organisms or was physically identical to the heat produced by a purely physical process, such as the burning of a candle. (For an analysis of the early history of this topic, see Mendelsohn, 1964.) Some of the earliest advocates of the idea that respiration and combustion were similar included Robert Boyle, Robert Hooke, and John Mayow in seventeenth-century England. However, the first measurements of rate of heat production in an animal (guinea pigs) were published a hundred years later by the English biologist Adair Crawford (1779) and by the

French scientists Antoine Lavoisier and Pierre Simon de Laplace (1780). The object of these studies was not to determine the guinea pigs' rate of energy expenditure so much as to show that animal heat required a substance, called oxygen, that was obtained from the atmosphere. Lavoisier and Laplace demonstrated that the amount of heat released relative to the amount of carbon dioxide produced was similar in respiring animals and in combusting candles. Crawford showed that respiration produces nearly the same amount of heat relative to the amount of oxygen consumed as combustion, which, as Kleiber (1961) noted, is a more accurate statement than that made by Lavoisier and Laplace. Crawford was also the first to show that heat production by mammals increased at low ambient temperatures.

Both sets of experiments used direct calorimetry. Lavoisier and Laplace, measured heat production, using the latent heat of fusion, by the amount of ice that melted around a chamber containing a guinea pig. The precision of this technique was reduced by the adherence of water to ice at the beginning of the experiment. Crawford improved the experimental design by abandoning the use of ice; he estimated heat production by the increase in the temperature of a water jacket surrounding the chamber.

Lavoisier and Crawford were the principal founders of the theory that animal heat originated from the combustion of oxygen with the production of carbon dioxide as a waste product, and thus is a process similar to combustion in a physical system. Unlike the vague speculations of Boyle, Hooke, and Mayow a century earlier, the conclusions of Lavoisier and Crawford were specific. Lavoisier and Crawford, however, erroneously argued that animal heat was generated at the site of gas exchange in the lungs. The view that heat was produced throughout the body was proposed by several physiologists at the end of the eighteenth century (e.g., Lagrange, in Hassenfratz, 1791) and was demonstrated in the beginning of the nineteenth century (Davy, 1838; Magnus, 1837), principally by comparing measurements of the amounts of oxygen and carbon dioxide in arterial and venous blood.

By the early nineteenth century, the collective work in England and France had finally established the physical similarity between combustion and metabolism (although, as we have come to understand, they are based on significantly different chemical systems). This work permitted later students of animal energetics to use indirect calorimetry, that is, to use rate of carbon dioxide production and rate of oxygen consumption as indices of an animal's rate of heat production.

The Nineteenth Century

The discovery of a correlation between rate of heat production and body size was the most influential observation made in the nineteenth century with respect to comparative mammalian energetics. This idea was presented as a theoretical relationship in 1838 at a meeting of the Royal Academy of Medicine in Paris by Sarrus, a professor of mathematics at Strasbourg, and Rameaux, a doctor of medicine and of science, and published the following year (Sarrus and Rameaux, 1839). The authors argued that (1) heat production is proportional to the amount of oxygen consumed; (2) heat production equals heat loss as long as body temperature is constant; (3) heat loss is proportional to the free surface area; and (4) oxygen consumption is therefore proportional to the ⅔ power of body mass. Apparently independent of Sarrus and Rameaux, the German biologist Carl Bergmann (1847) described the same relationship, as well as the propensity of small species to enter torpor, the use of heat generated by activity for temperature regulation at small masses, and his still-famous rule for body size relative to latitude. Bergmann was the first to conclude directly that "a gram of a large animal must, in general, respire less than a gram of a small animal." He also introduced the terms *homoiotherm* and *poikilotherm*, and seemingly was the first person to advocate use of "Newton's law of cooling" to analyze heat exchange between an organism and its environment. Carl Bergmann, consequently, appears to have been the father of comparative mammalian energetics and the first biologist to place energetics into an ecological and evolutionary context.

Experimental verification of the "surface law" was accomplished by various European biologists, including Bergmann and Leuckart (1852), Rubner (1883), and Richet (1885, 1889). Bergmann and Leuckart analyzed the extensive set of data on metabolism in domestic animals accumulated by Regnault and Reiset (1849), noting again the high mass-specific rates of small endotherms. The clearest demonstration of the importance of surface area for heat production in endotherms was produced by Max Rubner, who measured carbon dioxide production in domestic dogs. At masses that ranged from 3 to 31 kg, rates of heat production were approximately constant at 1000 kcal/(m^2 · day), whereas these rates expressed either on a total or mass-specific basis varied markedly over the mass range. He concluded that the constancy of surface-specific rates indicated that heat loss determined heat production. Richet demonstrated that small rabbits have rates of heat production similar to those of large rabbits when rates are expressed per unit surface area, but not when expressed per unit mass. The intraspecific conclusions of Rubner and Richet were confirmed in an interspecific comparison of domesticated animals by Voit (1901). A complication plagues these conclusions, however. Rubner and Richet used surface areas calculated with formulae developed by Meeh (1879). The inability to measure directly the surface area from which heat is lost, if only because it is subject to rapid modification, inherently diminishes the usefulness of a surface-specific expression for heat loss. Richet, in fact, was concerned with the appropriate surface area to use, especially given the large ears of rabbits.

The Twentieth Century

The Fall of the Surface Law

At the beginning of the twentieth century the validity of the surface law generally appeared to be taken for granted. Cracks in this view, however, soon appeared. Hoesslin, as early as 1888, showed on theoretical and experimental grounds that Rubner's heat-loss theory was inadequate. Francis G. Benedict, an American physiologist who was trained in Heidelberg and was director of the Nutrition Laboratory at the Carnegie Institution for thirty years, pointed out (Benedict, 1915) that neither body mass nor surface area alone could account for the variation in human basal rates of metabolism (BMRs). He suggested that many factors influence BMR in humans, including the proportion of active tissue (see also Le Breton, 1926) and the "nervous state."

Other problems with the surface law appeared soon thereafter. August Krogh (1916), a student of Christian Bohr at Copenhagen and winner of the 1920 Nobel Prize in physiology, demonstrated that total rate of metabolism increased with body mass in ectotherms in a manner similar to endotherms. If heat loss dictated rate of heat production in endotherms, as might be expected in species that actively maintain a temperature differential with the environment, how can a similar scaling function in ectotherms, which do not actively maintain such a differential, be explained? To compound the debate further, Lambert and Teissier (1927) made a dimensional analysis of various physiological functions in mammals, including basal rate of metabolism in relation to body mass and surface area. They concluded that heat production in thermoneutrality was, in fact, proportional to mass raised to the 0.67 power.

Early in the twentieth century mathematics was increasingly used to analyze the morphology and physiology of organisms. This intellectual climate, which reflected the growing view that organisms were elaborate physicochemical "machines," was encouraged by two books. They were D'Arcy W. Thompson's *On Growth and Form*, first published in 1917, and Julian Huxley's *On Relative Growth*, published in 1932. Unlike Thompson, who used a geometric description of growth, Huxley advocated use of power functions to describe the effect of body size on the structure of organisms. Because of the convenience of his approach, Huxley had a much greater long-term impact on comparative biology than Thompson.

Max Kleiber, a Swiss nutritionist transplanted to the University of California, and Samuel Brody, at the University of Missouri for over forty years, were the first to use regression analysis to show that the total basal rate of domestic endotherms was proportional to mass raised to a power greater than the 0.67 implied by the surface law. For example, Brody and Procter (1932) showed that the fitted power was 0.734 for domesticated animals, and Kleiber (1932) found this power to be 0.74 or 0.73, depending on whether ruminants were included. Given the variability in data, Kleiber rounded the power to 0.75 to facilitate the ease of calculation in an age when calculators and computers were not generally available. The use of 0.75, then, was as much for convenience as for "objective reality."

Two of the fundamental books of this period were written by Benedict and Brody. Benedict's book, *Vital Energetics* (1938), included data from mice and elephants, which led his summary to be known as the "mouse-to-elephant" curve. As valuable as were Benedict's data, his analysis was severely criticized by Kleiber (1961, p. 202) for arguing that the use of logarithms "completely masks metabolic differences [among] species." Benedict's analysis, although quantitative, represented a temporary retreat from a mathematical analysis of energetics. Brody published his monumental book *Bioenergetics and Growth* in 1945, summarizing the data on the energetics, work, growth, and production in domesticated mammals. He maintained that the best mass exponent for basal rate was 0.73.

All empirical, interspecific studies of BMR in domesticated endotherms suggested that the mass exponent for metabolism varied from 0.72 to 0.75. These powers are significantly greater than would be expected from the surface law. Early regressional analyses by Kleiber and

Brody, however, were biased by the use of repeated values for one or more species, depending simply on the number of measurements that had been made. Repeated values have the statistical effect of giving unequal value to the species used.

Following the work of Krogh, two of his students, the Danish biologists Eric Zeuthen (1953) and Axel Hemmingsen (1950, 1960), showed that the relationship between standard rate of metabolism and body mass is nearly the same in all organisms, including unicellular organisms, ectotherms, and endotherms. The general conformation of standard rates of metabolism in ectotherms and unicellular organisms to the "¾ rule," as it has been called, essentially destroyed the suggestion that heat loss is the principal factor dictating heat production.

By the middle of the twentieth century, Kleiber's relation was almost universally regarded as the standard in the study of mammalian energetics. Physiologists at this time were concerned with conformation to Kleiber's relation, not with the residual variation about the regression, probably because most of the work on mammals to 1950 had been restricted to domesticated species, which in terms of metabolism are relatively uniform (see McNab, 1988). No impetus existed to account for differences among species, except as they reflected differences in body mass. Because of the apparently successful description of the effect of body mass on rate of metabolism, many other properties of organisms were also described as functions of mass (see the extensive summaries in Peters, 1983; Calder, 1984; and Schmidt-Nielsen, 1984).

The conclusion that surface area was an inadequate explanation for the influence of body mass on the basal rate of endotherms led to a search for other factors that might influence the level of metabolism. This search was complicated because two slightly different questions were intertwined: what sets the *level* of metabolism, and what is the *scaling power* (i.e., mass exponent) for metabolism? Further work took three somewhat overlapping directions: (1) Can a (simple?) physical explanation for the ¾ rule be found, or is ¾ even the best description of the scaling power? (2) Is the level of metabolism determined by phyletic progression? and (3) Does the residual variation in basal rate have ecological correlates?

Physical Explanations

No one doubts that physics is an important determinant of heat loss and heat production in endotherms exposed to cold temperatures. The question, however, is whether such considerations are important in setting the basal rates of endotherms or standard rates of ectotherms. Although heat loss through the body surface appeared to be eliminated as the principal factor, at least for differences among species, other physical models have been constructed. MacMahon (1973), for example, using the elastic properties of skeletons and some assumptions on the geometric similarity of animal design, derive the conclusion that power output was proportional to mass raised to the 0.75 power. Similarly, Economos (1982) analyzed the conditions required for biological similarity to account for those species that do not conform to the 0.75 power.

Heusner (1982) suggested that the 0.75 power is a statistical artifact produced by lumping dissimilar species together; in his view, the true mass exponent for the scaling of metabolism is 0.67, which is often found intraspecifically. He argued that one cannot directly compare individuals belonging to different species because they are not structurally similar. Heusner concluded that interspecific regressions scale at about the 0.75 power because some additional factors that influence rate of metabolism are, themselves, correlated with body mass. Examples of these correlations are noted in McNab, 1988. Unfortunately, even though species-specific differences in structure are eliminated by examining intraspecific mass exponents, other factors are introduced, such as differences associated with ontogeny (e.g., the cost of growth at small masses). This problem necessarily occurs in endotherms because the only way that a sufficiently large range in mass can be obtained within a species is by including young and subadult individuals. The question of whether variation in standard energetics can be explained by physics alone remains unresolved, especially because all comparative sets of data show great variation in standard rates even at a fixed mass.

Phylogeny and Energetics

The influence of "biological" factors on the basal rate of mammals was hidden for a long time because nearly all measurements were made on domestic species. One of the first physiologists to compare the energy expenditures of wild mammals was Martin (1902). He reported data on the expenditures of monotremes, marsupials, and eutherians. Differences in mass alone could not account for the differences in basal rate among these mammals. Early comparative studies of mammalian energetics were also made by Gelineo (1934), Giaja (1938), Kalabuchov (1950), Eisentraut (1960), Kayser (1961), Slonim (1962), MacMillen and Nelson (1969), and Dawson and Hulbert (1970). These studies led to the suggestion that basal rate reflects phylogenetic affiliation. By implication, monotremes have low basal rates because they are monotremes, not because monotremes feed on invertebrates or because they live in tropical and temperate environments (i.e., ecological factors). They surely do not have a low basal rate because they have a small surface area. Recent analyses by Hayssen and Lacy (1985) and Elgar and Harvey (1987) restated the view that most variation in mammalian basal rates independent of body mass is related to phylogenetic status. However, all of these phylogenetic analyses assume that species constitute statistically independent observations, an assumption criticized by Felsenstein (1985), Pagel and Harvey (1989), and others. Because of the different levels of relatedness among species in an interspecific analysis, phylogenetic "causes" may be inferred incorrectly, which confounds a phylogenetic explanation for differences in BMR.

A difficulty with a phylogenetic "explanation" is that it avoids the question of why such differences in energetics evolved. It does not address the question of what factors influence (or set) rate of metabolism: it states only that rates are *correlated* with taxonomic affiliation. Many other characteristics, such as morphology, ecology, and behavior, also are correlated with taxonomic affiliation. Is it conceivable that a species will have a rate of metabolism dictated by phylogeny that is incompatable with its physical environment, behavior, or food habits? If not, we are left with a series of internal correlations that are exceedingly difficult to untangle.

Ecological Explanations

The inadequacy of mass alone to account for the variation in standard energetics led some biologists to examine the effect of environment and habit on mammalian energy expenditure. This search was stimulated by Pearson (1947) and Morrison (1948) in the United States and Hart (1950) in Canada, who measured rate of metabolism in wild mammals. Most influential were three papers published by Scholander, Irving, and co-workers (Scholander et al., 1950a, 1950b, 1950c).

Laurence Irving was pivotal in the examination of climatic correlates in mammalian (and avian) energetics. After obtaining his Ph.D. degree at Stanford and spending ten years at the University of Toronto, he spent 1937 to 1947 at Swarthmore College. He brought several Scandinavian biologists to Swarthmore, including August Krogh, who gave lectures in respiratory physiology that were published in 1939 by the University of Pennsylvania; Per Scholander, who came to Swarthmore in 1939 and stayed until 1949; and Knut and Bodil Schmidt-Nielsen, who were resident at Swarthmore from 1946 through 1948. Oliver P. Pearson and Peter R. Morrison were both undergraduates at Swarthmore, where they were influenced by Irving and Scholander. Their interest in energetics continued while they were graduate students along with C. P. Lyman and G. A. Bartholomew at Harvard University. Bartholomew, who was at the University of California, Los Angeles for over forty years, was a prolific contributor to the study of mammalian energetics, but his greatest contribution may be his training of graduate students, three of whom have contributed to this volume (MacMillen, Wunder, and French), as have two of his academic grandsons (Hill and Thompson).

The three papers published in 1950 by Scholander and Irving compared the energetics of tropical and arctic mammals and birds and concluded that adjustment to climate by endotherms was accomplished by modifications in insulation, not in basal rates, although the latter conclusion was technically limited to terrestrial mammals. One of the analytical tools used by these authors was Newton's law of cooling. Although such use was preceded more than a hundred years by Bergmann and sixteen years by Burton (1934), most recent work derives from the work of Scholander, Irving, and colleagues. Because Newton (1701) was concerned with the cooling of furnaces, not the increased heat production of endotherms exposed to cold temperatures, this model is best called the Scholander-Irving model of endothermy.

Irving and Krog (1954), Irving et al. (1955), Hart (1957), and Irving (1960) continued to focus interest on the influence of body mass, insulation, and climate on rate of metabolism in mammals and birds, but still maintained that the principal (only?) adaptation of energetics to climate was found in the insulation. Soon thereafter several studies, including those of McNab and Morrison (1963), Hulbert and Dawson (1974), and Shkolnik and Schmidt-Nielsen (1976), demonstrated a correlation of basal rate in mammals with climate. Additional studies of mammals have shown an association of basal rate with burrowing habits (McNab, 1966), food habits (McNab, 1969, 1986), nocturnal habits (Crompton et al., 1978), and, reminiscent of the suggestion of Benedict, body composition (McNab, 1978).

Recent Advances

Many of the studies summarized to this point were conducted under standard conditions in the laboratory. These conditions include no activity, postabsorption, and thermoneutral temperatures. Such studies are of great value because they permit different species to be compared under identical conditions. Mammals, however, usually are not living under standard conditions; they often are active—walking and running, gathering and digesting food, reproducing, and showing other behaviors. The energy expenditures associated with some of these behaviors have recently been measured.

One of the earliest attempts to measure the cost of locomotion was made by Wunder (1970) on chipmunks. He was followed by many others, especially Richard Taylor and his co-workers. Taylor et al. (1982) summarized a massive amount of data to show that the cost of movement can be predicted by knowledge of a mammal's mass and velocity. Tucker (1968) first measured the cost of flight in birds, work that was extended to bats by Thomas and Suthers (1972) and Carpenter (1975). Summaries on the comparative cost of various forms of locomotion were given by Tucker (1970) and Schmidt-Nielsen (1972).

Data on standard and activity energetics can be combined with field data on the proportion of time spent at various activities to estimate field energy budgets. The original attempt at such estimation was made by Pearson (1954) on an Anna's hummingbird. Synthetic energy budgets were first applied to mammals by McNab (1963). The difficulty with such estimates is that they have no confidence intervals, and therefore one does not know whether they are any better than just assuming that field expenditures are two to three times the basal rate.

Most recently, measurements of rate of metabolism, based on a doubly labeled water technique originally described by Nathan Lifson and his colleagues (1955), have been made in the field. This technique permits the rate of carbon dioxide production (and thus rate of energy expenditure) to be measured over a period of a few days in a free-living animal. It was first used for animals in the field by LeFebvre (1964) when he measured the metabolism of free-flying pigeons. Mullen (1970, 1971) was the first to apply this technique to mammals in the field. Many recent measurements of field expenditures of mammals are summarized in Nagy, 1987. The principal difficulty with these measurements is that they integrate all of the many factors that influence energy expenditure into a single number, which makes them nontransferable to other species and to other behavioral and environmental conditions (McNab, 1989). A recent analysis by Koteja (1991) showed that mass-independent field energy expenditures are correlated with basal rates in eutherians (r = 0.831) but not in marsupials (r = 0.066).

The best prospect for future studies of mammalian energetics appears to combine measurements of a species under standard conditions in the laboratory with field measurements so that the latter can be dissected into the fundamental components that constitute a field expenditure. Such analyses may allow us to examine the influence of different physical and biological conditions on the energy budget of a species. They will permit ecologically relevant hypotheses to be tested, such as the limitation to brood size imposed by energetics or the thermal bases for the limit to distribution.

ACKNOWLEDGMENTS

I thank S. D. Thompson, O. P. Pearson, T. Tomasi, B. Schmidt-Nielsen, and two anonymous reviewers for reviewing this manuscript. A. A. Heusner kindly brought several references to my attention.

Literature Cited

Benedict, F. G. 1915. Factors affecting basal metabolism. J. Biol. Chem., 20:263–299.

——. 1938. Vital Energetics. Carnegie Inst. Wash. Publ., 503:1–215.

Bennett, A. F. 1987. The accomplishments of ecological physiology. Pp. 1–8 *in* New Directions in Ecological Physiology (M. E. Feder, A. F. Bennett, W. W. Burggren, and R. B. Huey, ed.). Cambridge University Press, Cambridge. 364 pp.

Bergmann, C. 1847. Über die Verhältnisse der Wärmeökonomie der Tiere zu ihrer Grösse. Göttinger Studien, 1:1–395.

Bergmann, C., and R. Leuckart. 1852. Anatomisch-physiologische Übersicht des Tierreichs. J. B. Mueller, Stuttgart.

Brody, S. 1945. Bioenergetics and Growth. Reinhold, New York. 1023 pp.

Brody, S., and R. C. Procter. 1932. Relation between basal metabolism and mature body weight in different species of mammals and birds. M. Agric. Exp. Stn. Res. Bull., 166:89–101.

Burton, A. C. 1934. The application of the theory of heat flow to the study of energy metabolism. J. Nutr., 7:497–533.

Calder, W. A. 1984. Size, Function, and Life History. Harvard University Press, Cambridge. 431 pp.

Carpenter, R. E. 1975. Flight metabolism of bats. Pp. 883–890 *in* Swimming and Flying in Nature (T. Y.–T. Wu, C. J. Brokaw, and C. Brennen, eds.). Plenum, New York.

Crawford, A. 1779. Experiments and observations on animal heat, and the inflammation of combustible bodies. London.

Crompton, A. W., C. R. Taylor, and J. A. Jagger. 1978. Evolution of homeothermy in mammals. Nature, 272:333–336.
Davy, J. 1838. An account of some experiments on the blood in connexion with the theory of respiration. Philos. Trans. Soc. Lond., 128:283–299.
Dawson, T. J., and A. J. Hulbert. 1970. Standard metabolism, body temperature, and surface areas of Australian marsupials. Am. J. Physiol., 218:1233–1238.
Economos, A. C. 1982. On the origin of biological similarity. J. Theor. Biol., 94:25–60.
Elgar, M. A., and P. H. Harvey. 1987. Basal metabolic rates in mammals: allometry, phylogeny and ecology. Funct. Ecol., 1:25–36.
Eisentraut, M. 1960. Heat regulation in primitive mammals and in tropical species. Bull. Mus. Comp. Zool., 124:31–43.
Felsenstein, J. 1985. Phylogenies and the comparative method. Am. Nat., 125:1–15.
Gelineo, S. 1934. Influence du milieu thermique d'adaptation sur la thermogénèse des homéothermes. Ann. Physiol. Physiochim. Biol., 10:1083–1115.
Giaja, J. 1938. L'homéothermie et la thermorégulation. Herman et Cie., Paris.
Hart, J. S. 1950. Interrelations of daily metabolic cycle, activity, and environmental temperature of mice. Can. J. Res. Sect., D Zool. Sci. 28:293–307.
——. 1957. Climatic and temperature induced changes in the energetics of homeotherms. Rev. Can. Biol., 16:133–174.
Hassenfratz, J. 1791. Mémoire sur la combinaison de l'oxyène avec le carbone et l'hydrogène du sang, sur la dissolution de l'oxygène dans le sang, et sur la manière dont le calorique se dégage. Ann. Chim., 9:261–274.
Hayssen, V., and R. C. Lacy. 1985. Basal metabolic rates in mammals: taxonomic differences in the allometry of BMR and body mass. Comp. Biochem. Physiol., 81A:741–754.
Hemmingsen, A. 1950. The relation of standard (basal) energy metabolism to total fresh weight of living organisms. Rep. Steno Mem. Hosp., 4:7–58.
——. 1960. Energy metabolism as related to body size and respiratory surfaces, and its evolution. Rep. Steno Mem. Hosp. 9:7–110.
Heusner, A. A. 1982. Energy metabolism and body size. I. Is the 0.75 mass exponent of Kleiber's equation a statistical artifact? Respir. Physiol., 48:1–12.
Hoesslin, H. von. 1888. Über die Ursache der scheinbaren Abhängigkeit des Umsatzes von der Grösse der Körperoberfläche. Arch. Physiol. 1888:329–379.
Hulbert, A. J., and T. J. Dawson. 1974. Thermoregulation in perameloid marsupials from different environments. Comp. Biochem. Physiol., 47A:591–616.
Huxley, J. 1932. On Relative Growth. Methuen, London. 267 pp.
Irving, L. 1960. Birds of Anaktuvuk Pass, Kobuk, and Old Crow. U.S. Natl. Mus. Bull., 117:1–409.
Irving, L., and J. Krog. 1954. Temperature of skin in the Arctic as a regulator of heat. J. Appl. Physiol., 7:355–364.
Irving, L., J. Krog, and M. Monson. 1955. The metabolism of some Alaskan animals in winter and summer. Physiol. Zool., 28:173–185.
Kalabuchov, N. I. 1950. The Physiological-Ecological Peculiarities of Animals under Habitat Conditions (in Russian). Kharkov State University, Kharkov, USSR. 267 pp.
Kayser, C. 1961. The Physiology of Natural Hibernation. Pergamon Press, Paris. 325 pp.
Kleiber, M. 1932. Body size and metabolism. Hilgardia, 6:315–353.
——. 1961. The Fire of Life. John Wiley, New York. 454 pp.
Koteja, P. 1991. On the relation between basal and field metabolic rates in birds and mammals. Funct. Ecol., 5:56–64.
Krogh, A. 1916. The Respiratory Exchange of Animals and Man. Longmans, Green and Co., London.

——. 1939. The Comparative Physiology of Respiratory Mechanisms. University of Pennsylvania Press, Philadelphia. 173 pp.

Lambert, R., and G. Teissier. 1927. Théorie de la similitude biologique. Ann. Physiol. Physicochim. Biol., 4:212–246.

Lavoisier, A., and P. S. de Laplace. 1780. Mémoire sur la Chaleur. Mem. Acad. R. Sci., 1780:355–408.

Le Breton, E. 1926. Récherches sur la notion de "masse protoplasmique active." Ann. Physiol. Physicochim. Biol., 2:606–645.

LeFebvre, E. A. 1964. The use of D_2O^{18} for measuring energy metabolism in *Columba livia* at rest and in flight. Auk, 81:403–416.

Lifson, N., G. B. Gordon, and R. McClintock. 1955. Measurement of total carbon dioxide production by means of D_2O^{18}. J. Appl. Physiol., 7:704–710.

MacMahon, T. 1973. Size and shape in biology. Science, 179:1201–1204.

MacMillen, R. E., and J. E. Nelson. 1969. Bioenergetics and body size in dasyurid marsupials. Am. J. Physiol., 217:1246–1251.

Magnus, G. 1837. Über die im Blute enthaltenen Gase, Sauerstoff, Stickstoff und Kohlensäure. Ann. Phys. Chem., 40:583–606.

Martin, C. J. 1902. Thermal adjustment and respiratory exchange in monotremes and marsupials—a study in the development of homoeothermism. Philos. Trans. R. Soc. Lond. B. Biol. Sci., 195:1–37.

McNab, B. K. 1963. A model of the energy budget of a wild mouse. Ecology, 44:521–532.

——. 1966. The metabolism of fossorial rodents: a study of convergence. Ecology, 47:712–733.

——. 1969. The economics of temperature regulation in Neotropical bats. Comp. Biochem. Physiol., 31:227–268.

——. 1978. Energetics of arboreal folivores: physiological problems and ecological consequences of feeding on an ubiquitous food supply. Pp. 153–162 *in* The Ecology of Arboreal Folivores (G. G. Montogomery, ed.). Smithsonian Institution Press, Washington, D.C.

——. 1986. The influence of food habits on the energetics of eutherian mammals. Ecol. Monogr., 56:1–19.

——. 1988. Complications inherent in scaling the basal rate of metabolism in mammals. Q. Rev. Biol., 63:25–54.

——. 1989. Laboratory and field studies of the energy expenditure of endotherms: a comparison. Trends Ecol. Evol., 4:111–112.

McNab, B. K., and P. R. Morrison. 1963. Body temperature and metabolism in subspecies of *Peromyscus* from arid and mesic environments. Ecol. Monogr., 33:63–82.

Meeh, K. 1879. Oberflächenmessungen des menschlichen Körpers. Z. Biol., 15:425–458.

Mendelsohn, E. 1964. Heat and Life. Harvard University Press, Cambridge. 208 pp.

Morrison, P. R. 1948. Oxygen consumption in several mammals under basal conditions. J. Cell. Comp. Physiol., 31:281–291.

Mullen, R. K. 1970. Respiratory metabolism and body water turnover rates of *Perognathus formosus* in its natural environment. Comp. Biochem. Physiol., 32:259–265.

——. 1971. Energy metabolism and body water turnover rates of two species of free-living kangaroo rats, *Dipodomys merriami* and *Dipodomys microps*. Comp. Biochem. Physiol., 39A:379–390.

Nagy, K. A. 1987. Field metabolic rate and food requirement scaling in mammals and birds. Ecol. Monogr., 57:111–128.

Newton, I. 1701. Philos. Trans. R. Soc. Lond., 22:824–829.

Pagel, M. D., and P. H. Harvey. 1989. Comparative methods for examining adaptation depend on evolutionary models. Folia Primatol., 53:203–220.

Pearson, O. P. 1947. The rate of metabolism of some small mammals. Ecology, 28: 127–145.
——. 1954. The daily energy requirements of a wild Anna hummingbird. Condor, 56:317–322.
Peters, R. H. 1983. The Ecological Implications of Body Size. Cambridge University Press, New York. 329 pp.
Regnault, V., and J. Reiset. 1849. Récherches chimiques sur la respiration des animaux des diverses classes. Ann. Chim. Phys., 26:299–519.
Richet, C. 1885. Récherches de calorimetre. Arch. Physiol., 6:237–291.
——. 1889. La chaleur animale. Paris.
Rubner, M. 1883. Über den Einfluss der Körpergrösse auf Stoff- und Kraftwechsel. Z. Biol., 19:535–562.
Sarrus, and Rameaux. 1839. Application des sciences accessoires et principalement des mathématiques à la physiologie générale. Bull. Acad. R. Med., 3:1094–1100.
Schmidt-Nielsen, K. 1972. Locomotion: energy cost of swimming, flying, and running. Science, 177:222–228.
——. 1984. Scaling: Why Is Animal Size So Important? Cambridge University Press, New York. 241 pp.
Scholander, P. F., R. Hock, V. Walters, and L. Irving. 1950a. Adaptation to cold in arctic and tropical mammals and birds in relation to body temperature, insulation, and basal metabolic rate. Biol. Bull., 99:259–271.
Scholander, P. F., R. Hock, V. Walters, F. Johnson, and L. Irving. 1950b. Heat regulation in some arctic and tropical mammals and birds. Biol. Bull., 99:237–258.
Scholander, P. F., V. Walters, R. Hock, and L. Irving. 1950c. Body insulation of some arctic and tropical mammals and birds. Biol. Bull., 99:225–236.
Shkolnik, A., and K. Schmidt-Nielsen. 1976. Temperature regulation in hedgehogs from temperate and desert environments. Physiol. Zool., 49:56–64.
Slonim, A. D. 1962. Individual Physiological Ecology of Mammals (in Russian). USSR Acad. Sci., Moscow and Leningrad. 496 pp.
Taylor, C. R., N. C. Heglund, and G. M. O. Maloiy. 1982. Energetics and mechanics of terrestrial locomotion. J. Exp. Biol., 97:1–21.
Thomas, S. P., and R. A. Suthers, 1972. The physiology and energetics of bat flight. J. Exp. Biol., 57:317–335.
Thompson, D. W. 1917. On Growth and Form. (2d ed., 1942). Cambridge University Press, Cambridge. 1116 pp.
Tucker, V. A. 1968. Oxygen consumption of a flying bird. Science, 154:150–151.
——. 1970. Energetic cost of locomotion in animals. Comp. Biochem. Physiol., 34:841–846.
Voit, E. 1901. Über die Grösse des Energiebedarfs der Tiere im Hungerzustande. Z. Biol., 41:113–154.
Wunder, B. A. 1970. Energetics of running activity in Merriam's chipmunk, *Eutamias merriami*. Comp. Biochem. Physiol., 33:821–836.
Zeuthen, E. 1953. Oxygen uptake as related to body size in organisms. Q. Rev. Biol., 28:1–12.

Richard E. MacMillen and David S. Hinds

2. Standard, Cold-induced, Metabolism of Rodents and Exercise-induced

Much attention has been devoted in the past to relationships between standard, or basal, metabolic rate (we use these terms synonymously herein; see Bartholomew, 1982) and body mass in mammals (Brody and Proctor, 1932; Kleiber, 1932), and this continues today (Hayssen and Lacy, 1985; MacMillen and Garland, 1989; McNab, 1988). Unfortunately this attention has created the impression that standard metabolic rate (SMR) is the only thermoregulatory element subject to natural selection and adaptive change. Instead SMR is a readily measurable component of an entire suite of endothermic energetic events that are responsive to thermal influences, levels of activity, reproductive condition, and so on. Examples of this suite of events include SMR, maximal metabolism, insulation, lower and upper critical ambient temperatures, body temperature, and evaporative water loss. This suite of events, or the combination of all the components, yields to selection, resulting in adaptive patterns of energy exchange. While analyses of SMR might in themselves hint at adaptive responses, a much more thorough and definitive analysis should include additional features of the suite of endothermic energetic events that constitute organismal energy exchange with the environment. A recent holistic approach to this problem (Nagy, 1987) that integrates the total 24-h energy costs of animals operating under noncaptive field conditions (field metabolic rate, FMR) synthesizes what is known through doubly

Richard E. MacMillen: Department of Ecology and Evolutionary Biology, University of California, Irvine, California 92717. David S. Hinds: Department of Biology, California State University, Bakersfield, California 93311.

labeled water studies of FMR of various taxonomic, ecological, and dietary groups of endotherms. Such an approach, while extremely valuable, does not however reveal anything about the individual components that constitute energy exchange.

In this chapter we pay special attention to some of these components in rodents, in an attempt to establish their relative importances in contributing separately and collectively to energy exchange. The components we examine are (1) SMR of postabsorptive rodents at rest in thermally neutral conditions (their minimal metabolic rates), (2) maximal metabolic rates of rodents exposed to high levels of heat loss, and (3) maximal metabolic rates of rodents while exercising (running on a treadmill). These three major parameters of energy exchange are expressed in terms of oxygen consumption and therefore represent minimal and maximal aerobic capacities of these rodents.

Rodents lend themselves ideally to such investigations because of their diversity of species, distribution, and size. We report herein on metabolic patterns of 27 species, 21 genera, and 6 families of rodents from all of the world's major continents. These rodents vary in body mass from 7 g to 3 kg, thus facilitating allometric analyses, and more than half (N = 14) we measured ourselves, thereby ensuring standardized measurements (Table 2.1).

In addition, we examine in detail rates of oxygen consumption during treadmill running at various speeds in members of the rodent family Heteromyidae. Heteromyids provide ideal comparative material within a single rodent family because they include both quadrupedal and bipedal species and extend over a magnitude in body mass. They therefore provide the opportunity for size-matched comparisons of the metabolic patterns of quadrupedality and bipedality within a single taxonomic unit.

Relationships between Minimal and Maximal Metabolism of Rodents

Minimal Metabolism: Standard Metabolic Rate

The standard metabolic rates of the rodents for which we also have data for cold-induced or exercise-induced maximal metabolic rates or both, when expressed on a whole-animal basis (ml O_2/[animal · min]), are linearly and positively related to body mass when these two vari-

TABLE 2.1

Rodents from various geographic regions for which data exist on body mass, standard metabolic rates (SMR), and maximal metabolic rates induced by cold exposure and by exercise

Family and species	Body mass (g)	Metabolic rate (ml O_2/min)			Region of origin	Source
		SMR	Max. cold-induced	Max. exercise-induced		
Caviidae						
Cavia cobava	500.0	6.50			South America	Hill, 1959
	913.5			56.60		Lechner, 1978
Cricetidae						
Baiomys taylori	7.0	0.33	1.43	1.88	North America	Rosenmann & Morrison, 1974; Seeherman et al., 1981
Callomys callosus	48.0	0.93	5.34		South America	Rosenmann & Morrison, 1974
Clethrionomys rutilus	28.0	1.28	6.30		North America	Rosenmann et al., 1975
Dicrostonyx groenlandicus	47.0	1.54			North America	Casey et al., 1979
	61.0			7.50		Hart & Heroux, 1955
Mesocricetus auratus	98.0	2.47	9.90		Eurasia	Pohl, 1965
	113.0			11.20		Pasquis et al., 1970
Ondatra zibethicus	869.0	11.60			North America	Fish, 1979
	1,100.0		64.20			Hart, 1962
Peromyscus californicus	41.3	0.94	3.48		North America	Hulbert et al., 1985
P. eremicus	18.8	0.43	2.41		North America	Hulbert et al., 1985
P. leucopus	21.4	0.57			North America	Hayssen & Lacy, 1985
	25.0		4.96			Segram & Hart, 1967
Heteromyidae						
Dipodomys merriami	33.4	0.70	4.04		North America	Hulbert et al., 1985
D. ordii	46.8	1.07			North America	Hinds & MacMillen, 1985
	56.2			7.36	North America	Authors

D. panamintinus	65.6	1.24		7.41	North America	Hinds & MacMillen, 1985; Authors
Heteromys desmarestianus	75.8	1.65			Central America	Hinds & MacMillen, 1985
	83.0			8.60		Authors
Liomys salvini	45.1	0.83	3.34		Central America	Hulbert et al., 1985
	52.8			5.15		Authors
Microdipodops megacephalus	11.0	0.50			North America	Hinds & MacMillen, 1985
	13.5			2.97		Authors
Perognathus fallax	19.9	0.43	2.69	3.64	North America	Hulbert et al., 1985; Authors
Muridae						
Conilurus penicillatus	213.2	2.71	14.90		Australia	Authors
Mus musculus	32.0	0.88	5.46	5.30	Eurasia	Pasquis, et al., 1970; Rosenmann & Morrison, 1974
Notomys alexis	38.8	0.83	3.48		Australia	Authors
Rattus colletti	165.7	2.05	11.50		Australia	Authors
R. norvegicus	253.0	4.47			Eurasia	Rosenmann & Morrison, 1974
	205.0		15.70	19.80		Seeherman et al., 1981
R. villosissimus	250.6	2.43	14.50		Australia	Authors
Uromys caudimaculatus	812.0	9.51	44.30		Australia	Authors
Pedetidae						
Pedetes capensis	2,300.0	13.20			Africa	Muller et al., 1979
	3,000.0			291.00		Seeherman et al., 1981
Sciuridae						
Eutamias merriami	75.0	1.31		8.84	North America	Wunder, 1970
Tamias striatus	92.0	3.93	18.10	21.50		Randolph, 1980; Seeherman et al., 1981

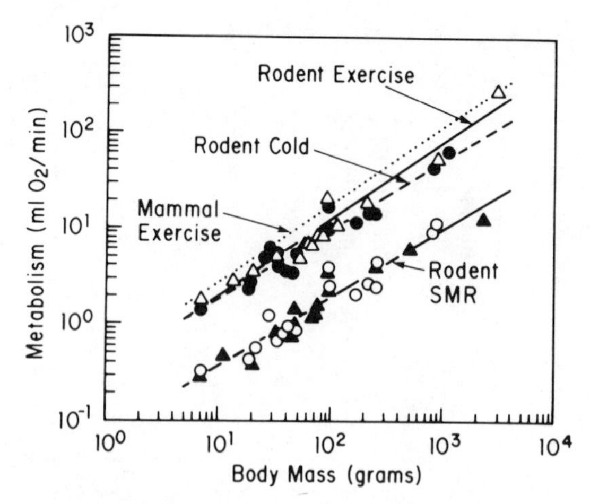

Fig. 2.1. The double logarithmic (log 10) relationships between body mass and standard metabolic rate (SMR), cold-induced maximal metabolic rate, and exercise-induced maximal metabolic rate of rodents. Each point represents the mean for a rodent species, and the lines are fit to the points by the method of least squares, after logarithmic transformation. Triangles represent species for which data are available for exercise-induced Vo_2 max; circles represent species for which data are available for cold-induced Vo_2 max. Unfilled triangles represent species means for exercise-induced Vo_2 max; filled circles for cold-induced Vo_2 max; filled triangles and unfilled circles are for SMR values (all species and data are identified in Table 2.1). The dotted line is the predicted relationship between body mass and exercise-induced Vo_2 max for mammals in general (from the equation of Taylor et al., 1981: ml O_2/min = 0.43 $g^{0.81}$).

ables are logarithmically transformed (Fig. 2.1). The regression equation for our data (ml O_2/min = 0.073 [$g^{0.707}$], N = 27 species; Table 2.2) does not differ significantly in slope or elevation from that for a more-extensive data set for rodents (ml O_2/min = 0.083 [$g^{0.669}$], N = 122 species: Hayssen and Lacy, 1985).

Cold-induced Maximal Metabolic Rate

The demonstration that endotherms contained in an atmosphere of approximately 79% helium and 21% oxygen will have greatly increased thermal conductances compared with those in air (Rosenmann and Morrison, 1974) has made the routine measurement of cold-induced maximal metabolic rate in endotherms an easy procedure, as well as one that does not subject them to traumatic cold stress. As a consequence many data have accrued for rodents (Table 2.1), making possible allometric analysis and comparison. These data for cold-induced maximal metabolic rates for rodents are depicted in Figure 2.1

TABLE 2.2

Statistical relationships between minimal (SMR) and maximal metabolic rates (V_{O_2}, ml O_2/min) and body mass in the rodent species identified in Table 2.1

Variable	No. of species	Regression equation*	S_{yx}	S_b	r^2	Mean mass (g)	V_{O_2} at 200 g
Minimal metabolism							
SMR	27	0.073 ($g^{0.707}$)	0.125	0.041	0.924	229.9	3.1
Maximal metabolism							
Cold-induced	19	0.340 ($g^{0.724}$)	0.133	0.055	0.912	172.3	15.8
Exercise-induced	15	0.308 ($g^{0.802}$)	0.122	0.049	0.953	319.3	21.6

*Slopes are all statistically indistinguishable ($P = 0.21–0.95$), while elevations are all statistically different ($P < 0.05$).

and are compared allometrically with SMR data and data for exercise-induced maximal metabolic rate in Figure 2.1 and Table 2.2.

Cold-induced maximal metabolic rate (expressed in ml O_2/min) parallels the pattern for SMR, but with an elevation that is about five times as high (Fig. 2.1, Table 2.2); it results in the regression equation ml O_2/min = 0.340 ($g^{0.724}$), which is statistically higher in elevation than that for SMR, but does not differ in slope.

Exercise-induced Maximal Metabolic Rate

The use of treadmills has made equally routine the measure of exercise-induced maximal metabolic rates in mammals (Seeherman et al., 1981). By running mammals on treadmills operating at varying rotational speeds, one can rather readily determine that point at which maximal oxygen consumption is reached. These data for rodents (Table 2.1) are depicted in Figure 2.1, with appropriate regression statistics identified in Table 2.2.

Like the pattern for cold-induced maximal metabolism, the pattern for exercise-induced metabolic rate parallels that for SMR, but is elevated about seven times (Fig. 2.1, Table 2.2). The regression equation for exercise-induced metabolism as a function of body mass is ml O_2/min = 0.308 ($g^{0.802}$). Its slope is statistically indistinguishable from the slopes for SMR and cold-induced maximal metabolism, but its elevation is statistically higher than for the other two variables (Table 2.2).

Discussion

When rodents' standard metabolism, cold-induced maximal metabolism, and exercise-induced maximal metabolism are regressed on body mass, each is elevated successively over the other, yet the scaling exponents of all three are statistically indistinguishable from each other (Fig. 2.1, Table 2.2). Thus the influence of body mass on each of these aerobic events is the same, but maximal aerobic capacities in response to thermoregulatory events are lower than those in response to activity. These similarities in slope and differences in elevation are clearly demonstrated in Table 2.2, in which we express rates of oxygen consumption for these three variables for a 200-g rodent (approximately the mean mass of the species employed in this analysis). The maximal metabolic rate induced by exercise in these rodents has a statistically indistinguishable scaling exponent from mammals in general, but the rodents have a significantly depressed maximal exercise metabolism (Fig. 2.1; Taylor et al., 1981; slopes, $F = 0.009$, $P = 0.92$; elevations, $F = 5.660$, $P < 0.022$). Furthermore, cold-induced metabolism in rodents is 73% of that induced by exercise, and this difference is significant (Table 2.2), while in mammals in general cold-induced metabolism is about 80% of exercise-induced metabolism (Weibel, 1984). We conclude, therefore, that rodents also have a significantly lower cold-induced metabolism, although there is not enough nonrodent mammalian data to statistically demonstrate this conclusion.

These seeming reductions in maximal aerobic capacities in rodents compared with other mammals when exposed to cold or high levels of activity, are probably related to the typically smaller mass and more-sheltered, sedentary lifestyles of rodents. The "mammals in general" employed by Weibel (1984) and Taylor et al. (1981) were generally large, cursorial species that rely to a much greater extent on sustained activity and physiological thermoregulation than do rodents, whose smaller sizes facilitate a substantial reliance on behavioral, rather than physiological, adjustments. The consistent elevation of exercise-induced maximal aerobic capacity over cold-induced maximal aerobic capacity both in mammals in general and in rodents indicates that it takes more energy to run fast than to keep warm. Exercise apparently calls into play either different mitochondrial populations or more oxygen-consuming organ systems than does thermoregulation in the cold. Another plausible hypothesis is that the increased maximal aerobic capacity of exercising mammals is merely a Q_{10} effect commensurate with the increased body temperatures that accompany exercise.

That the scaling exponents (slopes of the regression lines, Fig. 2.1) for minimal aerobic metabolism and for both cold-induced and exercise-induced maximal aerobic capacities are virtually identical (Table 2.2), lends credence to our suggestion that each is a representative endothermic energetic event that contributes to a suite of components constituting endothermic energy exchange. That the events are equally sensitive to body mass, yet differ in elevation, indicates further that realistic adaptational inferences can best be made if the patterns for at least these three events are known, rather than data on only a single event (such as SMR). An even fuller understanding of the energy exchanges of rodents (and other endotherms) could be obtained by integrating the patterns of minimal and maximal metabolic rates with the pattern for 24-h energy expenditures in noncaptive rodents (FMR; Nagy, 1987).

Patterns of Locomotion Metabolism in Heteromyid Rodents

The process of determining the relationships between running velocity and metabolic rate in mammals running on treadmills made it apparent that each individual of each species has some running speed (or level of sustained effort) at which aerobic metabolism becomes maximal (Seeherman et al., 1981). Beyond this level no further increase in oxygen consumption occurs, and progressive anaerobiosis sustains, in an additive manner, further effort. These readily recognizable maximal aerobic capacity plateaus are the same as those utilized in assessing exercise-induced maximal metabolic rates in the earlier part of this paper. Typically, the rate of oxygen consumption of mammals running at speeds below that at which maximal aerobic capacity is attained is linearly and positively related to running speed (Seeherman et al., 1981). Work by Dawson and Taylor (1973), however, indicated that an aerobic plateau is reached by the red kangaroo while hopping bipedally at intermediate speeds and at metabolic rates that are considerably below those predicted to represent maximal aerobic capacity for a mammal of that size. They interpreted this plateau to represent an energetically efficient mode of locomotion in which elastic storage in leg and tail tendons and in ligaments of energy generated while hopping decreased the energetic cost of hopping (Dawson and Taylor, 1973).

The thought of an energetically inexpensive aerobic plateau in hopping mammals was a very attractive idea, particularly to ecologists working with bipedal rodents, as there appeared to be linkages between bipedality and energetic efficiencies in granivorous desert rodents (MacMillen, 1983; Mares, 1983; Nikolai and Bramble, 1983; Reichman, 1981). There is, however, no indication of such a phenomenon in smaller (≤3 kg) mammalian bipeds, including rodents (Thompson et al., 1980). Further, MacMillen (1983) showed that the only plateau that was demonstrable in the bipedal Ord's kangaroo rat (*Dipodomys ordii*) while running at various speeds on a treadmill, was that achieved at maximal aerobic capacity; this plateau was accompanied by very high levels of blood lactic acid (>100 mg%) characteristic of incremental anaerobiosis as running speed increased above 3 km/h.

TABLE 2.3

Slopes and Y intercepts of relationships between observed Vo_2 (ml/[g · h]) and running speed (km/h) in quadrupedal and bipedal heteromyid rodents

Species	Mass (g)	Slope			Y intercept			
		Obs.	Pred.	Obs./Pred.	Obs.	Pred.	Obs./Pred.	Obs. r^2
Quadrupedal								
Perognathus fallax	20.2	0.44*	1.83	0.24	8.86*	3.52	2.52	0.15
Liomys salvini	50.9	0.56*	1.37	0.41	4.37*	2.66	1.64	0.34
Heteromys desmarestianus	84.1	0.60*	1.17	0.51	3.73*	2.29	1.63	0.54
Bipedal								
Microdipodops megacephalus	13.9	1.08*	2.06	0.52	9.45*	3.95	2.39	0.37
Dipodomys ordii	56.4	1.13	1.32	0.86	3.37	2.58	1.31	0.83
D. panamintinus	66.9	1.36	1.25	1.09	2.49	2.45	1.02	0.90

Note: Predicted values are based on body mass and the following equations in Calder, 1984: slope or transport cost, ml O_2/(g · km) = 4.73 $g^{-0.32}$; Y intercept or postural metabolism, ml O_2/(g · h) = 8.76 $g^{-0.30}$.

*Significantly different from predicated value at $P < 0.05$.

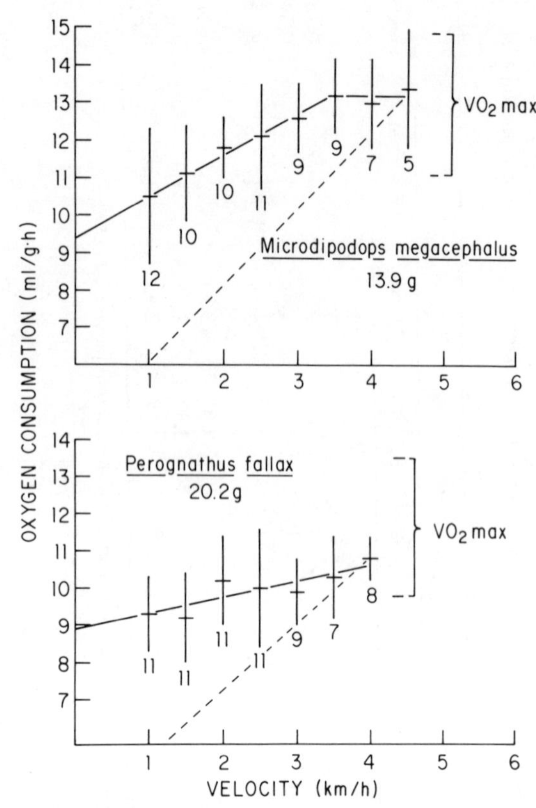

Fig. 2.2. The relationships between oxygen consumption and velocity of *Microdipodops megacephalus* and *Perognathus fallax* while running on a treadmill. The horizontal lines intersected by vertical lines represent the means ($\pm$SD) at a given running speed, and the numbers represent the number of individuals measured at that speed. The diagonal solid lines represent least squares fits to the data, and the brackets enclose the 95% confidence limits of exercise-induced maximal oxygen consumption for mammals as determined by Lechner (1978). The diagonal dashed lines represent predicted treadmill performance for mammals from the equation in Calder, 1984, converted to our units: ml $O_2/(g \cdot h) = 8.76\ g^{-0.30} + 4.73\ g^{-0.32}$ (km/h).

Because uncertainties still existed concerning metabolic patterns during locomotion in heteromyid rodents, we undertook an investigation in roughly size-matched species pairs of rates of oxygen consumption while running on a treadmill by quadrupedal (*Perognathus, Liomys, Heteromys*) and bipedal (*Microdipodops, Dipodomys*) heteromyids. The species employed and their body masses are identified in Table 2.3. Four of the six species are from North American desert localities, while the two largest quadrupeds (*Liomys salvini* and *Heteromys desmarestianus*) are from tropical habitats in Costa Rica (see MacMillen and

Hinds, 1983, for detailed descriptions of collecting localities). The methods that were employed in the measurements of oxygen consumption (Vo_2) while running on a treadmill are the same as those described by MacMillen (1983).

These relationships between Vo_2 and running speed for three bipedal and three quadrupedal heteromyids are depicted in Figures 2.2, 2.3, and 2.4 (Fig. 2.2, smallest-sized pair; Fig. 2.3, intermediate-sized pair; Fig. 2.4, largest-sized pair). The data that are depiced in these figures include mean (± 1 SD) Vo_2 at each running speed and a least-square regression line fitted to the individual Vo_2 measurements (solid line). The brackets enclose the 95% confidence limits for exercise-induced maximal aerobic capacity (Vo_2 max) predicted for a small mammal of that body mass (as determined by Lechner [1978]), and the dashed line is the predictive one for mammals in general relating Vo_2 to running

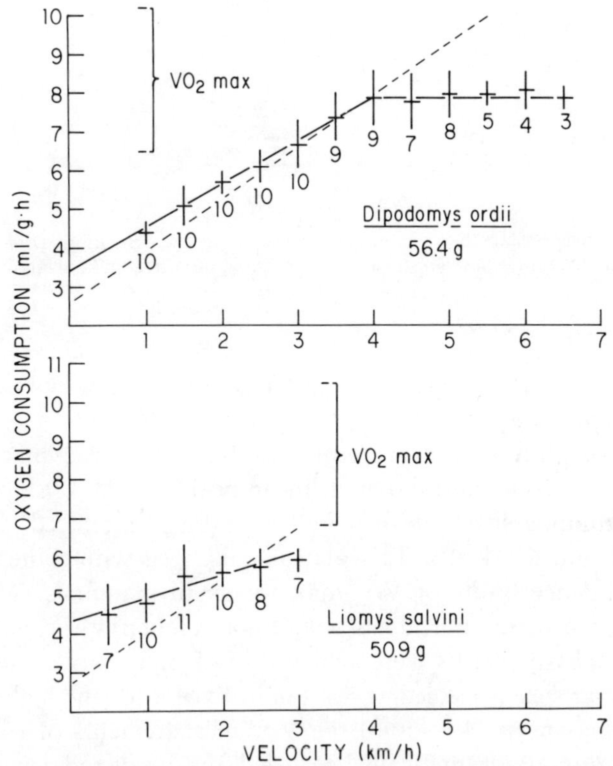

Fig. 2.3. The relationships between oxygen consumption and velocity of *Dipodomys ordii* and *Liomys salvini* while running on a treadmill. Symbols as in Figure 2.2.

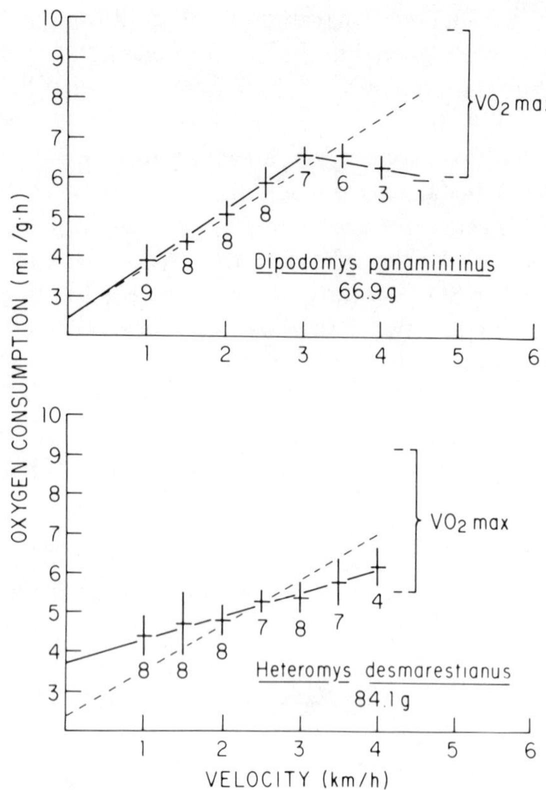

Fig. 2.4. The relationships between oxygen consumption and velocity of *Dipodomys panamintinus* and *Heteromys desmarestianus* while running on a treadmill. Symbols as in Figure 2.2.

speed (from Calder, 1984, p. 182, after various studies by C. R. Taylor and colleagues).

In the smallest pair of heteromyids (Fig. 2.2), the biped *Microdipodops megacephalus* shows a linear positive relationship between Vo_2 and running speed up to 3.5 km/h and a pronounced plateau between 3.5 and 4.5 km/h. This plateau falls well within the predicted 95% confidence limits of Vo_2 max for small mammals. Within this plateau region many individuals would not voluntarily run until steady-state Vo_2 measurements were achieved (3–5 min), and no individuals would run at speeds exceeding 4.5 km/h. Except for the highest voluntary running speed (4.5 km/h) the Vo_2 measurements of running *M. megacephalus* all dramatically exceeded the predicted levels. In the small quadruped *Perognathus fallax*, Vo_2 varied rather erratically with running speed (Fig. 2.2), and the regression line fitted to all data belies what may be a plateau in Vo_2 between 2.0 and 4.0 km/h. As in *M. megacephalus*, this plateau falls within predicted Vo_2 max limits for a small mammal of this size. Again, a pronounced attrition in voluntary running was observed at higher speeds, with no animal running faster than 4.0 km/h. Also, as in the case of *M. megacephalus*, Vo_2 max greatly exceeded predicted levels except at the highest running speed.

In the intermediate-sized pair (Fig. 2.3), the biped *Dipodomys ordii* conformed nearly precisely to predictions, with a linear positive relationship between Vo_2 and running speed that did not differ in slope or y intercept from the predicted relationship between 1.0 and 4.0 km/h (Table 2.3). There was a pronounced plateau between 4.0 and 6.5 km/h that corresponded with predicted levels of Vo_2 max; marked attrition in voluntary running again occurred within those higher speeds. The pattern relating Vo_2 to running speed in the intermediate-sized quadruped *Liomys salvini* (Fig. 2.3) was similar to that in the smaller quadruped *P. fallax* (Fig. 2.2) with a rather variable relationship exceeding the predicted levels at the lower running speeds (0.5 to 1.5 km/h). At higher speeds (2.0–3.0 km/h) there is some indication of a plateau that was accompanied by attrition in voluntary running. The highest values of Vo_2 measured were below the predicted limits of Vo_2 max for a mammal of this size.

In the largest pair of heteromyids observed (Fig. 2.4), the biped *Dipodomys panaminitinus* conformed nearly precisely to predicted patterns both in the relationship between Vo_2 and running speed, and in Vo_2 max, and with a very pronounced attrition in voluntary running while operating at Vo_2 max. Because individuals with higher rates of Vo_2 within the plateau region were the first to refuse to run there is an artifactual negative slope between 3.0 and 4.5 km/h. The largest quadruped (*Heteromys desmarestianus*) possessed the most positive relationship (by inspection) between Vo_2 and running speed (Fig. 2.4) of any of the quadrupeds, but again measures of Vo_2 at the lower running speeds were in excess of the predicted values. Measures of Vo_2 at the highest voluntary running speeds (3.5 and 4 km/h) fell within the predicted limits of Vo_2 max, with pronounced attrition of voluntary running within that range.

Although we have measured blood lactic acid concentration only in *Dipodomys ordii* (MacMillen, 1983), all of the data accrued on locomotion energetics in heteromyid rodents (MacMillen, 1983; Thompson, 1985; Thompson et al., 1980; Figs. 2.2, 2.3 and 2.4) argue per-

suasively against the occurrence of an energetically efficient aerobic plateau. While a plateau in Vo_2 does occur in heteromyids while running bipedally, it is undeniably an anaerobic plateau (Vo_2 max; Figs. 2.2, 2.3, and 2.4) and does not translate into energetic savings. It is interesting, however that the two bipedal *Dipodomys* species conform almost precisely to predicted locomotor performances, while the biped *Microdipodops* and the quadrupeds *Perognathus, Liomys*, and *Heteromys* generally have rates of Vo_2 while running that are well above the predicted rates, particularly at the slower running speeds (Figs. 2.2, 2.3, and 2.4; and Table 2.3). If these treadmill performances can be translated into field energetics, it does mean that at usual foraging speeds (0.5 to ca. 4.0 km/h within this size range: Thompson, 1985) the energetic costs of locomotion in *Dipodomys* are less than in the other three genera, but only because their costs conform to mammalian expectations while the costs of the quadrupeds exceed the expected. Thus our data indicate that among heteromyid rodents bipedal hopping is cheaper only because quadrupedal running is more expensive.

It is also interesting to note that the pattern of Vo_2 while running on a treadmill in the biped *Microdipodops* resembles much more closely the patterns of the quadrupeds *Perognathus, Liomys*, and *Heteromys* than those of the bipeds *Dipodomys ordii* and *D. panamintinus* (Figs. 2.2, 2.3, and 2.4; Table 2.3). These greater similarities in locomotor performance between *Microdipodops* and quadrupedal heteromyids are consistent with the recent demonstration by Djawdan and Garland (1988) that maximum burst speed in *Microdopodops megacephalus* matches much more closely that of quadrupedal *Perognathus* species than of bipedal *Dipodomys* species. A recent phylogenetic analysis of the Heteromyidae (Hafner and Hafner, 1983) suggests that *Microdipodops* genetically is more closely allied to *Dipodomys* than to any of the quadrupedal genera. A still more recent cladistic analysis based on myological characters (Ryan, 1989) agrees with this interpretation and concludes further that *Microdipodops* and *Dipodomys* did not evolve bipedalism independently. Our physiological data reveal substantial differences in bipedal locomotion energetics between *Microdipodops* and *Dipodomys*, suggesting that, if both are derived from common bipedal ancestry, the adaptational uses of bipedalism and their attendant energetic manifestations may well have diverged substantially between the two genera. The adaptational uses of bipedalism in these two genera remain to be clarified.

Lastly, we wish to comment on the rather variable (nonlinear) relationship between Vo_2 and running speed in especially *Microdipodops*, *Perognathus*, and *Liomys*, resulting in very low r^2 values (Table 2.3). While we did not quantify gait modes and transitions, these patterns are very reminiscent of those described by Hoyt and Taylor (1981) for horses running on a treadmill and as they changed gait. We suspect that the deviations from linearity observable for these genera in Figures 2.2 and 2.3 represent similar gait changes.

Conclusions

With respect to relationships between minimal and maximal aerobic metabolism and body mass, rodents appear to be energetically conservative. This conservatism is particularly apparent with regard to maximal aerobic capacities, in which exercise-induced Vo_2 max in rodents is significantly depressed below that predicted for a variety of mammals, dominated by larger, often cursorial, nonrodents (Fig. 2.1). In addition, rodents are apparently energetically conservative with respect to cold-induced Vo_2 max, which is significantly reduced below values observed in both rodents during locomotion and mammals in general in response to cold (Weibel, 1984). We suggest that this energetic conservatism is related to their generally small size, which allows them to spend more time in sheltered microhabitats and escape the macroenvironmental stresses to which larger mammals are more fully exposed.

This linkage between small size, lifestyle, and energetic performance results in an apparent dichotomy within the rodent family Heteromyidae, at least insofar as locomotion energetics is concerned. The highly mobile, bipedal *Dipodomys* species conform closely to locomotor patterns predictable for larger, highly mobile mammals. In contrast, species of the less-mobile genera all have relatively variable, energetically expensive patterns of locomotion either at all speeds (bipedal *Microdipodops* and quadrupedal *Perognathus*) or at all but the fastest speeds (quadrupedal *Liomys* and *Heteromys*).

ACKNOWLEDGMENTS

Many of the data used in this chapter were collected while we were supported variously by NSF grants DEB-7620116 and DEB-7923808, a grant under the Fulbright-Hays Act, and the Flinders University Research Committee. We are grateful to R. V. Baudinette and the Flinders University School of Biological Sciences for collegial and logistical support, and to the Coopers for exceptional aid.

Literature Cited

Bartholomew, G. A. 1982. Energy metabolism. Pp. 46–93 *in* Animal physiology—principles and adaptations (M. S. Gordon, G. A. Bartholomew, A. D. Grinnell, C. B. Jorgensen, and F. N. White, eds.). Macmillan, New York. 635 pp.

Brody, S., and R. C. Proctor. 1932. Relation between basal metabolism and mature body weight in different species of mammals and birds. Mo. Agric. Exp. Stn. Res. Bull., 166:89–101.

Calder, W. A., III. 1984. Size, function and life history. Harvard University Press, Cambridge. 431 pp.

Casey, T. M., P. C. Withers, and K. K. Casey. 1979. Metabolic and respiratory responses of arctic mammals to ambient temperate during the summer. Comp. Biochem. Physiol., 64A:331–342.

Dawson, T. J., and C. R. Taylor. 1973. Energetic cost of locomotion in kangaroos. Nature, 246:313–314.

Djawdan, M., and T. Garland, Jr. 1988. Maximal running speeds and bipedal and quadrupedal rodents. J. Mammal., 69:765–772.

Fish, F. E. 1979. Thermoregulation in the muskrat (*Ondatra zibethicus*): the use of regional heterothermia. Comp. Biochem. Physiol., 64A:391–397.

Hafner, J. C., and M. S. Hafner. 1983. Evolutionary relationships of heteromyid rodents. Great Basin Nat. Mem., 7:3–29.

Hart, J. S. 1962. Temperature regulation and adaptation to cold climates. Pp. 203–230 *in* Comparative Physiology of Temperature Regulation (J. P. Hannon and E. G. Vierick, eds.). Arctic Aeromedical Laboratory, Fort Wainwright, Alaska. 455 pp.

Hart, J. S., and O. Heroux. 1955. Exercise and temperature regulation and lemmings and rabbits. Can. J. Biochem. Physiol., 33:428–435.

Hayssen, V., and R. C. Lacy. 1985. Basal metabolic rates in mammals: taxonomic differences in the allometry of BMR and body mass. Comp. Biochem. Physiol., 81A:741–754.

Hill, J. R. 1959. The oxygen consumption of newborn and adult mammals. Its dependence on the oxygen tension in the inspired air and on the environmental temperature. J. Physiol., 149:346–373.

Hinds, D. S., and R. E. MacMillen. 1985. Scaling of energy metabolism and evaporative water loss in heteromyid rodents. Physiol. Zool., 58:282–298.

Hoyt, D. F., and C. R. Taylor. 1981. Gait and energetics of locomotion in horses. Nature, 292:239–240.

Hulbert, A. J., D. S. Hinds, and R. E. MacMillen. 1985. Minimal metabolism, summit metabolism and plasma thyroxine in rodents from different environments. Comp. Biochem. Physiol., 81A:687–693.

Kleiber, M. 1932. Body size and metabolism. Hilgardia, 6:315–353.

Lechner, A. J. 1978. The scaling of maximal oxygen consumption and pulmonary dimensions in small mammals. Respir. Physiol., 34:29–44.

MacMillen, R. E. 1983. Adaptive physiology of heteromyid rodents. Great Basin Nat. Mem., 7:65–76.

MacMillen, R. E., and T. Garland, Jr. 1989. Adaptive physiology of *Peromyscus*. Pp. 143–168 *in* Advances in the Study of *Peromyscus* (Rodentia) (G. L. Kirkland, Jr., and J. N. Layne, eds.). Texas Tech. University Press, Lubbock. 366 pp.

MacMillen, R. E., and D. S. Hinds. 1983. Water regulatory efficiency in heteromyid rodents: a model and its application. Ecology, 64:152–164.

Mares, M. A. 1983. Desert rodent adaptation and community structure. Great Basin Nat. Mem., 7:30–43.

McNab, B. K. 1988. Complications inherent in scaling the basal rate of metabolism in mammals. Q. Rev. Biol., 63:25–54.

Muller, E. F., J. M. Z. Kaman, and G. M. O. Maloiy. 1979. O_2 uptake, thermoregulation and heart rate in the springhare (*Pedetes capensis*). J. Comp. Physiol., 133:187–191.

Nagy, K. A. 1987. Field metabolic rate and food requirement scaling in mammals and birds. Ecol. Monogr., 57:111–128.

Nikolai, J. C., and D. M. Bramble. 1983. Morphological structure and function in desert heteromyid rodents. Great Basin Nat. Mem., 7:44–64.

Pasquis, P., A. Lacaisse, and P. Dejours. 1970. Maximal oxygen uptake in four species of small mammals. Respir. Physiol., 9:298–309.

Pohl, H. 1985. Temperature regulation and cold acclimation in the golden hamster. J. Appl. Physiol., 20:405–410.

Randolph, J. C. 1980. Daily energy metabolism of two rodents (*Peromyscus leucopus* and *Tamias striatus*) in their natural environments. Physiol. Zool., 53:70–81.

Reichman, O. J. 1981. Factors influencing foraging in desert rodents. Pp. 195–213 *in* Foraging Behavior (A. C. Kamil and T. D. Sargent, eds.). Garland STPM Press, New York and London.

Rosenmann, M., and P. R. Morrison. 1974. Maximum oxygen consumption and heat loss facilitation in small homeotherms by He-O_2. Am. J. Physiol., 226:490–495.

Rosenmann, M., P. Morrison, and D. Feist. 1975. Seasonal changes in the metabolic capacity of red-backed roles. Physiol. Zool., 48:303–310.

Ryan, J. R. 1989. Comparative myology and phylogenetic systematics of the Heteromyidae (Mammalia, Rodentia). Univ. Mich. Mus. Zool. Misc. Publ. No. 176:1–103.

Seeherman, H. J., C. R. Taylor, G. M. O. Maloiy, and R. B. Armstrong. 1981. Design of the mammalian respiratory system. II. Measuring maximum aerobic capacity. Respir. Physiol., 44:11–23.

Segram, N. D., and J. S. Hart. 1967. Oxygen supply and performance in *Peromyscus*. Comparison of exercise with cold exposure. Can. J. Physiol. Pharmacol., 45:543–549.

Taylor, C. R., G. M. O. Maloiy, E. R. Weibel, V. A. Langman, J. M. Z. Kamau, H. J. Seeherman, and N. C. Heglund. 1981. Design of the mammalian respiratory system. III. Scaling maximum aerobic capacity to body mass: wild and domestic mammals. Respir. Physiol., 44:25–37.

Thompson, S.D. 1985. Bipedal hopping and seed-dispersion selection by heteromyid rodents: the role of locomotion energetics. Ecology, 66:220–229.

Thompson, S. D., R. E. MacMillen, E. M. Burke, and C. R. Taylor. 1980. The energetic cost of bipedal hopping in small mammals. Nature, 287:223–224.

Weibel, E. R. 1984. The pathway for oxygen. Harvard University Press, Cambridge. 425 pp.

Wunder, B. A. 1970. Temperature regulation and the effects of water restriction on Merriam's chipmunk *Eutamis merriami*. Comp. Biochem. Physiol., 33:385–403.

Frank E. Fish

3. Aquatic Locomotion

The energetic cost of locomotion by mammals is expensive and therefore represents a major deficit to the available energy resources. For locomotion in water, the energetic demands and swimming performance of mammals are affected by the water-flow patterns around the swimmer dictated by the high density and viscosity of the aquatic medium (Daniel and Webb, 1987; Schmidt-Nielsen, 1972; Williams, 1987). The generation of propulsive forces (thrust) opposes resistive forces (drag) that increase the locomotor effort. In addition, energy expenditure for locomotion is hampered by periods of apnea during underwater swimming (Hochachka, 1980; Kooyman, 1985) and by a high thermal conductivity of the aquatic medium that potentially necessitates increased thermoregulatory demands (Fish, 1983; Hart and Fisher, 1964; Irving 1971; Nadel, 1977; Whittow, 1987). The large energetic demands of swimming mammals promoted the evolution of physiological, morphological, and behavioral adaptations that reduce energy consumption while allowing for effective locomotion. If aquatic mammals are adapted to swim in such a manner so as to minimize energy expenditure, there should be distinct metabolic and hydrodynamic advantages to swimming modes and morphologies employed by the most aquatically derived species. At varying times, mammals, such as cetaceans, sirenians, and pinnipeds, that spend all or most of their time inhabiting and locomoting in water, must generate adequate thrust for migration, high-speed swimming, and maneuverability. The evolution of such aquatic mammals represents the cul-

---

Department of Biology, West Chester University, West Chester, Pennsylvania 19383.

mination of a sequence of transitional stages from terrestrial quadrupeds to fully aquatic piscine-like morphologies and propulsive modes of high energetic efficiency (Barnes et al., 1985; Fish et al., 1988; Gaskin, 1982; Gingerich et al., 1983; Tedford, 1976). However, semiaquatic mammals must contend with swimming modes having metabolic and mechanical inefficiencies attributable to morphologies constrained by a compromise between movement on both land and water (Fish, 1984).

In this chapter, I discuss the energetics of swimming by mammals with particular regard to swimming modes and swimming strategies. Metabolic studies that estimate power input provide an indication of the energy potentially available for aquatic propulsion (Davis et al., 1985; DiPrampero et al., 1974; Fish, 1982, 1983; Williams, 1983a). However, because external work is performed in moving the body through a fluid, a full examination of the dynamics of swimming requires an estimate of the power output as the realized rate of energy use contributing to the performance of work (Fish, 1982). External work is manifest as a transfer of momentum between the animal and its environment (Daniel and Webb, 1987) resulting in a thrust force causing movement by the animal through the water. Therefore, studies concerning the energetics of swimming mammals should include examinations of hydrodynamics and biomechanics of the various locomotor modes to estimate power output as the rate of energy expended to produce thrust in addition to metabolic determinations of power input. Use of physiological, morphological, and hydrodynamic studies allows for an integrated approach to elucidate the dynamics of swimming by mammals.

Aquatic Mammal Diversity and Swimming Modes

The majority of mammalian orders have representatives that can be classified as semiaquatic or fully aquatic (Howell, 1930). With the possible exceptions of the giraffe and apes, all mammals have the ability to swim regardless of any specific adaptations for an aquatic existence (Dagg and Windsor, 1972). Perusal of the literature shows that cursorial (Bryant, 1919; Fregin and Nicholl, 1977), fossorial (Best and Hart, 1976; Hickman, 1983, 1984; Talmage and Buchanan, 1954), arboreal (Cole, 1922), and even aerial (Craft et al., 1958) mammals can swim.

TABLE 3.1
Mammalian swimming modes

Swimming mode	Principal appendage	Example	Reference
Drag-based oscillatory	Quadrupedal	Mink	Williams, 1983a
		Mole	Hickman, 1984
		Opossum	Fish, 1987
	Pectoral	Polar bear	Flyger & Townsend, 1968
	Pelvic	Muskrat	Fish, 1984
		Water opossum	Fish, 1987; Stein, 1981
		River otter	Tarasoff et al., 1972
Lift-based oscillatory	Pectoral	Sea lion	English, 1976; Feldkamp, 1987b
Lift-based undulatory	Pelvic	Phocid seal	Fish et al., 1988; Tarasoff et al., 1972
		Sea otter	Tarasoff et al., 1972
		Walrus	Gordon, 1981
	Caudal	Dolphin	Lang, 1966; Parry, 1949
		Manatee	Hartman, 1979

Terrestrial and semiaquatic mammals swim using mainly oscillatory propulsors (Table 3.1), which are either paired appendages that function as paddles or winglike hydrofoils (Webb and Blake, 1985). Paddling is associated with slow swimming and precise maneuverability (Webb, 1984) and generally is used in surface swimming. The paddling stroke cycle is composed of a power phase and a recovery phase. During the power phase, the paddle pushes posteriorly on the water, generating a drag force to its movement that produces thrust for the whole organism in the direction opposite the paddle movement (Fig. 3.1; Fish, 1984; Webb and Blake, 1985). The paddle is repositioned during the recovery phase without the generation of thrust.

The foreflippers, used by the Otariidae, act as oscillatory hydrofoils that generate thrust by the lifting-wing principle (Fig. 3.1; Webb and Blake, 1985) throughout the stroke cycle (Feldkamp, 1987b). This mechanism provides effective aquatic propulsion for otariid seals, which spend considerable amounts of time in the water foraging for food and undertaking long oceanic migrations (Ridgeway and Harrison, 1981), while maintaining a flipper structure that is adept at locomoting on land (English, 1976; Feldkamp, 1987a, 1987b).

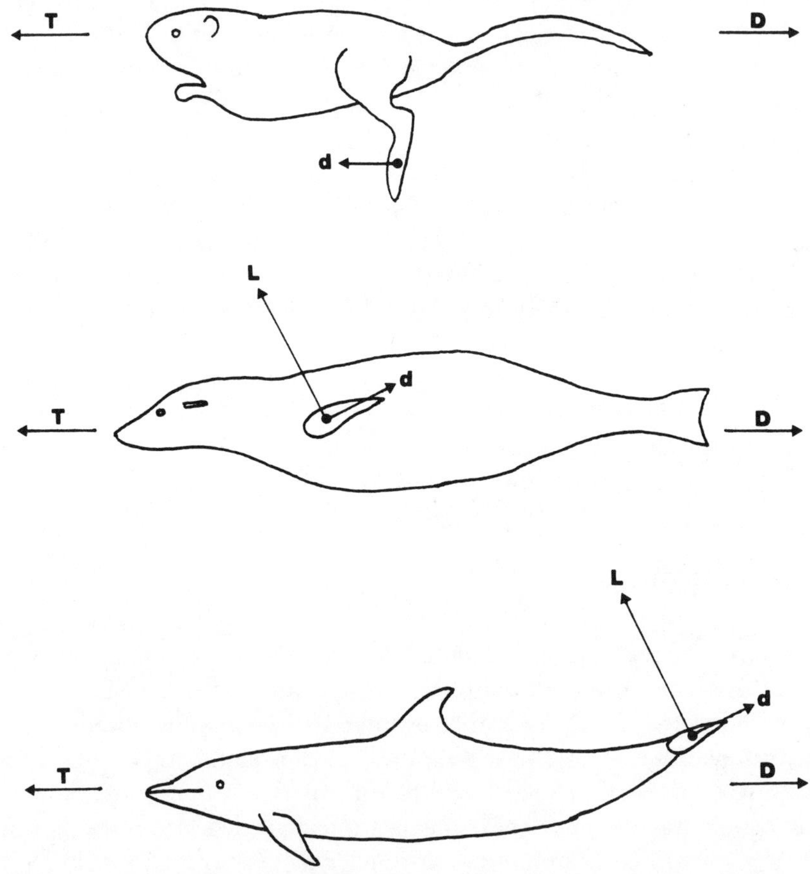

Fig. 3.1. Major forces associated with swimming, including (top) paddling (muskrat), (middle) lift-based oscillation (sea lion), and (bottom) lift-based undulation (dolphin). For each animal swimming in water, drag force (D) resists forward motion of the body and is opposed by an equal thrust force (T) generated by the propulsive appendages. The posterior movement of the hind feet of the muskrat acts as a paddle to generate a drag force on the appendage (d) that contributes to T. The foreflippers of the sea lion and the caudal fluke of the dolphin act as hydrofoils that produce an anteriorly directed lift force (L) that can be resolved to generate T.

The undulatory propulsive mechanism passes waves along the body or caudal fluke and is used by those mammals that are most restricted to the aquatic environment. Cetaceans and sirenians undulate the caudal fluke in a dorsoventral plane to effect propulsion (Hartman, 1979; Lang and Daybell, 1963; Nishiwaki and Marsh, 1985; Parry, 1949; Peterson, 1925; Slijper, 1961; Videler and Kamermans, 1985). Except for the orientation of propulsive movements and the asymmetry of the propulsive musculature, these mammals reflect the typical swimming pattern observed in many fish. Periodic motion of the caudal fluke generates thrust as a component of an anteriorly inclined lifting force (Fig. 3.1). The Phocidae and Odobenidae laterally undulate the posterior body to move the paired hind flippers in the horizontal plane, effecting a fishlike movement (Backhouse, 1961; Fish et al., 1988; Gordon, 1981; Ray, 1963; Tarasoff et al., 1972). By laterally undulating its compressed tail, the otter shrew (*Potamogale* spp.) may produce propulsion with the nonwebbed feet pressed against the body (Walker, 1975). In addition, fast, submerged swimming by otters is accomplished by the undulatory mode (Chanin, 1985; Kenyon, 1969; Tarasoff et al., 1972).

Undulatory swimming is a rapid and relatively high-powered propulsive mode (Webb, 1984). This mode is used for swimming durations from several seconds to weeks, as involved in cruising, sprints, patrolling, station holding, and migrations.

Swimming Speed

Reports on swimming speeds of various aquatic mammals have in many instances been anecdotal and often unreliable. The reason for questioning these reports is that estimates of swimming speeds based on observations from ships, airplanes, or shorelines have often been made without fixed reference points, information on currents, or accurate timing instruments. These observations have led to erroneous conclusions regarding swimming performance as exemplified for dolphins by Gray's Paradox (Gray, 1936; Parry, 1949). In this case, power output calculated from a simple hydrodynamic model for a dolphin swimming at 10 m/s for 7 s was greater than the power that was assumed to be developed by the muscles (Gray, 1936). The paradox is resolved when one considers that the dolphin was demonstrating a burst performance (Kooyman, 1989; Lang and Daybell, 1963), and the muscle power output used as a standard was based on the sustained performance of the dog and humans (Gray, 1936). Although controlled laboratory studies do allow for measurements of precise swimming speeds, these speeds may not reflect ecologically relevant levels of performance. The future use of microprocessors carried by freely swimming and diving mammals should provide accurate swimming speeds, as has already been done for otariids (Ponganis et al., 1990).

The range of swimming speeds varies markedly. Differences in swimming speeds of mammals are related to body size, swimming mode, and relation to the water surface. Large animals have higher cruising and maximum sprint speeds than smaller swimmers (Aleyev, 1977; Kooyman, 1989). This relationship holds up to a body length of 4.5 m and is due to allometric differences. Body surface area and resistance to movement in water are proportional to the square of body length, whereas muscle mass and power that may be developed for swimming are proportional to the cube of body length. Consideration of length-specific swimming speeds, however, shows that large mammals demonstrate lower performance levels, as exemplified by cetaceans (Webb, 1975a). Mammals that paddle are slower than either undulatory swimmers or lift-based oscillators. This pattern is most likely due to decreased net thrust production and efficiency of paddling compared with the other propulsive modes (see below). In addition, paddling mammals are usually surface swimmers whose speed is limited by interference from surface waves (Fish, 1982, 1984; Williams, 1983a, 1989). Paddlers, such as the muskrat (*Ondatra zibethicus*) and rice rat (*Oryzomys palustris*), maintain routine swimming speeds at or below the predicted speed of maximum wave resistance (Fish, 1984). Sea otters (*Enhydra lutris*) are restricted to sustained surface swimming speeds less than 0.8 m/s, but when submerged can swim at speeds from 0.6 to 1.39 m/s by undulation (Williams, 1989).

Hydrodynamic Drag

To propel itself through water at a constant velocity (U), a mammal needs to generate a thrust force (T) at the expense of metabolic energy that is equal to the sum of resistive drag forces (D), so that:

$$\eta M = TU = DU \tag{1}$$

where η is a dimensionless overall efficiency, M is the metabolic rate, and TU and DU represent thrust and drag power outputs, respectively. Because the rate of energy expended to overcome drag is related to the thrust power output and ultimately to the rate of metabolic energy expenditure, investigators have used estimates of drag in studies of swimming energetics. Complete explanations of hydrodynamics in biological systems can be found in publications by Webb (1975a), Vogel (1981), and Blake (1983b).

Drag consists of frictional and pressure force components that arise from the flow regime about the body (Blake, 1983b; Webb, 1978; Yates, 1983). The flow is divided into two regions: boundary layer and outer flow. The boundary layer arises as the result of water viscosity. Water particles adhere to the body surface, so there is no relative velocity difference, resulting in the no-slip condition. At a short distance from the body, water velocity approximates the outer flow. The velocity gradient between the body surface and outer flow develops because of shear stresses in the boundary layer caused by the fluid viscosity and accounts for the frictional drag component. The pressure component of drag arises from distortion of fluid around the body in the outer flow (Webb, 1978). Deflection of the outer flow due to the shape of the body produces a pressure gradient from varying flow velocities interacting with the boundary layer separating it from the body. This interaction translates into a wake where kinetic energy is lost in addition to a net pressure force that acts in opposition to forward motion (Webb, 1978).

The type of flows within the boundary layer and outer flow also influences frictional and pressure components of drag. Flows may be laminar, turbulent, or transitional. Flow type is determined by the size and speed of the animal and by the density and viscosity of the fluid medium. The influence of these parameters is represented by the nondimensional Reynolds number:

$$Re = \frac{UL}{\nu} \qquad (2)$$

where Re is the Reynolds number, L is body length, and ν is the kinematic viscosity of water (Blake, 1983b; Webb, 1978). Boundary flow about a submerged, rigid streamlined body is laminar up to a Re of approximately 5×10^5, turbulent above a Re of 5×10^6, and transitional between those values (Webb, 1975a; William, 1987). The onset of transitional flow that is partly laminar and partly turbulent occurs at the critical Re, which is influenced by disturbances in the outer flow, surface roughness, and pressure gradients opposite to the direction of flow (Webb, 1975a). The large size and high swimming speed of marine mammals in particular indicate a high Re of greater than 10^6 (e.g., *Enhydra lutris*, Re = 1.7×10^6: Williams, 1989; *Zalophus californianus*, Re = 8.4×10^6: Kooyman, 1989; *Lagenorhynchus obliguidens*, Re = 1.5×10^7: Lang and Daybell, 1963) and thus a transitional or fully turbulent boundary layer.

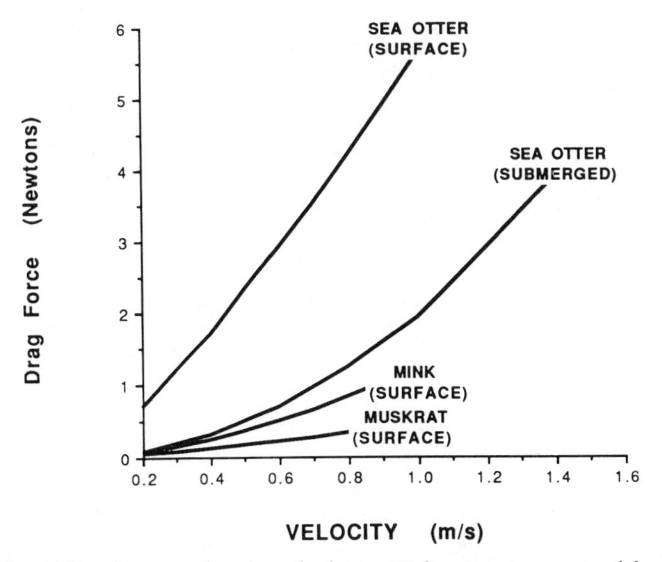

Fig. 3.2. Plot of drag force as a function of velocity (U) from measurements of dead-drag on muskrat (Fish, 1984), mink (Williams, 1983a), and sea otter (Williams, 1989). Drag measurements were obtained on muskrat, mink, and sea otter at the water surface and also on the sea otter in a submerged position.

Thrust and power output based on drag determinations were estimated for a variety of aquatic mammals by use of standard hydrodynamic equations (Au and Weihs, 1980; Gray, 1936; Hui, 1987; Parry, 1949), models (Aleyev, 1977; Purves et al., 1975), dead animals (Fish, 1984; Williams, 1983a, 1989), and towing or coasting (Feldkamp, 1987a; Innes, 1984; Lang and Daybell, 1963; Lang and Pryor, 1966; Williams and Kooyman, 1985). In all cases the bodies are rigid or assumed to be analogous to a flat plate with an equivalent surface area. These rigid-body analogies for aquatic mammals demonstrate that drag and power output increase curvilinearly with increasing velocity (Fig. 3.2), but the magnitude of the drag force differs with the size of the animal. Because swimming speed varies with the size of the animal, comparisons of the energetics of aquatic mammals over a 10^7-fold range of body mass are difficult.

A convenient method of estimating the resistance of a body moving through water is by computation of the dimensionless drag coefficient, C_D:

$$C_D = \frac{D}{0.5 \rho S U^2} \quad (3)$$

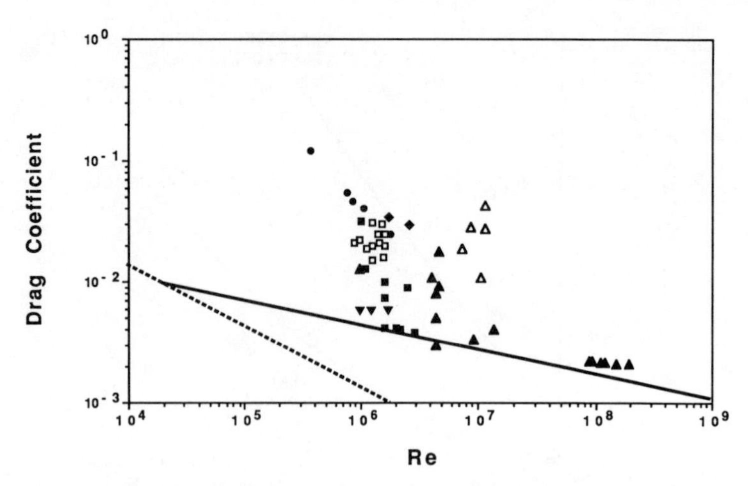

Fig. 3.3. Plot of the drag coefficient (C_D) as a function of the Reynolds number (Re). Symbols represent cetaceans (▲ and Δ: Aleyev, 1977; Kermack, 1948; Lang and Daybell, 1963; Purves et al., 1975; Videler and Kamermans, 1985; Webb, 1975a; Yates, 1983), pinnipeds (■ and □: Feldkamp, 1987a; Fish et al., 1988; Williams and Kooyman, 1985), sea otter (▼: Williams, 1989), beaver (●: Kurbatov and Mordvinov, 1974), and human (♦: Williams and Kooyman, 1985). Closed symbols represent estimates of C_D based on rigid-body analogies; open symbols represent values determined from hydrodynamic thrust-based models. The solid line represents the minimum C_D assuming turbulent boundary conditions; the broken line is for C_D assuming laminar conditions.

where ρ is water density, S is wetted surface area, and U is velocity. C_D depends on Re and the flow conditions about the body (Webb, 1978); C_D is positively related to drag force, which is best measured directly from animals (Williams, 1987). In addition, C_D can be compared with a reference drag coefficient representing the theoretical minimum based on a flat plate with equivalent surface area and Re (Fish et al., 1988).

C_Ds computed from gliding and towing drag measurements for submerged otariid and phocid seals and odontocete cetaceans range from 0.003 to 0.018 at Re of 10^6–10^7 (Feldkamp, 1987a; Lang and Daybell, 1963; Mordvinov and Kurbatov, 1972; Videler and Kamermans, 1985; Williams and Kooyman, 1985). The values of C_D for these marine mammals is greater or equivalent to the minimum drag coefficient for a turbulent boundary layer (Fig. 3.3). The blue whale (*Balaenoptera musculus*) has the lowest calculated value of C_D at 0.0022 for Re = 1.9×10^8, assuming a turbulent boundary layer (Kermack, 1948). Although C_D for the blue whale is low, the whale's large surface area and high swimming speed result in the largest drag force reported for a swimming mammal. Calculated power output for the blue whale is 4×10^5 W (Kermack, 1948), which is 88.3 times as great as the highest power output measured for a dolphin (Lang and Pryor, 1966).

Values of C_D for aquatic mammals indicating turbulent boundary conditions contradict the assertion of drag reduction by maintenance of a laminar boundary layer as an answer to Gray's Paradox (Gray, 1936; Parry, 1949). On the basis of the calculated drag of a dolphin swimming with a turbulent boundary layer, Gray (1936) predicted a higher power output than could be developed by the locomotor muscle mass. Although Gray (1936) underestimated the power generated by a dolphin swimming at burst speed by assuming equivalence with estimates of muscle power output of sustained activity, drag on the dolphin was believed to be reduced by maintenance of laminar flow within the boundary layer.

Attempts to reconcile Gray's Paradox by a mechanism that maintains a laminar boundary layer have focused on hydrodynamic characteristics of the integument of marine mammals. A compliant skin that could dampen turbulence and maintain laminar boundary conditions by active or passive mechanisms was viewed as a possible resolution to the paradox (Kramer, 1960a, 1960b, 1965; Sokolov, 1962). The structure of cetacean skin is similar to humanmade compliant surfaces (Kramer, 1960b, 1965; Yurchenko and Babenko, 1980). Mobile skin folds observed on swimming dolphins (Essapian, 1955) were thought to absorb energy through elastic deformation and to dampen turbulence in the boundary layer, resulting in a reduction of the total drag (Aleyev, 1977; Kramer, 1965; Sokolov, 1960; Yurchenko and Babenko, 1980). In addition, dermal ridges in the skin, the infusion of desquamated epidermal cells into the boundary layer, secretions from the dolphin eye, and heating by the skin to change boundary layer viscosity were hypothesized to retain laminar flow and reduce drag (Lang, 1966; Purves, 1963; Sokolov et al., 1969).

Mordvinov and Kurbatov (1972) reported that the body hair of phocid seals dampens turbulent eddies, thus reducing drag. Most studies on swimming mammals, however, have found little evidence promoting a drag reduction mechanism by laminarization of boundary flow due to properties of the integument. Experiments using naked women towed through water as analogues to swimming dolphins show that mobile skin folds represent a parasitic feature that does not improve drag reduction (Aleyev, 1977). Results reported by Lang and Daybell (1963) refute the assumption of drag reduction by a laminar boundary layer in dolphins. In a study on a live dolphin in which turbulence was induced over its surface, the drag was the same as when no turbulence was induced, indicating a normally turbulent flow (Lang and Daybell, 1963; Webb, 1975a). Flow visualization experiments in mammals show that the majority of the body surface has turbulent flow conditions (Fish, 1984; Kurbatov and Mordvinov, 1974; Mordvinov, 1974; Purves et al., 1975; Rosen, 1963; Williams and Kooyman, 1985). Although it produces a higher frictional drag component than laminar flow, a turbulent boundary layer will generate a smaller pressure drag (Webb, 1975a). The difference in magnitudes of drag components ultimately produces a smaller total drag for an animal with turbulent flow compared with laminar boundary conditions. The high energy content of the turbulent boundary layer prevents separation of the boundary layer from the body into the outer flow. Separation results in increased pressure and total drag.

Significant drag reduction in aquatic mammals is largely dependent on body shape. Highly aquatic mammals, such as cetaceans and pinnipeds, have streamlined bodies and appendages that incur low drag (Williamson, 1972; Feldkamp, 1977a; Fish et al., 1988). In comparison, less-aquatic mammals, such as humans and beavers, have high values of C_D (Kurbatov and Mordvinov, 1974; Williams and Kooyman, 1985), which are at least 75% as great as the minimum turbulent C_D at equivalent Re (Fig. 3.3).

The fineness ratio (FR = body length/maximum body diameter), which serves as a crude indicator of the streamlining of a body (Williams, 1987), has an optimal value of 4.5 for the lowest drag, where volume is maximized for a minimum surface area (Hertel, 1966; Webb, 1975a). Cetaceans and phocids maintain a range of FR (3.2–5.6) that spans the optimal FR (Aleyev, 1977; Fish et al., 1988; Hertel, 1966; Mordvinov, 1972; Williams and Kooyman, 1985), although phocids are represented at the lower end of this range. The sea lion (*Zalophus californianus*) and sea otter (*Enhydra lutris*) have moderately elongate body forms with FRs of 5.5 (Feldkamp, 1987a) and 5.8 (Williams, 1989), respectively. Extremes of FR that indicate increased pressure drag are displayed by semiaquatic paddlers: muskrat (*Ondatra zibethicus*) and beaver (*Castor canadensis*) have relatively nonstreamlined bodies with FR values of 2.5 and 3.0, respectively (Kurbatov and Mordvinov, 1974), while the FR of mink (*Mustela vison*) is 9.1 (Williams, 1983a).

Increased drag and energy loss from surface swimming result from the formation of waves that augment drag up to five times (Hertel, 1966), increasing metabolic expenditure and limiting swimming speed when compared with submerged swimming (Fish, 1984). The effect of surface waves is negated when the animal is submerged below a depth of three times the body diameter (Hertel, 1966). Periods of submerged swimming by wild minks and sea otters indicate a behavioral strategy to reduce drag (Williams, 1983a, 1989). Dolphins, however, utilize the pressure field of surface waves in bow riding to minimize their locomotor effort (see review in Hertel, 1969). It has been suggested recently that even large whales can save energy by extracting up to 33% of their propulsive power from ocean waves (Bose and Lien, 1990).

Although submerged swimming reduces drag compared with surface swimming, air-breathing mammals can not indefinitely avoid the water surface and its enhanced drag. An alternate strategy is to leap from the water and become airborne. Porpoising consists of serial leaps in which the animal leaves the water and thus reduces drag and energy cost (Au, 1980; Au and Weihs, 1980; Blake 1983a, 1983b). At high swimming speeds, the energy required to leap a given distance is less than the energy to swim an equivalent distance at the water surface. Williams (1987) observed the minimum porpoising speed for adult harbor seals ranged from 2.5 to 3.0 m/s, which agreed with the predicted porpoising speed based on Blake's (1983a) model. This behavior for energy savings may be limited in harbor seals, because porpoising is mainly confined to excited males during the mating season (King, 1983). In addition, Hui (1989) found the emergence angle of 36.9° for free-ranging dolphins (*Delphinus delphis* and *Stenella attenuata*) when porpoising was not the angle predicted for maximum energy savings. Porpoising for energy conservation predicts an emergence angle of 45° for maximum leap distance (Au and Weihs, 1980; Blake, 1983a) or an angle of 30° as a compromise for maximum leap distance and maximum forward speed (Gordon, 1980).

Thrust-based Models

Power outputs based on drag measurements from rigid-body analogies provide only a minimum estimate of the energy expenditure of swimming, because such models do not account for movements of the body or appendages, gross flow effects, interactions, and drag-reducing mechanisms (Webb, 1975a). The propulsive undulatory and oscillatory movements of the body and appendages incur increased energy loss due to increased drag and inertial forces from accelerations of the propulsor and the water (Daniel, 1984; Fish, 1984; Fish et al., 1988; Lighthill, 1971). Consequently, hydrodynamic models based on thrust calculated from the kinematics of swimming mammals provide a better estimate of power output than drag determinations. Calculated estimates of power output from such models are 3–16 times greater than values calculated for equivalent rigid bodies (Fish, 1984; Fish et al., 1988; Webb, 1975a; Yates, 1983). A further benefit of hydrodynamic models is that comparisons of thrust generation and mechanical efficiency can be evaluated for the different swimming modes used by mammals.

A model to examine paddling locomotion was developed by Blake (1979, 1980) and employed by Fish (1984, 1985) to investigate the energetics of surface-swimming semiaquatic rodents. The power phase of the paddling stroke of the muskrat and rice rat is characterized by a posterior acceleration of the hindfoot generating a thrust force due to the drag on the foot (Figs. 3.1, 3.4; Fish, 1984). Maximum thrust is realized when the hindfoot is oriented 90° to the horizontally inclined body (Fish, 1984). The jointed hindlimbs of the muskrat allow a paddle angle close to 90° for a large portion of the power phase (Webb and Blake, 1985). Thrust production is enhanced by an increase in plantar

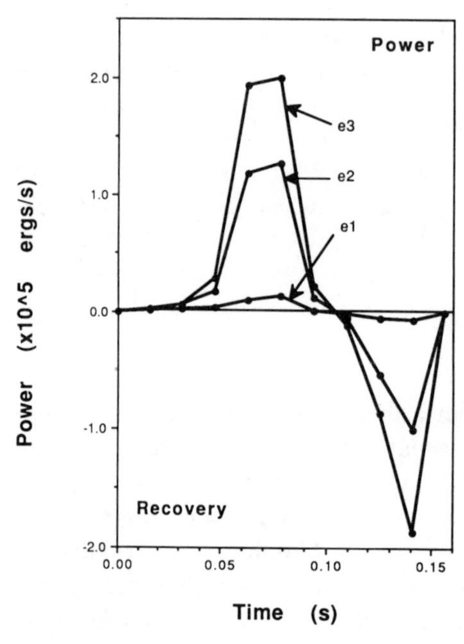

Fig. 3.4. Power output during a complete paddling cycle for the rice rat (*Oryzomys palustris*). Thrust power generated during the power phase and drag power expended during the recovery phase were computed from the model by Blake (1979, 1980). Power produced was determined for proximal (e1), middle (e2), and distal (e3) segments of the paddling hindfoot.

surface area of the feet by elongation of bony elements and addition of lateral fringe hairs on each digit or interdigital webbing (Fish, 1984; Howell, 1930; Mordvinov, 1976). The increase in effective surface area of the feet provides an economical generation of thrust, because the mass of water being worked on is also increased (Alexander, 1983). Mink, which have relatively unmodified feet for a semiaquatic mammal, compensate for deficiencies in thrust production by using long stroke lengths, high stroke frequencies, and quadrupedal swimming (Williams, 1983a). Long stroke lengths and high frequencies, however, are not economical; it is more economical to generate thrust by accelerating a large fluid mass to a small velocity than the converse (Alexander, 1983).

In contrast to the power phase, drag on the foot during the recovery phase is minimized. This action prevents a reduction of net thrust production as the foot is repositioned. In the muskrat, mean power loss attributed to the recovery phase represents 20–39% of thrust power generated during the power phase (Fish, 1984). This small power loss is accomplished by configural changes that reduce foot area by 55% and temporal changes that reduce the relative velocity, thereby minimizing the drag on the foot.

Additional energy losses accrue from paddling because of inertial effects of accelerating and decelerating the mass of the paddle and the added mass (Blake, 1979; Fish, 1984). The added mass is the mass of water entrained with the paddle as it moves (Blake, 1983b). Inertial effects are a major source of energy loss, particularly in small paddlers (Fish, 1984, 1985). In rice rats, acceleration of the paddle and added mass accounts for 31–53% of the total energy necessary for paddling.

Because of large energy expenditures incurred from the recovery phase and inertial effects, efficiency of the paddling mode is low (Webb and Blake, 1985). Mechanical or propeller efficiency (η_p) is calculated as the ratio of the energy utilized for thrust generation to the total energy expended during the paddling stroke. Maximum η_p values for paddling by muskrats and rice rats are 0.33 and 0.25, respectively (Fish, 1984, 1985). The higher η_p for the muskrat is due to enhanced thrust production by its modified hindfeet.

Unlike paddling mammals, which use the drag-based mechanism, otariids, cetaceans, and phocids use lift-based mechanisms to effect propulsion (Feldkamp, 1987a, 1987b; Fish et al., 1988; Lighthill, 1969; Webb, 1975a; Yates, 1983). This mechanism generates a large thrust force as a component of a lift force (Fig. 3.1) produced from the oscillatory movements of the hydrofoil (i.e., pectoral and pelvic flippers, caudal flukes). Shape and movements of the hydrofoil provide increased efficiency due to a high lift-to-drag ratio and continual generation of thrust throughout the entire stroke cycle.

Power output of the lift-based oscillatory swimming in mammals has not been computed by hydrodynamic models, but the kinematics and efficiency of the swimming mode for sea lions have been examined (English, 1976; Feldkamp, 1987a, 1987b; Godfrey, 1985). Feldkamp (1987b) found that *Zalophus* uses a combination of paddling and lift-based propulsion. The paddling component of the stroke occurs at the beginning of the recovery phase with the winglike, high aspect ratio (AR = span/chord) foreflippers. The remainder of the recovery phase (upstroke) and the power phase (downstroke) generate forces that resolve into downward and upward lift forces and anteriorly directed thrust forces. This mechanism has been likened to the stroke of flying birds (Feldkamp, 1987b).

Although up- and downstrokes produce thrust and represent the majority of the stroke cycle of sea lions, the paddling phase generates the greatest amount of thrust (Feldkamp, 1987b). Therefore, the three-phase system generates a large thrust force by paddling that is enhanced by winglike movements that incur less drag in a recovery phase. Additional drag reduction is affected by the morphology of the foreflippers. The high aspect ratio (AR = 7.9) reduces the formation of vortices, reducing the induced drag component caused by movements of the foreflippers (Feldkamp, 1987b). Consequently, the lift-to-drag ratio is improved, and thrust is generated more efficiently. Feldkamp (1987a) calculated a maximum η_p of 0.80 for the lift-based mode of the sea lion when swimming at the highest speeds.

In the undulatory lift-based mode of cetaceans and phocid seals, thrust is generated solely by a caudal hydrofoil represented by a caudal fluke or alternating hind flippers. The swimming motions of cetaceans and phocids are analogous to the thunniform mode of fish (Aleyev, 1977; Fish et al., 1988; Lindsey, 1978; Webb, 1975a), which use "lunate tail" propulsion (Lighthill, 1969). As in the thunniform mode, the presence of a double-jointed system at the hydrofoil base allows the angle of inclination of the hydrofoil to be adjusted throughout the stroke cycle, maintaining nearly continuous maximum thrust (Fish et al., 1988; Lindsey, 1978). In addition, moderate to high aspect ratio (AR = 3.4–5.5), low sweep-back angle, and flexibility of the hydrofoil enhance reduced drag with high thrust and efficiency.

Parry (1949) developed a model of undulatory swimming for dolphins based on quasistatic flow (Webb, 1975a). Using this model and data from Norris and Prescott (1961) and Lang and Daybell (1963), Webb (1975a) calculated the thrust power for three species of dolphin. The model thrust power was 6.3–16.0 times greater than the theoretical frictional drag power assuming turbulent conditions (Fig. 3.3). Webb (1975a) assumed that the calculated thrust power was not unreasonable if the dolphins were swimming near the surface, where drag is high.

A model developed by Lighthill (1970) uses unsteady wing theory to calculate the total thrust and power output by a rigid hydrofoil. Lighthill's (1970) model was used by Webb (1975a) to calculate a thrust power of 4030 W for a dolphin (*Lagenorhynchus obliguidens*) swimming at 5.5 m/s (Lang and Daybell, 1963). This value is 65% of the thrust power for the same dolphin calculated from Parry's (1949) model and 10.2 times the theoretical frictional drag power (Webb, 1975a). A revision of Lighthill's model (Chopra and Kambe, 1977) used for the dolphin (Chopra and Kambe, 1977; Yates, 1983) and phocids (Fish et al., 1988) also predicts thrust power greater than drag power at equivalent Re (Fig. 3.3).

Mechanical efficiencies of lunate tail propulsion are the highest for any swimming mode in mammals. Under optimal conditions, efficiency may be as high as 99% (Wu, 1971). For the dolphin, Webb (1975a) and Yates (1983) calculated efficiencies of 0.77 and 0.92, respectively, whereas phocids have efficiencies of 0.85 (Fish et al., 1988).

Power Input

Power input represents the rate of energy use that is potentially available to do work; it is limited proximately by metabolic capacities and ultimately by the availability of food resources (Hui, 1987). Power input for swimming mammals can be determined from estimates of metabolic rate and therefore is related to thrust and drag power outputs by Equation 1. Williams (1987) has reviewed the methodology for measuring swimming metabolism.

Active metabolic rates have been determined from measurements of oxygen consumption for both oscillatory and undulatory swimming mammals including muskrat (Fish, 1982; 1983), mink (Williams, 1983a), sea otter (Williams, 1989), sea lion (Costello and Whittow,

1975; Feldkamp, 1987a; Kruse, 1975), phocid seal (Craig and Pasche, 1980; Davis et al., 1985; Innes, 1984; Øritsland and Ronald, 1975; Williams et al., 1991), cetaceans (T. M. Williams, pers. comm., 1991; Worthy et al., 1987), and human (DiPrampero et al., 1974; Holmer, 1972; Nadel et al., 1974). Metabolic rate is directly related to swimming speed and increases linearly (Fish, 1982; Innes, 1984; Nadel et al., 1974) or curvilinearly (Davis, et al., 1985; Feldkamp, 1987a; Holmer, 1972; Nadel et al., 1974; Williams, 1983a). A steep increase in metabolic rate with swimming speed is associated with the high resistance caused by drag, because drag power output increases as U^3.

Feldkamp (1987a) noted that the cost of swimming for sea lions was less than for comparatively sized mammalian runners at the same speeds. He argued the reduced cost of swimming was attributed to the buoyant effect of water, which removed any energy expenditure for maintaining posture. However, metabolic studies of running and swimming mink showed the converse (Williams, 1983a, 1983b). At equivalent speeds of 0.7 m/s, the mass-specific metabolic rate for swimming mink was 1.6 times that for running and represented the maximum metabolic rate. Running mink attained the maximum rate at a speed nearly 1.0 m/s faster than swimming. This difference is not unexpected, because, despite their aquatic habits, mink are mainly terrestrial in design. Surface paddling by mink, as a compromise for amphibious behaviors, has limitations because of substantial energy losses due to surface effects (see below) and inefficiencies of the propulsive mode (Williams, 1983a).

Increased metabolic effort by anaerobic mechanisms above the aerobic capacity in swimming mammals has been suggested, although information on these mechanisms has been gathered only in experiments on diving (Kooyman, 1987, 1989). Fish (1982) suggested that increased power input to generate thrust at high surface speeds is supplied by anaerobic metabolism. Hui (1987) estimated that an 11 m/s burst of less than 2 s by a dolphin represents a 166-fold increase of the metabolism over resting rates when including the anaerobic contribution.

Extra energy expenditures are required during swimming to cope with thermoregulatory demands (Fish, 1983; MacArthur, 1984; Nadel et al., 1974). For muskrat, a 5°C decrease in water temperature below thermoneutrality can account for a 22% higher metabolic rate at the same swimming speed (Fish, 1983). Such an increase is due to the interaction of the conductivity of the water and convective effect from the velocity. Williams (1986) found the thermal conductance of the mink to increase with increasing swimming speed that resulted in heat loss exceeding metabolic heat production and a drop in core body temperature. A decrease in body temperature by as much as 3.1°C in free-swimming Weddell seals during both short-duration and long, exploratory dives (Kooyman, 1989) may indicate that heat loss to the water increases because of convection (Whittow, 1987). An elevated metabolic rate has been suggested as a mechanism to maintain homeothermy in aquatic mammals (Hampton and Whittow, 1976; Irving, 1971; Kanwisher and Sundnes, 1966; Whittow, 1987), but this assertion recently has been disputed. Seals and whales have been reported to have basal metabolic rates equivalent to rates predicted for terrestrial mammals (Lavigne et al., 1986; Worthy and Edwards, 1990; Yasui and Gaskin, 1986). The blubber layer provides sufficient insulation for homeothermy in water without an elevated metabolism. Indeed, overheating may be more of a problem in highly active large cetaceans (Brodie, 1975; Worthy and Edwards, 1990).

The conflicting energetic demands of diving with anoxic conditions and exercise suggest metabolic adjustments. Physiological responses by seals are graded according to dive mode (Guppy et al., 1986; Kooyman, 1987). Short feeding dives are considered to be aerobic, whereas longer exploratory dives display the classical dive response of energy conservation and anaerobic metabolism (Castellini et al., 1985; Guppy et al., 1986; Hochachka and Guppy, 1987). Typical dives are short in duration, so that aquatic mammals are within the aerobic dive limits (Dolphin, 1987; Estes, 1989; Feldkamp et al., 1989; LeBoeuf et al., 1986). This behavior comes into conflict during active swimming in that swimming near the surface encumbers increased drag and energy requirements (see above). Harbor seals can stay within their aerobic dive limits, remaining submerged for 82–92% of the time, when swimming under 1.2 m/s (Williams et al., 1991), but they decrease their submergence time during higher and more strenuous swimming speeds to maintain an aerobic, fat-based metabolism (Davis et al., 1991; Williams et al., 1991).

During diving, submerged swimming may be very energy-efficient when compared with surface swimming (Castellini, 1988; Hochachka and Guppy, 1987; Kooyman et al., 1973; Whittow, 1987). Field and laboratory studies of diving seals show depressed oxygen consumption rates compared with sustained exercise (Castellini et al., 1985). Low metabolic rates allow diving mammals to increase their dive time (Fedak et al., 1988; Kooyman et al., 1981). Costello and Whittow

(1975) concluded that the need to conserve oxygen during diving was larger than the high energetic demands of swimming. Although the sea otter displays an oxygen consumption during submerged swimming that is 41% lower than when surface swimming, hypometabolism is unlikely because these animals maintain their metabolism with oxygen supplied from the enlarged lungs without initiating the diving response (Kooyman, 1973; Williams, 1989). The lower metabolism of diving marine mammals differs from semiaquatic mammals such as the muskrat in which diving and underwater exercise incur an increase in the energetic expenditure above the resting metabolic rate and approach the energy expended during surface swimming (MacArthur and Krause, 1989).

The metabolic demands of swimming have been calculated using estimates of the hydrodynamic power output. Hui (1987) computed the total power input for a dolphin of the *Stenella-Delphinus* morphology based on the assumptions of a rigid-body analogy. His estimates of dolphin power input for routine and maximum swimming speeds were 1.0–3.4 and 13.4 times the resting metabolic rate, respectively. This result compares favorably with activity levels of aquatic and terrestrial mammals. However, similar calculations for phocid and otariid seals underestimated the power input when compared with measurements of oxygen consumption (Lavigne et al., 1982).

Association between metabolic rate and swimming speed is important in consideration of the ecology of aquatic mammals. Limitations due to hydrodynamics and energy metabolism will influence the swimming performance and behavior of the animal. Muskrats reach a limit in their aerobic capacity at 0.6 m/s and routinely swim at a slightly lower velocity (Fish, 1982). This behavior ensures that the muskrat can economically locomote without invoking an anaerobic metabolism and its associated oxygen debt. The low metabolic effort for harbor seals at routine swimming speed (1.4 m/s) would be advantagous during diving when energy conservation is critical, whereas the more economical but higher maximum range speed (2.2–2.3 m/s) is believed to be used in migrations or movement to food patches (Williams, 1987).

Aerobic Efficiency and Cost of Transport

Aerobic efficiency (η_a) is calculated as the ratio of power output to aerobically supplied power input and relates the thrust power to the active metabolic rate of a swimming animal. The η_a for swimming mammals is lower than the maximum value of 0.22 reported for fish (Webb, 1975b). Both otariid and phocid seals have the highest values of aerobic efficiency (0.12–0.30) for aquatic mammals, probably because of the streamlined pinniped body form and swimming modes (Feldkamp, 1987a; Innes, 1984; Williams et al., 1991). Because η_a, however, was computed using drag estimates that represent the minimum thrust required, values for seals could be higher if the power output was measured using thrust-based models. Peak values of η_a for surface paddlers show these animals to be less efficient than the submerged lift-based propulsion of seals. The η_a for muskrats (Fish, 1984), minks (Williams, 1983a), and humans (DiPrampero et al., 1974) are 0.046, 0.014, and 0.052, respectively.

The cost of transport (CT) is also used as a means of assessing the metabolic efficiency of locomotion (Schmidt-Nielsen, 1972; Tucker, 1970). CT is defined as the metabolic energy required to transport a unit mass a unit distance. Figure 3.5 shows minimum CT for various aquatic mammals as a function of body mass. The values of mammalian CT are compared with the regression for fish, which represents the lowest minimum CT for any animal or method of locomotion.

All aquatic mammals have values of minimum CT that are higher than for fish of equivalent sizes. Although higher maintenance costs associated with homeothermy could account for the difference of CT, high drag due to surface swimming and locomotory modes additionally would produce an elevated CT (Fish, 1982).

Surface-paddling muskrats (Fish, 1982), mink (Williams, 1983a), sea otters (Williams, 1989), and humans (P. E. DiPrampero, pers. comm., 1979) have the highest minimum CT for swimming mammals at 10–25 times the CT of fish of equivalent sizes. A lower minimum CT is attained by mammals that swim submerged and use more-efficient propulsive modes. The amount of oxygen used by humans to swim a unit distance is six to nine times that used by harbor seals (Craig and Pasche, 1980). The sea otter has a 40% lower minimum CT when swimming in a submerged undulatory mode than when surface paddling (Williams, 1989). Further decreases in the minimum CT are attained for lift-based oscillators and undulators that are only 1.9–4.6 times the minimum CT for a similar-sized fish (Costello and Whittow, 1975; Davis et al., 1985; Feldkamp, 1987a; Innes, 1984; Kruse, 1975). Schmidt-Nielsen (1972) predicted from hydrodynamic data and estimates of muscular efficiency that the minimum CT for a dolphin

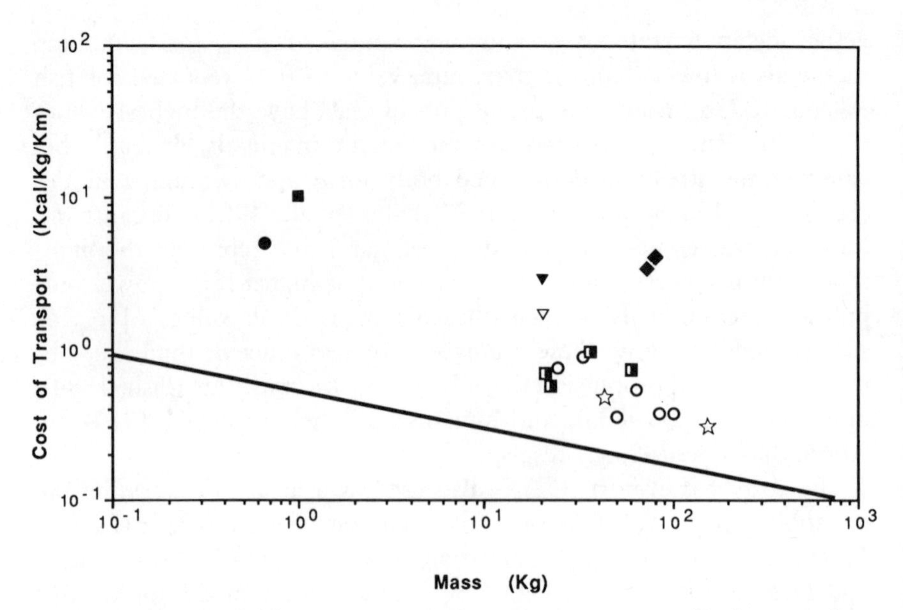

Fig. 3.5. Comparison of the minimum cost of transport over a range of body masses. Symbols represent mink (■: Williams, 1983), muskrat (●: Fish, 1982), sea otter (▼ and ∇: Williams, 1984), human (◆: P. E. DiPrampero, pers. comm., 1979), sea lion (◨: Costello and Whittow, 1975; Feldkamp, 1987a; Kruse, 1975), phocid seals (○: Davis et al., 1985; Innes, 1984), and dolphins (☆: T. M. Williams, pers. comm., 1991; Worthy et al., 1987). Closed symbols represent surface paddlers, open symbols represent submerged undulatory propulsors, and half symbols represent lift-based oscillators. The solid line is the minimum cost of transport extrapolated from data on fish (Davis et al., 1985).

should fall on the line for fish. However, there has been no controlled study that confirms this assertion. The CT calculated from the average metabolic rate of a 41.5-kg harbor porpoise swimming at approximately 2 m/s (Worthy et al., 1987) is still 2.5 times the minimum value, although the single estimate may not reflect the minimum CT of the porpoise. A similar assertion of low minimum CT for the sea lion (Lavigne et al., 1982; Luecke et al., 1975) also represents an underestimate from the metabolically derived CT (Feldkamp, 1987a).

Regardless of swimming mode, the minimum CT for many aquatic mammals has been found to coincide with their routine swimming speeds. During foraging bouts or migrations, this behavior would be economically advantagous because it minimizes energy expenditure while maximizing distance traveled. Fish (1982) found that muskrats swim at a speed within aerobic limits at the minimum CT. Dive velocities of otariids were observed to be equal to or less than the minimum CT velocity (Ponganis et al., 1990). Thus, for the available oxygen stores fur seals and sea lions are able to cover the greatest distance during the foraging dives and remain within the aerobic dive limits. Migrating gray whales (*Eschrichtius robustus*) have a minimum CT based on breathing rate that occurs at the mean velocity of 2.0 m/s (Sumich, 1983). Economical travel would be paramount for extending the stored energy reserves over a migration of 15,000–20,000 km, during which the whales fast.

Conclusions

Swimming performance (speed, acceleration, and endurance) is directly related to the ability to effectively use available energy resources in response to hydrodynamic requirements in the balance of thrust and drag (Weihs and Webb, 1983). For mammals, the evolutionary transition from terrestrial to semiaquatic and ultimately fully aquatic habits has allowed for increased swimming performance by abandoning the water surface and adopting low-drag body forms with changes in propulsive mode (Fig. 3.6). Metabolic and biomechanic studies of swimming show lift-based modes to have higher efficiency and higher performance levels than the paddling mode. The semiaquatic nature of most paddlers restricts the development of aquatic specialization for increased efficiency and performance. The vastly different environments for these mammals dictates concessions of locomotor agility. Highly aquatic mammals use efficient lift-based modes, which reduce drag and increase thrust. The oscillatory pectoral mode of otariids affords these animals greater high-speed maneuverability with a constant generation of thrust throughout the stroke cycle. Although different in orientation, the undulatory modes of cetaceans and phocids are analogous to the modes of piscine vertebrates for high efficiency, rapid propulsion. The convergence of swimming mode and low-drag body form indicate the importance of energetics in highly aquatic mammals. However, ecological and historical constraints limit the evolution of optimal designs, so that aquatic mammals represent compromises between form, function, and phylogeny.

The above review on swimming mammals, although demonstrating the energy relationship between active metabolic rate and power output for mammals and elucidating the physical causations of energy loss, reflects a largely incomplete picture. The size of most aquatic mammals, their availability, and problems associated with data collec-

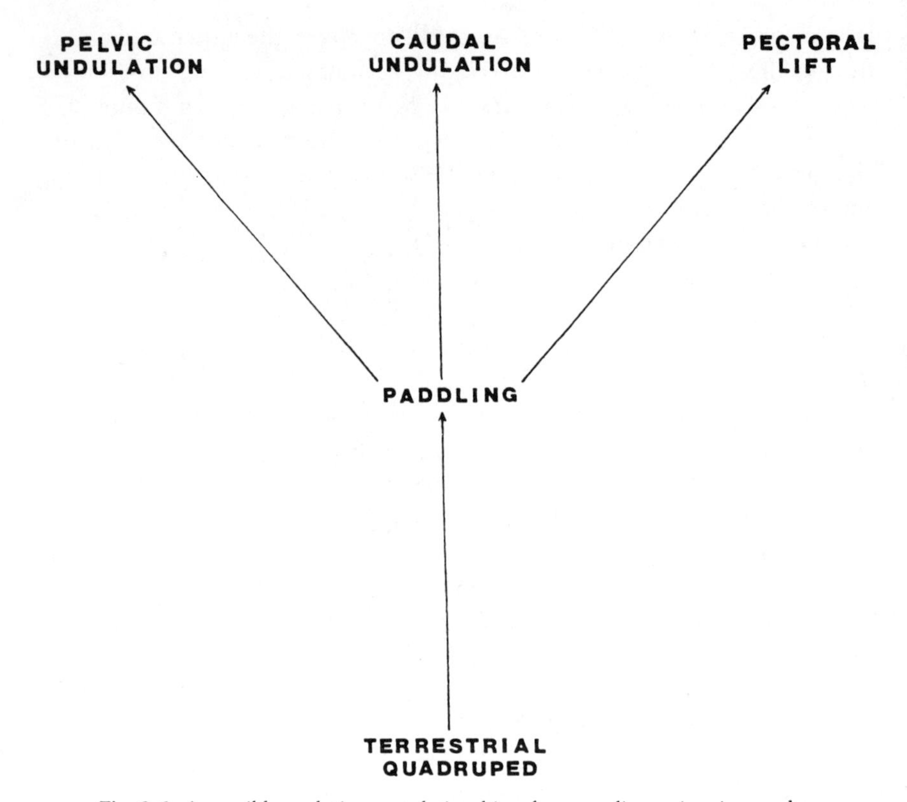

Fig. 3.6. A possible evolutionary relationship of mammalian swimming modes.

tion in water both in the field and laboratory have prevented a full examination of the evolutionary and ecological diversity over a complete range of performance levels. The basis for all future studies on swimming mammals must be reliable measurements and observations of performance, including speeds, accelerations, and maneuverability. Ecological and evolutionary questions can be addressed by examining a greater diversity of aquatic mammals than has previously been investigated. This diversity should include considerations of size, swimming mode, population variation, and variation between closely related species. Specifically, the cetaceans and sirenians should be targeted for examination because of their importance as highly derived aquatic mammals and the paucity of information on their active metabolism and mechanism of thrust production. The convergence of similar body designs, despite different activity levels and diets, may provide an understanding of similar physical constraints and evolutionary pathways of these two phylogenetically different groups. Finally, investigations of the evolutionary transition from terrestrial to semiaquatic to fully aquatic mammals should employ an experimental design in which direct comparisons of performance are measured in terrestrial and aquatic situations. Examinations of this type would provide insight to the compromises inherent in physiological and morphological adaptations that operate in two different physical environments.

ACKNOWLEDGMENTS

I am grateful to J. T. Beneski, H. L. Skelton, T. E. Tomasi, and an anonymous reviewer for comments on the manuscript. I also thank C. A. Hui, P. W. Webb, and T. M. Williams for their comments and discussions over the years, which have benefited this review.

Literature Cited

Alexander, R. McN. 1983. Animal Mechanics. Blackwell, Oxford. 301 pp.

Aleyev, Yu. G. 1977. Nekton. Junk, The Hague. 435 pp.

Au, D. 1980. Leaping dolphins. Letter to the editor. Nature, 287:759.

Au, D., and D. Weihs. 1980. At high speeds dolphins save energy by leaping. Nature, 284:548–550.

Backhouse, K. M. 1961. Locomotion of seals with particular reference to the forelimb. Symp. Zool. Soc. Lond., 5:59–75.

Barnes, L. G., D. P. Domning, and C. E. Ray. 1985. Status of studies on fossil marine mammals. Mar. Mamm. Sci., 1:15–53.

Best, T. L., and E. B. Hart. 1976. Swimming ability of pocket gophers (Geomyidae). Tex. J. Sci., 27:361–366.

Blake, R. W. 1979. The mechanics of labriform locomotion. I. Labriform locomotion in the angelfish (*Pterophyllum eimekei*): an analysis of the power stroke. J. Exp. Biol., 82:255–271.

——. 1980. The mechanics of labriform locomotion. II. An analysis of the recovery stroke and the overall fin-beat cycle propulsive efficiency in the angelfish. J. Exp. Biol., 85:337–342.

——. 1983a. Energetics of leaping dolphins and other aquatic animals. J. Mar. Biol. Assoc. U.K., 63:61–70.

——. 1983b. Fish locomotion. Cambridge University Press, London. 208 pp.

Bose, N., and J. Lien. 1990. Energy absorption from ocean waves: a free ride for cetaceans. Proc. R. Soc. Lond. B, 240:591–605.

Brodie, P. F. 1975. Cetacean energetics, an overview of intraspecific size variation. Ecology, 56:152–161.

Bryant, H. C. 1919. The coyote not afraid of water. J. Mammal, 1:87–88.

Castellini, M. A. 1988. Visualizing metabolic transitions in aquatic mammals: does apnea plus swimming equal "diving"? Can. J. Zool., 66:40–44.

Castellini, M. A., B. J. Murphy, M. Fedak, K. Ronald, N. Gofton, and P. W. Hochachka. 1985. Potentially conflicting metabolic demands of diving and exercise in seals. J. Appl. Physiol., 58:392–399.

Chanin, P. 1985. The Natural History of Otters. Facts on File Publ., New York. 179 pp.
Chopra, M. G., and T. Kambe. 1977. Hydrodynamics of lunate-tail swimming propulsion. Part 2. J. Fluid Mech., 79:49–69.
Cole, L. J. 1922. Red squirrels swimming a lake. J. Mammal., 3:53–54.
Costello, R. R., and G. C. Whittow. 1975. Oxygen cost of swimming in a trained California sea lion. Comp. Biochem. Physiol., 50:645–647.
Craft, T. J., M. I. Edmondson, and R. Agee. 1958. A comparative study of the mechanics of flying and swimming in some common brown bats. Ohio J. Sci., 58:245–249.
Craig, A. B., Jr., and A. Pasche. 1980. Respiratory physiology of freely diving harbor seals (*Phoca vitulina*). Physiol. Zool., 53:419–432.
Dagg, A. I., and D. E. Windsor. 1972. Swimming in northern terrestrial mammals. Can J. Zool., 50:117–130.
Daniel, T. L. 1984. Unsteady aspects of aquatic locomotion. Am. Zool., 24:121–134.
Daniel, T. L., and P. W. Webb. 1987. Physical determinants of locomotion. Pp. 343–369 *in* Comparative Physiology: Life in Water and on Land (P. Dejours, L. Bolis, C. R. Taylor, and E. R. Weibel, eds.). Liviana Press, New York. 556 pp.
Davis, R. W., M. A. Castellini, T. M. Williams, and G. L. Kooyman. 1991. Fuel homeostasis in the harbor seal during submerged swimming. J. Comp. Physiol. B, 160:627–635.
Davis, R. W., T. M. Williams, and G. L. Kooyman. 1985. Swimming metabolism of yearling and adult harbor seals *Phoca vitulina*. Physiol. Zool., 58:590–596.
DiPrampero, P. E., D. R. Pendergast, D. W. Wilson, and D. W. Hennie. 1974. Energetics of swimming in man. J. Appl. Physiol., 37:1–5.
Dolphin, W. F. 1987. Dive behavior and estimated energy expenditure of foraging humpback whales in southeast Alaska. Can. J. Zool., 65:354–362.
English, A. W. 1976. Limb movements and locomotor function in the California sea lion (*Zalophus californianus*). J. Zool. (Lond.), 178:341–364.
Essapian, F. S. 1955. Speed-induced skin folds in the bottle-nosed porpoise, *Tursiops truncatus*. Breviora Mus. Comp. Zool., 43:1–4.
Estes, J. A. 1989. Adaptations for aquatic living by carnivores. Pp. 242–282 *in* Carnivore Behavior, Ecology, and Evolution (J. L. Gittleman, ed.). Cornell University Press, Ithaca, New York. 624 pp.
Fedak, M. A., M. R. Pullen, and J. Kanwisher. 1988. Circulatory responses of seals to periodic breathing: heart rate and breathing during exercise and diving in the laboratory and open sea. Can. J. Zool., 66:53–60.
Feldkamp, S. D. 1987a. Swimming in the California sea lion: Morphometrics, drag and energetics. J. Exp. Biol., 131:117–135.
——. 1987b. Foreflipper propulsion in the California sea lion, *Zalophus californianus*. J. Zool. (Lond.), 212:43–57.
Feldkamp, S. D., R. L. DeLong, and G. A. Antonelis. 1989. Diving patterns of California sea lions, *Zalophus californianus*. Can. J. Zool., 67:872–883.
Fish, F. E. 1982. Aerobic energetics of surface swimming in the muskrat *Ondatra zibethicus*. Physiol. Zool., 55:180–189.
——. 1983. Metabolic effects of swimming velocity and water temperature in the muskrat (*Ondatra zibethicus*). Comp. Biochem. Physiol., 75:397–400.
——. 1984. Mechanics, power output and efficiency of the swimming muskrat (*Ondatra zibethicus*). J. Exp. Biol., 110:183–201.
——. 1985. Swimming dynamics of a small semi-aquatic mammal. Am. Zool., 25:13A.
——. 1987. Swimming mode changes associated with terrestrial-semiaquatic transition in mammals. Am. Zool., 27:86A.
Fish, F. E., S. Innes, and K. Ronald. 1988. Kinematics and estimated thrust production of swimming harp and ringed seals. J. Exp. Biol., 137:157–173.

Flyger, V., and M. R. Townsend. 1968. The migration of polar bears. Sci. Am., 218:108–116.

Fregin, G. F., and T. Nicholl. 1977. Swimming: its influences on heart rate, respiration rate, and some hematological values in the horse. J. Equine Med. Surg., 1:288–293.

Gaskin, D. E. 1982. The Ecology of Whales and Dolphins. Heinemann, London. 459 pp.

Gingerich, P. D., N. A. Wells, D. E. Russell, and S. M. I. Shah. 1983. Origin of whales in epicontinental remnant seas: new evidence from the early Eocene of Pakistan. Science, 220:403–406.

Godfrey, S. J. 1985. Additional observations of subaqueous locomotion in the California sea lion (*Zalophus californianus*). Aquat. Mamm., 11:53–57.

Gordon, C. N. 1980. Leaping dolphins. Letter to the editor. Nature, 287:759.

Gordon, K. R. 1981. Locomotor behaviour of the walrus (*Odobenus*). J. Zool. (Lond.). 195:349–367.

Gray, J. 1936. Studies in animal locomotion. VI. The propulsive powers of the dolphin. J. Exp. Biol., 13:192–199.

Guppy, M., R. D. Hill, R. C. Schneider, J. Qvist, G. C. Liggins, W. M. Zapol, and P. W. Hochachka. 1986. Microcomputer-assisted metabolic studies of voluntary diving of Weddell seals. Am. J. Physiol., 250:R175–R187.

Hampton, I. F. G., and G. C. Whittow. 1976. Body temperature and heat exchange in the Hawaiian spinner dolphin, *Stenella longirostris*. Comp. Biochem. Physiol., 55A:195–197.

Hart, J. S., and H. D. Fisher. 1964. The question of adaptations to polar environments in marine mammals. Fed. Proc., 23:1207–1214.

Hartman, D. S. 1979. Ecology and behavior of the manatee (*Trichechus manatus*) in Florida. Spec. Publ. Am. Soc. Mammal. No. 5. 153 pp.

Hertel, H. 1966. Structure, Form, Movement. Reinhold, New York. 251 pp.

——. 1969. Hydrodynamics of swimming and wave-riding dolphins. Pp. 31–63 *in* The Biology of Marine Mammals (H. T. Andersen, ed.). Academic Press, New York. 511 pp.

Hickman, G. C. 1983. Burrows, surface movement, and swimming of *Tachyoryctes splendens* (Rodentia: Rhizomyidae) during flood conditions in Kenya. J. Zool. (Lond.), 200:71–82.

——. 1984. Swimming ability of talpid moles, with particular reference to the semiaquatic *Condylura cristata*. Mammalia, 43:505–513.

Hochachka, P. W. 1980. Living without oxygen. Harvard University Press, Cambridge. 181 pp.

Hochachka, P. W., and M. Guppy. 1987. Diving mammals and birds. Pp. 36–56 *in* Metabolic Arrest and the Control of Biological Time (P. W. Hochachka and M. Guppy, eds.). Harvard University Press, Cambridge. 227 pp.

Holmer, I. 1972. Oxygen uptake during swimming in man. J. Appl. Physiol., 33:502–509.

Howell, A. B. 1930. Aquatic Mammals. Charles C Thomas, Springfield, Ill. 338 pp.

Hui, C. 1987. Power and speed of swimming dolphins. J. Mammal, 68:126–132.

——. 1989. Surface behavior and ventilation in free-ranging dolphins. J. Mammal, 70:833–835.

Innes, H. S. 1984. Swimming energetics, metabolic rates and hind limb muscle anatomy of some phocid seals. Ph.D. dissertation, University of Guelph, Ontario.

Irving, L. 1971. Aquatic mammals. Pp. 47–96 *in* Comparative Physiology of Thermoregulation. Vol. 3 (G. C. Whittow, ed.). Academic Press, New York. 278 pp.

Kanwisher, J., and G. Sundnes. 1966. Thermal regulation in cetaceans. Pp. 397–407 *in* Whales, Dolphins and Porpoises (K. S. Norris, ed.). University of California Press, Berkeley. 789 pp.

Kenyon, K. W. 1969. The sea otter in the eastern Pacific Ocean. N. Am. Fauna, 68:1–352.

Kermack, K. A. 1948. The propulsive powers of blue and fin whales. J. Exp. Biol., 25:237–240.

King, J. E. 1983. Seals of the World. Cornell University Press, Ithaca. 240 pp.

Kooyman, G. L. 1973. Respiratory adaptations of marine mammals. Am. Zool., 13:457–468.

——. 1985. Physiology without restraint in diving mammals. Mar. Mamm. Sci., 1:166–178.

——. 1987. A reappraisal of diving physiology: seals and penguins. Pp. 459–469 *in* Comparative Physiology: Life in Water and on Land. Vol. 9 (P. Dejours, L. Bolis, C. R. Taylor, E. R. Weibel, eds.). Fidia Research Series, Liviana Press, Padova. 556 pp.

——. 1989. Diverse Divers. Springer-Verlag, Berlin. 200 pp.

Kooyman, G. L., M. A. Castellini, and R. W. Davis. 1981. Physiology of diving in marine mammals. Annu. Rev. Physiol., 43:343–356.

Kooyman, G. L., D. H. Kerem, W. B. Campbell, and J. J. Wright, 1973. Pulmonary gas exchange in freely diving Weddell seals (*Leptonychotes weddelli*). Respir. Physiol., 17:283–290.

Kramer, M. O. 1960a. Boundary layer stabilization by distributing damping. J. Am. Soc. Nav. Eng., 72:25–33.

——. 1960b. The dolphin's secret. New Sci., 7:1118–1120.

——. 1965. Hydrodynamics of the dolphin. Pp. 111–130 *in* Advances in Hydroscience. Vol. 2 (V. T. Chow, ed.). Academic Press, New York.

Kruse, D. H. 1975. Swimming metabolism of California sea lions, *Zalophus californianus*. M.S. thesis, San Diego State University.

Kurbatov, B. V., and Yu. E. Mordvinov. 1974. Hydrodynamic resistance of semiaquatic mammals. Zool. Zh., 53:104–110.

Lang, T. G. 1966. Hydrodynamic analysis of cetacean performance. Pp. 410–432 *in* Whales, Dolphins and Porpoises (K. S. Norris, ed.). University of California Press, Berkeley. 789 pp.

Lang, T. G., and D. A. Daybell. 1963. Porpoise performance tests in a seawater tank. Nav. Ord. Test Stn., China Lake, Calif. NAVWEPS Rept. 8060. NOTS Tech. Publ. 3063.

Lang, T. G., and K. Pryor. 1966. Hydrodynamic performance of porpoises (*Stenella attenuata*). Science, 152:531–533.

Lavigne, D. M., W. Barchard, S. Innes, and N. A. Oritsland. 1982. Pinniped bioenergetics. FOA Fisheries Series No. 5, Mammals in Seas, 4:191–235.

Lavigne, D. M., S. Innes, G. A. J. Worthy, K. M. Kovacs, O. J. Schmitz, and J. P. Hickie. 1986. Metabolic rates of seals and whales. Can. J. Zool. 64:279–284.

LeBoeuf, B. J., D. P. Costa, A. C. Huntley, G. L. Kooyman, and R. W. Davis. 1986. Pattern and depth of dives in Northern elephant seals, *Mirounga angustirostris*. J. Zool. (Lond.), 208:1–7.

Lighthill, M. J. 1969. Hydrodynamics of aquatic animal propulsion. Annu. Rev. Fluid Mech., 1:413–446.

——. 1970. Aquatic animal propulsion of high hydromechanical efficiency. J. Fluid Mech., 44:265–301.

——. 1971. Large-amplitude elongated-body theory of fish locomotion. Proc. R. Soc. Lond. Ser. B. Biol. Sci., 179:125–138.

Lindsey, C. C. 1978. Form, function, and locomotory habits in fish. Pp. 1–100 *in* Fish Physiology: Locomotion. Vol. 7 (W. S. Hoar and D. J. Randall, eds.). Academic Press, New York. 576 pp.

Luecke, R. H., V. Natarajan, and F. E. South. 1975. A mathematical biothermal model of the California sea lion. J. Therm. Biol., 1:35–45.

MacArthur, R. A. 1984. Aquatic thermoregulation in the muskrat (*Ondatra zibethicus*): energy demands of swimming and diving. Can. J. Zool., 62:241–248.
MacArthur, R. A., and R. E. Krause. 1989. Energy requirements of freely diving muskrats (*Ondatra zibethicus*). Can J. Zool., 67:2194–2200.
Mordvinov, Yu. E. 1972. Some hydrodynamic parameters of body shape in Pinnipedia. Hydrobiol. J. 8:81–84.
——. 1974. The character of boundary layer in the process of swimming in the muskrat (*Ondatra zibethica*) and mink (*Mustela lutreola*). Zool. Zh., 53:430–435.
——. 1976. Locomotion in water and the indices of effectiveness of propelling systems for some aquatic mammals. Zool. Zh., 55:1375–1382.
Mordvinov, Yu. E., and B. V. Kurbatov. 1972. Influence of hair cover in some species of Phocidae upon the value of general hydrodynamic resistance. Zool. Zh., 51:242–247.
Nadel, E. R. 1977. Thermal and energetic exchanges during swimming. Pp. 91–119 *in* Problems with Temerature Regulation during Exercise (E. R. Nadel, ed.). Academic Press, New York. 141 pp.
Nadel, E. R., I. Holmer, U. Bergh, P. O. Astrand, and A. J. A. Stolwij. 1974. Energy exchanges of swimming man. J. Appl. Physiol., 36:465–471.
Nishiwaki, M., and H. Marsh. 1985. Dugong. Pp. 1–31 *in* Handbook of Marine Mammals. Vol. 3 (S. H. Ridgeway and R. Harrison, eds.). Academic Press, New York. 362 pp.
Norris, K. S., and J. H. Prescott. 1961. Observations on Pacific cetaceans of California and Mexican waters. Univ. Calif. Publ. Zool., 63:291–402.
Øritsland, N. A., and K. Ronald. 1975. Energetics of the free diving harp seal (*Pagophilus groenlandicus*). Rapp. P.-V. Reun. Cons. Int. Explor. Mer., 169:451–454.
Parry, D. A. 1949. The swimming of whales and a discussion of Gray's paradox. J. Exp. Biol., 26:24–34.
Petersen, C. G. J. 1925. The motion of whales during swimming. Nature, 116:327–329.
Ponganis, P. J., E. P. Ponganis, K. V. Ponganis, G. L. Kooyman, R. L. Gentry, and F. Trillmich. 1990. Swimming velocities in otariids. Can. J. Zool., 68:2105–2112.
Purves, P. E. 1963. Locomotion in whales. Nature, 197:334–337.
Purves, P. E., W. H. Dudok van Heel, and A. Jonk. 1975. Locomotion in dolphins. Part I. Hydrodynamic experiments on a model of the bottle-nosed dolphin, *Tursiops truncatus*, (Mont.). Aquat. Mamm., 3:5–31.
Ray, G. C. 1963. Locomotion in pinnipeds. Nat. Hist. (N.Y.), 72:10–21.
Ridgeway, S. H., and R. J. Harrison. 1981. Handbook of Marine Mammals. Vol. 1. The walrus, sea lions, fur seals and sea otter. Academic Press, London. 235 pp.
Rosen, M. W. 1963. Flow visualization experiments with a dolphin. Nav. Ord. Test. Stn., China Lake, Calif. NAVWEPS Rept. 8062, NOTS Tech. Publ. 3065.
Schmidt-Nielsen, K. 1972. Locomotion: energy cost of swimming, flying, and running. Science, 117:222–228.
Slijper, E. J. 1961. Locomotion and locomotory organs in whales and dolphins (Cetacea). Symp. Zool. Soc. Lond., 5:77–94.
Sokolov, V. 1960. Some similarities and dissimilarities in the structure of the skin among the members of the suborders Odontoceti and Mystacoceti (Cetacea). Nature, 185:745–747.
——. 1962. Adaptations of the mammalian skin to the aquatic mode of life. Nature, 195:464–466.
Sokolov, V., I. Bulina, and V. Rodionov. 1969. Interaction of dolphin epidermis with flow boundary layer. Nature 222:267–268.

Stein, B. R. 1981. Comparative limb myology of two opossums, *Didelphis* and *Chironectes*. J. Morphol., 169:113–140.

Sumich, J. L. 1983. Swimming velocities, breathing patterns, and estimated costs of locomotion in migrating gray whales, *Eschrichtius robustus*. Can. J. Zool., 61:647–652.

Talmage, R. V., and G. D. Buchanan. 1954. The armadillo (*Dasypus novemcinctus*). A review of its natural history, ecology, anatomy and reproductive physiology. Rice Inst. Pamph. Monogr. Biol., 41:1–135.

Tarasoff, F. J., A. Bisaillon, J. Pierard, and A. P. Whitt. 1972. Locomotory patterns and external morphology of the river otter, sea otter, and harp seal (Mammalia). Can. J. Zool., 50:915–929.

Tedford, R. H. 1976. Relationship of pinnipeds to other carnivores (Mammalia). Syst. Zool., 25:363–374.

Tucker, V. A. 1970. Energetic cost of locomotion in animals. Comp. Biochem. Physiol., 34:841–846.

Videler, J., and P. Kamermans. 1985. Differences between upstroke and downstroke in swimming dolphins. J. Exp. Biol., 119:265–274.

Vogel, S. 1981. Life in Moving Fluids. Willard Grant Press, Boston. 352 pp.

Walker, E. P. 1975. Mammals of the World. Johns Hopkins University Press, Baltimore. 1500 pp.

Webb, P. W. 1975a. Hydrodynamics and energetics of fish propulsion. Bull. Fish. Res. Board Can., 190:1–159.

——. 1975b. Efficiency of pectoral-fin propulsion of *Cymatogaster aggregata*. Pp. 573–584 *in* Swimming and Flying in Nature. Vol. 2 (T. Y. Wu, C. J. Brokaw, and C. Brennen, eds.). Plenum Press, New York. 1005 pp.

——. 1978. Hydrodynamics: nonscombrid fish. Pp. 189–237 *in* Fish Physiology: Locomotion. Vol. 7 (W. S. Hoar and D. J. Randall, eds.). Academic Press, New York. 576 pp.

——. 1984. Body form, locomotion and foraging in aquatic vertebrates. Am. Zool., 24:107–120.

Webb, P. W., and R. W. Blake. 1985. Swimming. Pp. 110–128 *in* Functional Vertebrate Morphology (M. Hildebrand, D. M. Bramble, K. F. Liem, and D. B. Wake, eds.). Harvard University Press, Cambridge. 429 pp.

Weihs, D., and P. W. Webb. 1983. Optimization of locomotion. Pp. 339–371 *in* Fish Biomechanics (P. W. Webb and D. Weihs, eds.). Praeger, New York. 398 pp.

Whittow, G. C. 1987. Thermoregulatory adaptations in marine mammals: interacting effects of exercise and body mass. A review. Mar. Mamm. Sci., 3:220–241.

Williams, T. M. 1983a. Locomotion in the North American mink, a semi-aquatic mammal. I. Swimming energetics and body drag. J. Exp. Biol., 103:155–168.

——. 1983b. Locomotion in the North American mink, a semi-aquatic mammal. II. The effect of an elongate body on running energetics and gait patterns. J. Exp. Biol., 105:283–295.

——. 1986. Thermoregulation of the North American mink during rest and activity in the aquatic environment. Physiol. Zool., 59:293–305.

——. 1987. Approaches for the study of exercise physiology and hydrodynamics in marine mammals. Pp. 127–145 *in* Approaches to Marine Mammal Energetics (A. C. Huntley, D. P. Costa, G. A. J. Worthy, and M. A. Castellini, eds.). Spec. Publ. Soc. Mar. Mammal., 1:1–253.

——. 1989. Swimming by sea otters: adaptations for low energetic cost locomotion. J. Comp. Physiol. A, 164:815–824.

Williams, T. M., and G. I. Kooyman. 1985. Swimming performance and hydrodynamic characteristics of harbor seals *Phoca vitulina*. Physiol. Zool., 58:576–589.

Williams, T. M., G. L. Kooyman, and D. A. Croll. 1991. The effect of submergence on heart rate and oxygen consumption of swimming seals and sea lions. J. Comp. Physiol. B, 160:637–644.
Williamson, G. R. 1972. The true body shape of rorqual whales. J. Zool. (Lond.), 167:277–286.
Worthy, G. A., and E. F. Edwards. 1990. Morphometric and biochemcial factors affecting heat loss in a small temperate cetacean (*Phocoena phocoena*) and a small tropical cetacean (*Stenella attenuata*). Physiol. Zool., 63:432–442.
Worthy, G. A. J., S. Innes, B. M. Braune, and R. E. A. Stewart. 1987. Rapid acclimation of cetaceans to an open-system respirometer. Pp. 115–126 *in* Approaches to Marine Mammal Energetics (A. C. Huntley, D. P. Costa, G. A. J. Worthy, and M. A. Castellini, eds.). Spec. Publ. Soc. Mar. Mammal., 1:1–253.
Wu, T. Y. 1971. Hydrodynamics of swimming propulsion. Part 2. Some optimum shape problems. J. Fluid Mech., 46:521–544.
Yasui, W. Y., and D. E. Gaskin. 1986. Energy budget of a small cetacean, the harbour porpoise, *Phocoena phocoena* (L.). Ophelia, 25:183–197.
Yates, G. T. 1983. Hydromechanics of body and caudal fin propulsion. Pp. 177–213 *in* Fish Biomechanics (P. W. Webb and D. Weihs, eds.). Praeger, New York. 398 pp.
Yurchenko, N. F., and V. V. Babenko. 1980. Stabilization of the longitundinal vortices by skin integuments of dolphins. Biophysics, 25:309–315.

Thomas E. Tomasi and Alan S. Gleit

4. The Allometry of Thyroxine Utilization Rates

The allometric relationship between basal metabolic rate (BMR) and body mass has received much attention; it was popularized by Kleiber (1932), expanded by proponents such as Brody (1945), Pearson (1947), Stahl (1967), and Dawson and Hulbert (1970), and recently reevaluated by Heusner (1982, 1985), Hayssen and Lacy (1985), Elgar and Harvey (1987), McNab (1988), and MacMillen and Hinds (this volume). The metabolic rates of some species, and of most animals for some time in their lives, do not fit the basic allometric relationship (because most measurements are actually resting metabolic rates, RMR is used for the rest of this chapter). The search for ecological explanations for the evolution of this residual variation in metabolism is well under way (McNab, this volume). In fact, several of the chapters in this volume address this variation, from an ecological point of view. Whatever the physical and evolutionary forces involved, however, selective pressures must work through the proximal (physiological) causes of these metabolic differences. The physiological determinants that mediate these metabolic differences have not been widely studied and have yet to be critically and comprehensively reviewed.

The primary physiological factor regulating RMR, at least intraspecifically, is the thyroid gland and its hormones, thyroxine (T_4) and triiodothyronine (T_3). Thyroid function, then, would be expected to be correlated with metabolism. Unfortunately, while a myriad of measurable parameters have been used to estimate thyroid function, most do

Thomas E. Tomasi: Department of Biology, Southwest Missouri State University, Springfield, Missouri 65804. Alan S. Gleit: Department of Computer Informations Systems, Southwest Missouri State University, Springfield, Missouri 65804.

not directly assess the magnitude of the thyroid hormone effect. A couple of techniques, however, provide reasonable measures of the effect of thyroid hormone (specifically T_4) on peripheral tissues, expressed as the rate of hormone use (ng/day). These are (1) the administration of exogenous T_4 to inhibit thyroidal secretion (Grosvenor and Turner, 1958) and (2) the radiothyroxine turnover technique (Gregerman and Crowder, 1963). The former technique requires certain assumptions as to hypothalamic regulation and has been reported to yield higher values in intraspecific comparisons (Gregerman and Crowder, 1963; Tomasi and Horwitz, 1987; Wills and Schindler, 1970). Therefore, it is of interest to ascertain whether or not there is a systematic and statistical difference between the two techniques. However, at least initially, both are included in our analysis under the term *thyroxine utilization rate* (T_4U).

Thyroxine utilization rates have been studied extensively in laboratory animals and in humans, but only infrequently in wild mammals under "natural" conditions. In the only previous studies to evaluate the relationship between the RMR and T_4U of wild mammals, Yousef and Johnson (1975) and Tomasi (1984) concluded that these parameters are correlated.

The purposes of this chapter are to review the available literature on thyroxine utilization rates of mammals and to analyze these data with respect to the allometric relationship between T_4U and body mass. In conjunction with this relationship, we reexamine the correlation between RMR and T_4U. Finally, we examine the available data for the methodological differences discussed above and for taxonomic patterns of T_4U.

T_4U Data

We carefully searched all available literature for studies with sufficient data to calculate T_4U (103 T_4U values from 49 studies). These data typically took one of two forms. In the "inhibition of T_4 secretion" studies ($N = 54$), data were already expressed as units of T_4 per day. Often, data were given per unit of body mass, which was negated by multiplying by body mass. When no body mass was given, a mean mass for the species was used as an estimate (making the assumption that the "per mass" T_4U is a reliable value for the entire mass range of the given group of animals; an implied assumption when the authors expressed their data in this manner).

In the "radiothyroxine turnover" studies (N = 49), data for serum T_4 concentration, volume of thyroxine distribution, and fractional thyroxine turnover were required to calculate T_4U. These data were often expressed as metabolic clearance and serum T_4 or as thyroxine pool size and fractional turnover. No distinction was made between studies in which these parameters were all measured on the same individuals and studies in which they were measured on different animals that were all exposed to the same treatment. As in the above case, data expressed per body mass were converted to per animal.

All T_4U values were converted to units of nanograms of thyroxine used per day (ng/day) for comparison. For studies in which animals received surgical, medicinal, or environmental treatments, data from the control animals were used in our interspecific comparisons. When seasonal data were given, summer values were taken to best approximate control values.

All control T_4U values judged to be scientifically sound are given in Table 4.1. Note that while 37 species representing seven mammalian orders are included, most of the measurements were made on laboratory or domestic animals. Few of the data are on species for which natural selection might be operating to optimize T_4U.

Analysis of Data

The relationship between control T_4U and body mass was assessed by means of linear regressions after variously transforming the data. The analysis in which the data best fit the regression, with a random symmetrical error, was taken to represent the mathematical relationship between T_4U and body mass. This data analysis assumes that a *single* underlying principle determines the T_4U for all mammals and that this relationship can be represented mathematically as a straight line (after transforming the data). We did not assume a priori that this is the case, but undertook the following interspecific comparison of T_4U rates to determine if such a relationship could be found that fit the available data.

Once the form of this regression was determined, we then weighted each datum, on the basis of sample size and variance reported in the respective studies (generalized least squares), to improve our estimate of regression parameters.

TABLE 4.1

Thyroxine utilization rates (T_4U) of mammals

Species	No. of individuals	Body mass (g)	T_4U (ng/day)	SE	Technique[a]	Reference
Marsupials						
Didelphis virginiana	4	3,000[b]	35,700	7,200	S	Bauman & Turner, 1966
Macropus eugenii	4	5,000[b]	22,850	1,400	T	Kaethner & Good, 1975
	4	5,000[b]	22,500	2,200	T	Kaethner & Good, 1975
Insectivores						
Sorex vagrans		5	251		T	Tomasi, 1984
Suncus murinus	5	23	437	39	S	Dryden et al., 1969
	5	34	544	78	S	Dryden et al., 1969
	16	35	536	25	S	Dryden et al., 1969
Scalopus aquaticus	2	83	1,452		S	Leach et al., 1962
	3	87	1,522		S	Leach et al., 1962
	6	109	2,136		S	Leach et al., 1962
Primates						
Macaca mulatta	4	3,500[b]	7,084	629	T	Stolte et al., 1966
	4	6,500[b]	13,475	1,400	T	Stolte et al., 1966
	7	4,600	6,950	437	T	Dowling et al., 1961
Homo sapiens	5	60,000[b]	82,800	9,000	T	Inada et al., 1975
	9	70,000[b]	82,544	8,470	T	Ingbar & Freinkel, 1955
	5	70,000[b]	136,000	6,708	T	Pittman et al., 1980
	7	70,000[b]	140,000	6,300	T	Suda et al., 1978
	6	70,000[b]	82,400	10,492	T	Pittman et al., 1971
	13	72,300	101,400	6,222	T	Bianchi et al., 1983
Rodents						
Ammospermophilus leucurus	13	76	813	91	S	Yousef & Johnson, 1975
Spermophilus tereticaudus	11	91	673	64	S	Yousef & Johnson, 1975
S. lateralis	13	170	2,295	170	S	Yousef & Johnson, 1975

TABLE 4.1—*Continued*

Species	No. of individuals	Body mass (g)	T_4U (ng/day)	SE	Technique[a]	Reference
S. tridecemlineatus	15	200[b]	3,020	300	T	Bauman & Anderson, 1970
S. richardsoni		451	965		T	Demeneix & Henderson, 1978
Perognathus longimembris	3	11	83	18	S	Yousef & Johnson, 1975
P. formosus	6	15	95	14	S	Yousef & Johnson, 1975
P. intermedius	3	18	144	25	S	Yousef & Johnson, 1975
Dipodomys merriami	14	39	347	94	S	Yousef & Johnson, 1975
D. microps	13	52	510	42	S	Yousef & Johnson, 1975
D. deserti	14	98	941	314	S	Yousef & Johnson, 1975
Reithrodontomys megalotis		12	56		T	Tomasi, 1984
Peromyscus maniculatus		18	162		T	Tomasi, 1984
	12	25[b]	362		S	Eleftheriou & Zarrow, 1962
	12	25[b]	188		S	Eleftheriou & Zarrow, 1962
	5	26	346	29	S	Yousef & Johnson, 1975
P. eremicus	6	21	218	21	S	Yousef & Johnson, 1975
Neotoma lepida	14	99	990	89	S	Yousef & Johnson, 1975
Mesocricetus auratus	46	90[b]	522	7	S	Bauman et al., 1968
	42	90[b]	522	7	S	Bauman et al., 1968
	10	119	260	11	T	Tomasi & Horwitz, 1987
Hamster (species?)		150[b]	930		S	Premachandra, 1962
Microtus montanus		29	424		T	Tomasi, 1984
Rattus norvegicus	8	134	1,890	190	T	Gregerman & Crowder, 1963
	10	159	2,230	260	T	Gregerman & Crowder, 1963
	13	187	2,250	180	T	Gregerman & Crowder, 1963
	10	197	2,000	140	T	Gregerman & Crowder, 1963
	10	266	2,800	140	T	Gregerman & Crowder, 1963
	14	320	3,850	340	T	Gregerman & Crowder, 1963
	22	326	3,750	130	T	Gregerman & Crowder, 1963

	10	335	4,040	190	T	Gregerman & Crowder, 1963
	41	339	5,460	390	T	Gregerman & Crowder, 1963
	15	453	4,410	220	T	Gregerman & Crowder, 1963
	6	187	1,470	258	S	Djososoebagio & Turner, 1964
	12	203	1,181	164	S	Djososoebagio & Turner, 1964
	6	207	1,834	236	S	Djososoebagio & Turner, 1964
	12	210	1,754	120	S	Djososoebagio & Turner, 1964
	12	216	2,030	199	S	Djososoebagio & Turner, 1964
	6	222	2,204	111	S	Djososoebagio & Turner, 1964
	12	222	1,605	209	S	Djososoebagio & Turner, 1964
		200[b]	2,280	80	T	Balsam & Sexton, 1975
		200[b]	5,460		S	Heroux & Brauer, 1965
	8	214	4,800	707	T	Yousef & Johnson, 1968
	8	424	3,400	283	T	Yousef & Johnson, 1968
	8	519	4,300	601	T	Yousef & Johnson, 1968
	54	215[b]	2,494	94	S	Bauman & Turner, 1967
	12	500[b]	5,000	159	S	Bauman & Turner, 1967
		220[b]	2,332	330	T	Thompson et al., 1977
	20	280[b]	3,640	308	S	Grosvenor & Turner, 1958
	25	300[b]	2,910	180	S	Anderson et al., 1961
	34	400[b]	3,720	200	S	Anderson et al., 1961
	5	379	2,822	151	T	DiStefano et al., 1982
	5	532	3,168	322	T	Jang & DiStefano, 1985
	7	387	1,990	300	T	Heroux & Petrovic, 1969
		587	4,901	153	T	Balsam & Leppo, 1974
Mus musculus	9	25	200	15	T	Wills & Schindler, 1970
	6	25	115	15	T	Wills & Schnidler, 1970
	6	27	186	11	T	Wills & Schindler, 1970
	16	30	414	51	S	Pipes et al., 1960
	26	30[b]	378	21	S	Wada et al., 1959
	10	30[b]	360	33	S	Wada et al., 1959
	9	30[b]	399	42	S	Wada et al., 1959
	14	35[b]	420	53	S	Wada et al., 1959
	13	35[b]	417	40	S	Wada et al., 1959
Cavia sp.		1,300[b]	6,240		S	Premachandra, 1962

TABLE 4.1—*Continued*

Species	No. of individuals	Body mass (g)	T_4U (ng/day)	SE	Technique[a]	Reference
Carnivores						
Canis familiaris		12,000[b]	29,400		T	Belshaw et al., 1974
Procyon lotor	18	10,000[b]	13,000	9,000	S	Bauman et al., 1965
	4	10,000[b]	8,000	8,000	S	Bauman et al., 1965
Felis domesticus	6	5,000[b]	28,000	2,450	T	Broome et al., 1987
Perissodactyls						
Equus caballus	17	453,600[b]	490,000	17,000	T	Irvine, 1967
E. asinus	2	106,000	240,000	13,500	T	El-Nouty et al., 1978
Artiodactyls						
Lama glama	2	115,000	250,800	15,100	T	El-Nouty et al., 1978
Rangifer tarandus		70,000[b]	178,700		T	Yousef & Luick, 1971
Bos taurus	37	350,000[b]	2,079,000		S	Pipes et al., 1963
	5	365,000[b]	1,533,000		S	Premachandra et al., 1958
	9	400,000[b]	1,760,000		S	Premachandra et al., 1958
	2	400,000[b]	1,320,000		S	Premachandra et al., 1958
	3	450,000[b]	1,305,000		S	Premachandra et al., 1958
	17	595,000	1,904,000	178,500	T	Yousef & Johnson, 1967
Capra hircus	4	33,400	422,000		T	Abdullah & Falconer, 1977
	4	35,000[b]	136,150	10,150	S	Flamboe & Reineke, 1959
	5	40,000[b]	87,200	21,200	S	Flamboe & Reineke, 1959
Ovis aries	4	51,800[b]	42,500		S	Henneman et al., 1955
	3	66,100	291,000	36,000	T	Fisher et al., 1972

[a]See text for explanation of techniques: S, secretion; T, turnover.
[b]Estimated body mass.

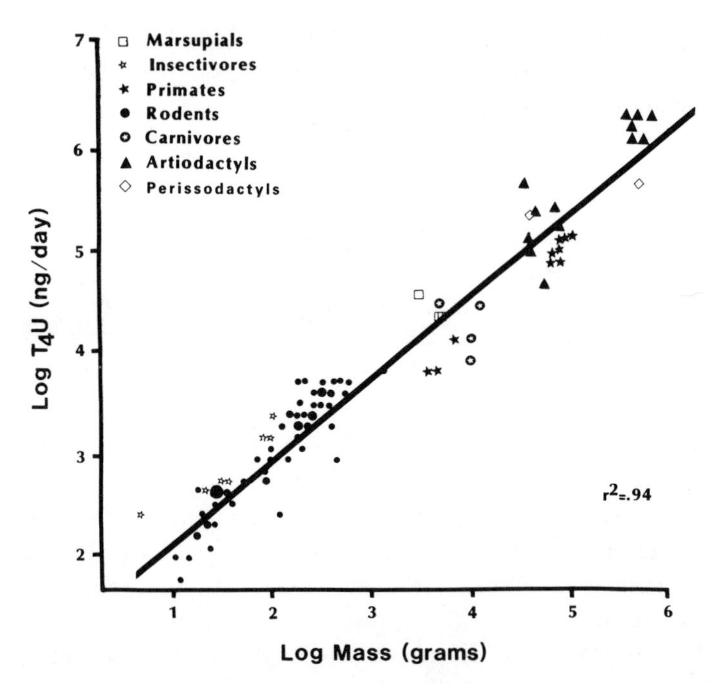

Fig. 4.1. Allometric relationship between thyroxine utilization rate (T_4U) and body mass of wild and domestic mammals (N = 103). Larger dots represent multiple data points. For the simple linear regression: $T_4U = 22.3(\text{mass})^{0.81}$ (or $\log T_4U = 0.81[\log M] + 1.348$: $r^2 = 0.94$). However, note that separate regressions drawn for small and large animals would differ significantly in both slope and intercept, indicating that these data are not truly linear despite the good fit of the linear model. This curvilinearity can be linearized by raising log M values to the 0.83 power, yielding the best mathematical equation: $\log T_4U = 1.183(\log M)^{0.83} + 0.886$ ($r^2 = 0.94$): see text for more details.

After the overall regression was calculated using 102 of the 103 values (because of the extremely high T_4U value for its body mass, *Sorex vagrans* was omitted as it would have had a disproportionately large effect on the regression), various subsets of the data were analyzed separately for comparison using analysis of covariance (ANCOVA).

Control data for all species (excluding *S. vagrans*) are shown in Figure 4.1 (log-log) and the regression for these data is:

$$\text{Log } T_4U = 0.81 (\log M) + 1.348$$

or

$$T_4U = 22.301 (M)^{0.81}$$

with mass (M) in grams ($r^2 = 0.94$). When adjusted for the sample size and variance of each datum (generalized least squares), the parameters for this equation become:

$$\text{Log } T_4U = 0.84 (\log M) + 1.202$$

or

$$T_4U = 15.921 (M)^{0.84}$$

Although the fit to this log-log regression is good ($r^2 = 0.94$), the data appear slightly curvilinear; for each X value, the data are not randomly distributed above and below the regression line. This curvilinearity can be most easily demonstrated by dividing the data into "large" and "small" species (> or < 250 g) and calculating the two separate regressions. If the entire data set yielded a truly linear regression, the two smaller regressions would yield approximately the same equation. However, in this case, the regression coefficient for large species (0.81: $r^2 = 0.91$) is significantly different ($P < 0.05$) from that for the small species (0.97: $r^2 = 0.80$). An additional transformation can be performed to remove this curvilinearity. In this case, raising the value (log M) to the power 0.83 yields the best linearization:

$$\text{Log } T_4U = 1.183 (\log M)^{0.83} + 0.886$$

Although the fit of the data to this regression ($r^2 = 0.94$) is no better than to the first regression, errors are now more randomly distributed.

This equation, adjusted for sample size and variance (generalized least squares), becomes:

$$\text{Log } T_4U = 1.227 (\log M)^{0.83} + 0.704$$

This mathematical expression best describes the available data.

When looking for taxonomic differences in these T_4U data, we had predicted that insectivores, which tend to have elevated metabolic rates, and heteromyid rodents, which have characteristically depressed RMR, would have relatively high and low T_4U rates, respectively. In addition, the T_4U data for primates appear to be relatively low. For these three groups, the mass range is sufficient to calculate and compare separate regressions (ANCOVA). When this is done, primates are indeed found to have significantly lower T_4U rates (unadjusted $P < 0.001$), and insectivores have significantly higher T_4U rates ($P < 0.002$) than the other orders taken as a group. The T_4U data for heteromyids approach but are not significantly different ($0.05 < P < 0.10$) from the rest of the data.

When data are separated by technique (inhibiting thyroidal secretion vs. radiothyroxine turnover), no significant difference is found either when primates and insectivores are excluded ($P > 0.5$) or included ($P > 0.3$). Even when considering only those species that were actually measured using both techniques ($N = 7$), no difference exists ($P > 0.5$).

For 32 of the species in Table 4.1, published values for RMR (BMR) are available (Table 4.2). Whenever possible, RMR values were taken from the same study from which the T_4U values were obtained. Otherwise, data were taken from recent reviews by Hayssen and Lacy (1985) and McNab (1988) or from Kleiber (1932) or Bruhn (1934). Each RMR was adjusted to the body mass used for T_4U determination (Table 4.1). If more than one line for that species was available in Table 4.1, the one most closely matching the mass reported in the RMR source was used. When log RMR (in milliliters of oxygen per hour) is regressed with log mass for these 32 species, the regression coefficient obtained is 0.77, with 98.75% of the variance in RMR explained by log mass. For these 32 species, 94.56% of the variance in log T_4U (in nanograms per day) is also explained by log mass. When log RMR is regressed with log T_4U, the regression is highly significant (Fig. 4.2; $P < 0.001$: $r^2 = 0.95$). The regression coefficient (0.90) is significantly less than 1.0 ($P < 0.05$). However with the exclusion of one species, *Bos taurus*, this difference is lost.

Despite the statistical significance of the log RMR and log T_4U linear regression, a multiple regression using both log mass and log T_4U as independent variables does not explain a significantly greater portion of the error variance in log RMR; r^2 is improved from 0.9875 to 0.9879. In addition, the residuals of log T_4U and log RMR (each regressed with log mass) are not statistically correlated ($r^2 = 0.09$: $0.05 < P < 0.10$).

The data analyzed in this chapter are limited to control conditions. However, since reproduction is an important theme in this volume, a few observations of how T_4U rates are affected by reproductive status are in order (Tomasi, 1991). In rodent studies, T_4U was elevated during pregnancy (Wills and Schindler, 1970) and lactation (Grosvenor

TABLE 4.2
Mean resting metabolic rates (RMR) of mammals for which throxine utilization rates are known, adjusted to the body masses in Table 4.1

Species	RMR (ml O_2/h)	Reference
Marsupials		
Didelphis virginiana	1,140	Hayssen & Lacy, 1985
Macropus eugenii	1,415	Hayssen & Lacy, 1985
Insectivores		
Sorex vagrens	20	Tomasi, 1984
Suncus murinus	64	Hayssen & Lacy, 1985
Scalopus aquaticus	137	Hayssen & Lacy, 1985
Primates		
Macaca mulatta	2,355	Bruhn, 1934
Homo sapiens	13,192	Kleiber, 1932
Rodents		
Ammospermophilus leucurus	79	Yousef & Johnson, 1975
Spermophilus tereticaudus	60	Yousef & Johnson, 1975
S. lateralis	242	Youself & Johnson, 1975
S. tridecemlineatus	114	McNab, 1988
S. richardsoni	289	Hayssen & Lacy, 1985
Perognathus longimembris	16	Yousef & Johnson, 1975
P. formosus	18	Yousef & Johnson, 1975
P. intermedius	24	Yousef & Johnson, 1975
Dipodomys merriami	42	Yousef & Johnson, 1975
D. microps	55	Yousef & Johnson, 1975
D. deserti	85	Yousef & Johnson, 1975
Reithrodontomys megalotis	23	Tomasi, 1984
Peromyscus maniculatis	31	Tomasi, 1984
P. eremicus	27	Yousef & Johnson, 1975
Neotoma lepida	74	Yousef & Johnson, 1975
Mesocricetus auratus	102	Tomasi & Horwitz, 1987
Microtus montanus	55	Tomasi, 1984
Rattus norvegicus	259	Kleiber, 1932
Cavia porcellus	715	McNab, 1988
Carnivores		
Canis familiaris	3,510	Kleiber, 1932
Perissodactyls		
Equus asinus	38,096	El-Nouty et al., 1978
Artiodactyls		
Lama glama	26,496	El-Nouty et al., 1978
Rangifer tarandus	21,980	Hayssen & Lacy, 1985
Bos taurus	50,018	Kleiber, 1932
Ovis aries	11,967	Kleiber, 1932

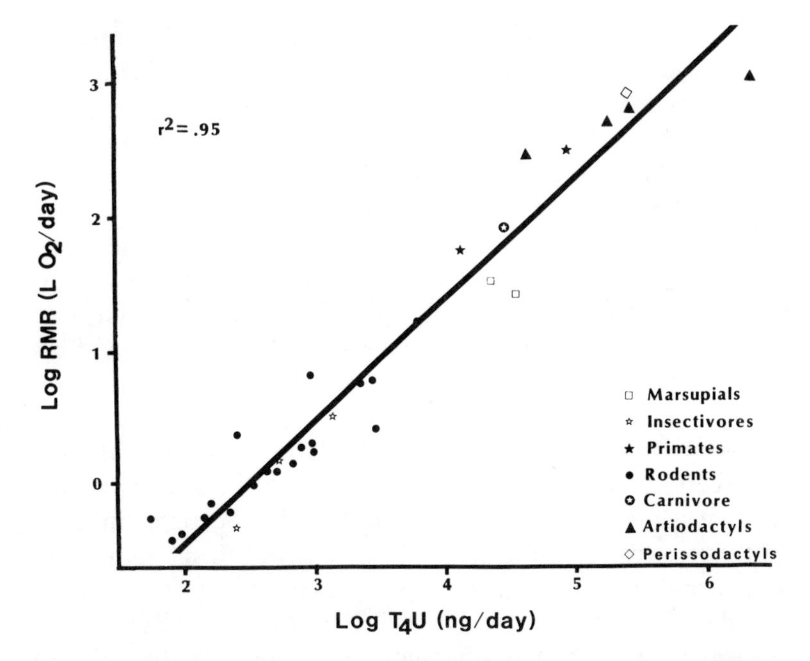

Fig. 4.2. Allometric relationship between resting metabolism (RMR) and thyroxine utilization (T_4U) of mammals for which both are known (N = 32). RMR = $0.25(T_4U)^{0.90}$.

and Turner, 1958) but was depressed by the removal of gonads (Wada et al., 1959). Pregnant or lactating sheep tended to have elevated rates of T_4U, but these data are confounded by seasonal changes in T_4U (Henneman et al., 1955). Conversely, lactating goats had either depressed rates of T_4U or showed no change, once the effect of season could be eliminated (Flamboe and Reineke, 1959). In pregnant monkeys, T_4U was elevated in only one of three groups (Dowling et al., 1961; Stolte et al., 1966). Obviously, more studies are needed before we can determine whether or not there is a "basic mammalian pattern" for T_4U that corresponds to reproductive status and the associated energy demands.

Interpretation of Data

When data derived from studies using the radiothyroxine turnover technique were compared with data from studies that inhibited thyroidal secretion, no significant difference was found. This finding validates the pooling of data from both techniques for allometric analysis. However, residual variation in the data and the small number of species measured using both techniques make this conclusion only tentative.

Since metabolic data are traditionally expressed allometrically (i.e., as a power function of body mass) this approach was also used for the T_4U data. In this way, similarities between metabolic and T_4U data might become evident, and a relationship between these two parameters may be suggested using the entire T_4U data set. Given the wide range of species, techniques, and experimental conditions used by various investigators, the fit of T_4U values to the allometric regression ($r^2 = 0.94$) is very good. Even more satisfying is that the allometric relationship (regression coefficient = 0.81) is similar (slightly higher) to that reported for the allometry of RMR. (While it is appropriate to apply a generalized least squares analysis to the data, in order to weight values from the literature on the basis of their sample size and variance, the weighted regression parameters are not used when comparing the allometric equations for T_4U and RMR because the generalized least squares analysis has not been applied to the latter when those data have been reviewed.) This finding is consistent with the hypothesis that T_4U rate is the primary physiological determinant of RMR in mammals and extends the conclusions of Yousef and Johnson (1975) and Tomasi (1984).

The best *mathematical* fit of the T_4U and mass data, however, is not a simple power function; the allometric relationship is not truly linear. When the data are linearized, the resultant equation describes the data better, but takes a more complicated form. It is not known what, if any, *biological* significance can be attributed to this equation, but it has predictive value because it is based on empirical data. Similarly, the RMR/mass allometric equation is often used for prediction, without any consensus on whether the nature of this relationship has any biological meaning.

Since body mass does not explain all of the variation in RMR or T_4U, other variables may be important in explaining the residual variation. Several groups of mammals have metabolic rates that do not adhere to the general relationship with mass (Dawson and Hulbert, 1970; McNab, 1966, 1979, 1980, 1983, 1988; Vogel, 1980; Yousef and Johnson, 1975), and factors such as taxonomic status, fossoriality, diet, morphology, and use of torpor have been suggested as *ecological* factors to explain some of the residual variation in RMR. On the basis of our analysis, at least taxonomic status appears to be a factor in T_4U as well. The analysis of other ecological effects on T_4U is not within the scope of this chapter.

Yousef and Johnson (1975) concluded, using 12 rodent species from the western United States, that a low T_4U served as an adaptation to living in the desert, presumably related to the low RMR usually noted in desert mammals (Hayssen and Lacy, 1985; McNab, 1979; Yousef and Johnson, 1975). Tomasi (1984) also noted a correlation between RMR and T_4U. When we analyze the 32 mammal species that have had both RMR and T_4U measured, the same pattern is evident (Fig. 4.2). Since the regression coefficient (log-transformed data) is indistinguishable from unity, and the relationship is expected to include the origin, these two parameters (untransformed data) appear to be proportional. Although more data are needed, the observation of a direct regression for these 32 species is consistent with the conclusion based on separate regressions of these two parameters with body mass. It would appear that "hypometabolic" and "hypermetabolic" species tend to be "hypothyroid" and "hyperthyroid," respectively, as measured by T_4U. Put another way, the species that deviate from the RMR allometric relationship seem to do so (physiologically) because their T_4U is elevated or depressed.

However, the analysis of residuals (log RMR and log T_4U on log mass) indicates that there is no direct relationship between T_4U and RMR in these 32 species, once the effect on mass is removed. Despite this statistical suggestion that T_4U explains RMR differences only because both are correlated with mass, this conclusion must be viewed as tentative for two reasons. First, in our analysis, RMR is so highly correlated with mass ($r^2 = 0.9875$) that there is little variance left to be explained by T_4U. In data sets with more residual variation in RMR, this may not be the case. McNab (1988) and Hayssen and Lacy (1985) reported that the variation in metabolism explained by body mass is only 0.78 and 0.75, respectively, when the data set is not limited as in our analysis ($N = 321$ and 293, respectively). Our 32 species are apparently not representative of mammals in general with respect to interspecific differences in metabolism. This may be due to the preponderance of laboratory and domestic species in our analysis, the same problem that beset early analyses of basal metabolism (see McNab, this volume).

Second, myriad intraspecific studies have shown that changes in thyroid function can be associated with changes in RMR, independent of mass. For example, Tomasi and Horwitz (1987) showed an increase in T_4U and RMR for hamsters acclimated to the cold, despite a decrease in body mass.

The observation that the Insectivora appear to possess rates of T_4U that are higher than other mammals (Fig. 4.1) is not unexpected as they (especially the shrews) have relatively high metabolic rates as well (Hayssen and Lacy, 1985; Vogel, 1980). For their T_4U rates, their RMRs are not excessive (Fig. 4.2). Similarly, while the heteromyid T_4U rates are not statistically distinct (Table 4.1), they are low enough that the low RMRs seen in this group of rodents appear to be appropriate (Fig. 4.2).

Conversely, primates (including humans) have lower T_4U rates than one would expect on the basis of their mass (Fig. 4.1). Since (simian) primates are not hypometabolic (Hayssen and Lacy, 1985; Muller 1985) their RMRs appear to be somewhat high relative to their T_4U (Fig. 4.2). With the exception of cattle (*Bos taurus*), ungulates (Perissodactyla and Artiodactyla) also appear to have slightly higher RMR values than would be expected on the basis of their T_4U rates (Fig. 4.2). Their RMRs also tend to be somewhat elevated relative to mass (Hayssen and Lacy, 1985). The two marsupials in this study appear to have relatively low RMRs for their T_4U rates, although their T_4U is appropriate for their mass (Fig. 4.1). From this, it follows that marsupial mass-specific RMRs would be depressed, and this has indeed been documented (Hayssen and Lacy, 1985). Although sample sizes are small, these latter three groups of mammals do not support the correlation between T_4U and RMR and no doubt contribute to the conclusion from residuals analysis that T_4U and RMR are not directly correlated, once the effect of body mass is removed.

Several explanations can be hypothesized for why the expected (and "apparent") correlation between T_4U and RMR was not obtained. Statistically, our interspecific analysis assumes that each species represents an independent observation. Yet there are varying degrees of relatedness between species, especially if there are several levels of taxonomic differences present (Felsenstein, 1985). Therefore, the assumption of independent observations is violated, and this may affect the outcome of the analyses.

Physiologically, tissue sensitivity to thyroxine may vary between species; primates and ungulates may have a greater cellular response to a given amount of thyroid hormone, while marsupials may be less responsive. Alternatively, these groups could have some (unknown) factor, other than thyroxine, that is contributing to the setting of RMR.

A fourth possible explanation for the lack of a direct correlation between rates of T_4U and RMR is that the fraction of thyroxine that is activated in vivo may differ. Thyroxine that is not activated is destroyed (wasted), but still is included in the calculated "utilization." This activation is performed in certain peripheral tissues via the deiodination of thyroxine (T_4) to triiodothyronine (T_3). As T_3 is actually the "active" thyroid hormone, the utilization rate of T_3 (T_3U) is a better indicator of thyroid physiology (Tomasi and Horwitz, 1987). T_4U values are good estimates only to the extent that the fraction of T_4 converted to T_3 is constant for all animals. Unfortunately, T_3U rates have been published for only eight species (including humans) to date (Tomasi, 1991); only three are wild species (El-Nouty et al., 1978). Until such time as sufficient values are available for wild species, a detailed allometric analysis of T_3U cannot be performed. We hope that sometime in the not-too-distant future, an analysis similar to that presented above will be possible for T_3.

AKNOWLEDGMENTS

We are grateful to M. Yousef for reading an earlier draft and encouraging us to pursue this review, and to D. Mitchell for assistance in figure and manuscript preparation.

Literature Cited

Abdullah, R., and I. R. Falconer. 1977. Responses of thyroid activity to feed restriction in the goat. Aust. J. Biol. Sci., 30:207–215.

Anderson, R. R., J. A. Grossie, and C. W. Turner. 1961. Effect of adrenalectomy and successive hydrocortisone replacement on thyroid secretion rate in male and female rats. Proc. Soc. Exp. Biol. Med., 107:571–573.

Balsam, A., and L. E. Leppo. 1974. Augmentation of the peripheral metabolism of L-triiodothyronine and L-thyroxine after acclimation to cold. J. Clin. Invest., 53:980–987.

Balsam, A., and F. C. Sexton. 1975. Increased metabolism of iodothyronines in the rat after short-term cold adaptation. Endocrinology, 97:385–391.

Bauman, T. R., and R. R. Anderson. 1970. Thyroid activity of the ground squirrel (*Citellus tridecemlineatus*) using a cannula technique. Gen. Comp. Endocrinol., 15: 369–373.

Bauman, T. R., R. R. Anderson, and C. W. Turner. 1968. Thyroid hormone secretion rates and food consumption of the hamster (*Mesocricetus auratus*) at 25.5° and 4.5°C. Gen. Comp. Endocrinol., 10:92–98.

Bauman, T. R., F. W. Clayton, and C. W. Turner. 1965. The L-thyroxine secretion rate, L-triiodothyronine equivalent, and biological half-life ($t_{1/2}$) of L-thyroxine-^{131}I in the raccoon (*Procyon lotor*). Gen. Comp. Endocrinol., 5:261–266.

Bauman, T. R., and C. W. Turner. 1966. L-thyroxine secretion rates and L-triiodothyronine equivalents in the opossum (*Didelphis virginianus*). Gen. Comp. Endocrinol., 6:109–113.

——. 1967. The effect of varying temperatures on thyroid activity and the survival of rats exposed to cold and treated with L-thyroxine or corticosterone. J. Endocrinol., 37:355–359.

Belshaw, B. E., M. Barandes, D. V. Becker, and M. Berman. 1974. A model of iodine kinetics in the dog. Endocrinology, 95:1078–1093.

Bianchi, R., G. Mariani, N. Molea, F. Vitek, F. Cazzuola, A. Carpi, N. Mazzuca, and M. G. Toni. 1983. Peripheral metabolism of thyroid hormones in man. I. Direct measurement of the conversion rate of thyroxine to 3,5,3′-triiodothyronine (T_3) and determination of the peripheral and thyroidal production of T_3. J. Clin. Endocrinol. & Metab., 56:1152–1163.

Brody, S. 1945. Bioenergetics and Growth. Reinhold, New York. 1023 pp.

Broome, M. R., M. T. Hays, and J. M. Turrel. 1987. Peripheral metabolism of thyroid hormones and iodide in healthy and hyperthyroid cats. Am. J. Vet. Res., 48:1286–1289.

Bruhn, J. M. 1934. The respiratory metabolism of infrahuman primates. Am. J. Physiol., 110:477–484.

Dawson, T. J., and A. J. Hulbert. 1970. Standard metabolism, body temperature, and surface areas of Australian marsupials. Am. J. Physiol., 218:1233–1238.

Demeneix, B. A., and N. E. Henderson. 1978. Thyroxine metabolism in active and torpid ground squirrels, *Spermophilus richardsoni*. Gen. Comp. Endocrinol., 35:86–92.

DiStefano, J. J., III, T. K. Malone, and M. Jang. 1982. Comprehensive kinetics of thyroxine distribution and metabolism in blood and tissue pools of the rat from only six blood samples: dominance of large, slowly exchanging tissue pools. Endocrinology, 111:108–117.

Djojosoebagio, S., and C. W. Turner. 1964. Effects of parathyroid extract, calciferol, hytakerol and dihydrotachysterol upon thyroid secretion rate in normal female rats. Proc. Soc. Exp. Biol. Med., 116:1099–1102.

Dowling, J. T., D. L. Hutchinson, W. R. Hindle, and C. R. Kleeman. 1961. Effects of pregnancy on iodine metabolism in the primate. J. Clin. Endocrinol. & Metab., 21:779–791.

Dryden, G. L., T. R. Bauman, C. H. Conaway, and R. R. Anderson. 1969. Thyroid hormone secretion rate and biological half-life ($t_{1/2}$) of L-thyroxine-^{131}I in the musk shrew (*Suncus murinus*). Gen. Comp. Endocrinol., 12:536–540.

Eleftheriou, B. E., and M. X. Zarrow. 1962. Seasonal variation in thyroid gland activity in deermice. Proc. Soc. Exp. Biol. Med., 110:128–131.

Elgar, M. A., and P. H. Harvey. 1987. Basal metabolic rates in mammals: allometry, phylogeny and ecology. Funct. Ecol., 1:25–36.

El-Nouty, F. D., M. K. Yousef, A. B. Magdub, and H. D. Johnson. 1978. Thyroid hormones and metabolic rate in burros, *Equus asinus*, and llamas, *Lama glama*: effects of environmental temperature. Comp. Biochem. Physiol., 60A:235–237.

Felsenstein, J. 1985. Phylogenies and the comparative method. Am. Nat., 125:1–15.

Fisher, D. A., I. J. Chopra, and J. H. Dussault. 1972. Extrathyroidal conversion of thyroxine to triiodothyronine in sheep. Endocrinology, 91:1141–1144.

Flamboe, E. E., and E. P. Reineke. 1959. Estimation of thyroid secretion rates in dairy goats and measurement of I^{131} uptake and release with regard to age, pregnancy, lactation, and season of the year. J. Anim. Sci., 18:1135–1148.

Gregerman, R. I., and S. E. Crowder. 1963. Estimation of thyroxine secretion rate in the rat by the radioactive thyroxine turnover technique: influences of age, sex and exposure to cold. Endocrinology, 72:382–392.

Grosvenor, C. E., and C. W. Turner. 1958. Effect of lactation upon thyroid secretion rate in the rat. Proc. Soc. Exp. Biol. Med., 99:517–519.

Hayssen, V., and R. C. Lacy. 1985. Basal metabolic rates in mammals: taxonomic differences in the allometry of BMR and body mass. Comp. Biochem. Physiol., 81A: 741–754.

Henneman, H. A., E. P. Reineke, and S. A. Griffin. 1955. The thyroid secretion rate of sheep as affected by season, age, breed, pregnancy, and lactation. J. Anim. Sci., 14:419–434.

Heroux, O., and R. Brauer. 1965. Critical studies on determination of thyroid secretion rate in cold-adapted animals. J. Appl. Physiol., 20:597–606.

Heroux, O., and V. M. Petrovic. 1969. Effects of high and low bulk diets on the thyroxine turnover rate in rats with acute and chronic exposure to different temperatures. Can. J. Physiol. Pharmacol., 47:963–968.

Heusner, A. A. 1982. Energy metabolism and body size. II. Dimensional analysis and energetic non-similarity. Respir. Physiol., 48:13–25.

——. 1985. Body size and energy metabolism. Annu. Rev. Nutr., 5:267–293.

Inada, M., K. Kasagi, S. Kurata, Y. Kazama, H. Takayama, K. Torizuka, M. Fukase, and T. Soma. 1975. Estimation of thyroxine and triiodothyronine distribution and of the conversion rate of thyroxine to triiodothyronine in man. J. Clin. Invest., 55: 1337–1348.

Ingbar, S. H., and N. Freinkel. 1955. Simultaneous estimation of rates of thyroxine degradation and thyroid hormone synthesis. J. Clin. Invest., 34:808–819.

Irvine, C. H. G. 1967. Thyroxine secretion rate in the horse in various physiological states. J. Endocrinol., 39:313–320.

Jang, M., and J. J. DiStefano, III. 1985. Some quantitative changes in iodothyronine distribution and metabolism in mild obesity and aging. Endocrinology, 116:457–468.

Kaethner, M. M., and B. F. Good. 1975. Seasonal thyroid activity in the tammar wallaby, *Macropus eugenii* (Desmarest). Aust. J. Zool., 23:363–369.

Kleiber, M. 1932. Body size and metabolism. Hilgardia, 6:315–353.

Leach, B. J., T. R. Bauman, and C. W. Turner. 1962. Thyroxine secretion rate of Missouri Valley mole, *Scalopus aquaticus*. Proc. Soc. Exp. Biol. Med., 110:681–682.

McNab, B. K. 1966. The metabolism of fossorial rodents: a study of convergence. Ecology, 47:712–733.

——. 1979. Climatic adaptation in the energetics of heteromyid rodents. Comp. Biochem. Physiol., 62A:813–820.

——. 1980. Food habits, energetics, and the population biology of mammals. Am. Nat., 116:106–124.

——. 1983. Energetics, body size, and the limits to endothermy. J. Zool., 199:1–29.

——. 1988. Complications inherent in scaling the basal rate of metabolism in mammals. Q. Rev. Biol., 63:25–54.

Muller, E. F. 1985. Basal metabolic rates in primates—the possible role of phylogenetic and ecological factors. Comp. Biochem. Physiol., 81A:707–711.

Pearson, O. P. 1947. The rate of metabolism of some small mammals. Ecology, 28:127–145.

Pipes, G. W., T. R. Bauman, J. R. Brooks, J. E. Comfort, and C. W. Turner. 1963. Effects of season, sex and breed on the thyroxine secretion rate of beef cattle and a comparison with dairy cattle. J. Anim. Sci., 22:476–480.

Pipes, G. W., J. A. Grossie, and C. W. Turner. 1960. Effect of underfeeding on thyroxine secretion rate of female mice. Proc. Soc. Exp. Biol. Med., 104:491–492.

Pittman, C. S., J. B. Chambers, Jr., and V. H. Read. 1971. The extrathyroidal conversion rate of thyroxine to triiodothyronine in normal man. J. Clin. Invest., 50:1187–1196.

Pittman, C. S., T. Shimizu, A. Burger, and J. B. Chambers, Jr. 1980. The nondeiodinative pathways of thyroxine metabolism 3,5,3′,5′-tetraiodothyroacetic acid turnover in normal and fasting human subjects. J. Clin. Endocrinol. & Metab., 50:712–716.

Premachandra, B. N. 1962. Thyroxine secretion rate in hamster and the guinea pig. Abstract 469. 22nd International Congress of Physiological Sciences.

Premachandra, B. N., G. W. Pipes, and C. W. Turner. 1958. Variation in the thyroxine secretion rate of cattle. J. Dairy Sci., 41:1609–1615.

Stahl, W. R. 1967. Scaling of respiratory variables in mammals. J. Appl. Physiol., 22:453–460.

Stolte, L., H. Kock, H. van Kessel, and L. Kock. 1966. Thyroxine utilization in nonpregnant, steroid-induced pseudopregnant, and pregnant monkeys. Acta Endocrinol., 52:383–390.

Suda, A. K., C. S. Pittman, T. Shimizu, and J. B. Chambers, Jr. 1978. The production and metabolism of 3,5,3′-triiodothyronine and 3,3′,5′-triiodothyronine in normal and fasting subjects. J. Clin. Endocrinol. & Metab., 47:1311–1319.

Thompson, M. E., G. P. Orczyk, and G. A. Hedge. 1977. In vivo inhibition of thyroid secretion by indomethacin. Endocrinology, 100:1060–1067.

Tomasi, T. E. 1984. Shrew metabolic rates and thyroxine utilization. Comp. Biochem. Physiol., 78A:431–435.

——. 1991. Utilization rates of thyroid hormones in mammals. Comp. Biochem. Physiol., 100A:503–516.

Tomasi, T. E., and B. A. Horwitz. 1987. Thyroid function and cold acclimation in the hamster, *Mesocricetus auratus*. Am. J. Physiol., 252:E260–E267.

Vogel, P. 1980. Metabolic levels and biological strategies in shrews. Pp. 170–180 *in* Comparative Physiology of Primitive Mammals (K. Schmidt-Nielsen, L. Bolis, and C. R. Taylor, eds., Cambridge University Press, Cambridge. 322 pp.

Wada, H., R. V. Berswordt-Wallrabe, and C. W. Turner. 1959. Thyroxine secretion rate of mice. Proc. Soc. Exp. Biol. Med., 102:608–609.

Wills, P. I., and W. J. Schindler. 1970. Radiothyroxine turnover studies in mice: effect of temperature, diet, sex, and pregnancy. Endocrinology, 86:1272–1280.

Yousef, M. K., and H. D. Johnson. 1967. A rapid method for estimation of thyroxine secretion rate of cattle. J. Anim. Sci., 26:1108–1112.

——. 1968. Effects of heat and feed restriction during growth on thyroxine secretion rate of male rats. Endocrinology, 82:353–358.

——. 1975. Thyroid activity in desert rodents: a mechanism for lowered metabolic rate. Am. J. Physiol., 229:427–431.

Yousef, M. K., and J. R. Luick. 1971. Estimation of thyroxine secretion rate in reindeer, *Rangifer tarandus*: effects of sex, age, and season. Comp. Biochem. Physiol., 40A: 789–795.

Bruce A. Wunder

5. Morphophysiological Indicators of the Energy State of Small Mammals

To exist, endothermic vertebrates must use energy. Consequently, they must find and process energy. These statements, by themselves, are tautological. Nonetheless, the study of accumulation and allocation of energy as it relates to life history tactics is valuable, for it gives us insight into what animals can and cannot do and how they are limited by when certain activities can occur. Most vertebrates that live in seasonal environments do not breed continuously nor are their foods homogeneous throughout the year. Rowsemitt et al. (1975) suggested that certain sorts of molt patterns do not occur simultaneously with breeding of beach voles (*Microtus breweri*). These observations suggest that the ability of small mammals to accumulate energy and to allocate that energy to reproduction (fitness) is limited, and certain other energy-requiring processes (e.g., molt or growth) cannot occur simultaneously with reproduction. Stenseth et al. (1980) developed a model with which they suggest that energy may limit reproduction in small mammals. Thus, there is an underlying tenet in small mammal ecology that the ability to accumulate energy and the avenues of allocation are important in determining what species can live where, when that species can breed, and how many young it can produce. These may be tied to various life history parameters such as food habits (McNab, 1981, 1986).

For those reasons energetics has been important to the study of small mammal ecology (Grodzinski and Wunder, 1975), especially the ecology of herbivores. They eat foods that are less energy-dense than those

Department of Biology, Colorado State University, Fort Collins, Colorado 80523.

of carnivores and that are generally more difficult to digest. Also, the quality and abundance of their food vary seasonally. Plants grow in summer and die back in fall and winter, translocating nutrients to roots (Wunder et al., 1977).

Much of what I discuss in this chapter is limited to small herbivores. I do this because most of the data are limited to such forms and because I believe they are limited more by energetics than are larger forms. Although less food is needed to maintain small mammals, they must process that food quickly and cannot withstand periods of deprivation for very long.

Energy Balance Model

In a discussion of how small herbivores might make adjustments in their balance of energy accumulation and allocation to meet different

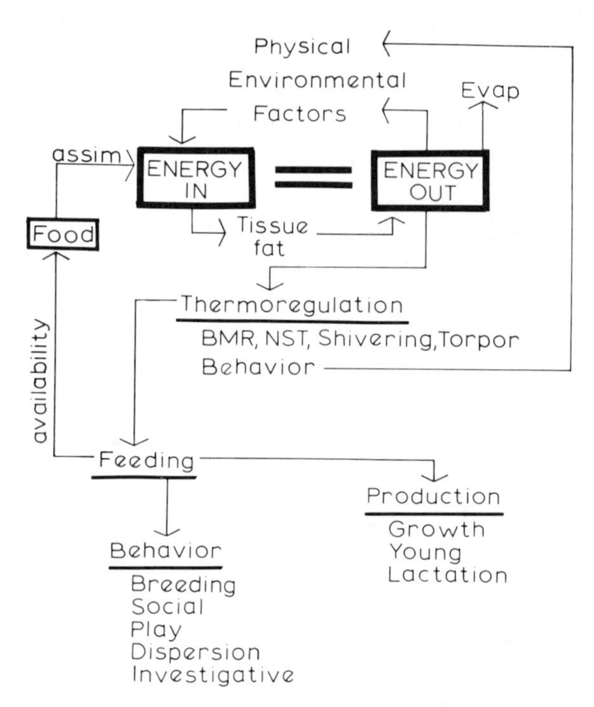

Fig. 5.1. A conceptual model of energy balance for a small mammal indicating a priority cascade for energy allocation. Lines represent both total energy flow and rate functions. (From Wunder, 1978a, courtesy of Pymatuning Laboratory of Ecology, University of Pittsburgh.) BMR, basal metabolic rate; NST, nonshivering thermogenesis; Evap, evaporation; assim, assimilation.

stressors in their environments, it is valuable to consider the pathways for balance. For that reason I present a descriptive model (Fig. 5.1) that I have discussed before (Wunder, 1978a, 1984). The thesis is quite simple. Energy expended by a small herbivore must equal that accumulated or the animal will either (1) heat or cool (imbalance of thermal energy) or (2) fatten or starve (imbalance of chemical energy). I depict the energy flow as a cascade because small homeotherms must allocate energy to different processes in a priority fashion (Wunder, 1978a). If they do not thermoregulate, they can do nothing else because they cannot behave. Second, they must gather energy by feeding because this is the feedback to balance their needs. When those first two needs, thermoregulation and feeding, are met then they can do other things. I suggest that when such a balance cannot be met or is tenuous they cannot breed; and, hence, timing for reproduction is limited. Increased needs for thermoregulation or reproduction or both will drive energy costs up and, hence, necessitate that the animals gather more energy per unit time, draw upon stores of chemical energy, or both.

To meet increased demands small herbivores must accumulate more energy. Ecologists tend to think of food accumulation as a process that consists primarily of finding and handling food. This usually means getting food to the mouth. However, in herbivores, especially those that feed on material with substantial cell wall component, digestion may be limiting (Demment and Van Soest, 1985; White, 1978; Wunder, 1985).

It is well known to animal scientists, but not always to ecologists, that herbivores that feed on leafy vegetation tend to keep the gut full processing the hard-to-digest food (Van Soest et al., 1983). It is also known that digestibility is related to how long food is kept in the gut (retention time). If energy needs increase, more food must be processed through the gut per unit time. If gut size remains constant, retention time may decrease, and, hence, digestibility could decrease, reducing the net energy gained per unit time (Sibly, 1981; Wunder 1985). Limits to passage rates of material through the gut and factors that affect those rates are discussed by Warner (1981). Recently it has been found that small birds (reviewed by Sibly, 1981) and mammals (Gross et al., 1985) respond to changes in energy need by changing gut size.

Application

Given this brief overview of energy balance of small herbivores, it is frustrating to note that there are few insights and even fewer data on how animals actually interpret that world in which they operate. We presume that they are cold-stressed during winter and must spend extra energy on thermoregulation while eating dead vegetation with relatively high cell wall content, which takes longer to digest. But we have few data to substantiate these suppositions. That it is cold is indicated by the fact that insulation usually changes seasonally (Bartholomew, 1982; Irving, 1964; Wunder, 1984). Given such changes in insulation to buffer the animals against a cold environment, it is not known whether they must continuously expend extra energy on thermoregulation or whether the insulation alone counteracts the cold. There are few measurements of energy expenditure by animals in the field. Probably the best are those of Nagy and colleagues (Nagy, 1987; Nagy and Martin, 1985; Nagy and Suckling, 1985), but they have not looked at seasonal patterns or animals found in cold, high alpine areas.

Since 1975 I have studied ways in which small herbivores adjust to seasonal environments. I have been particularly interested in how small mammals can change their capacity for heat production, what cuing mechanisms are involved (Wunder, 1979, 1984), and, more recently, how they change gut size to allow assimilation of more energy from the small intestine per unit time or for more fermentation of plant cell wall in the cecum per unit time (Gross et al., 1985). For these studies, it would be valuable to have some measures of how small herbivores actually interpret the world in which they live and how they respond to that world.

Shvarts (1975), Vorontsov (1967), and Bashenina (1984) pioneered in the use of morphophysiological characteristics to indicate to what habitat animals may be adapted and how those animals adjust to certain environmental stressors. Some indicators that they used allow insights into the energy states of small mammals. Recently Schieck and Millar (1985), building on earlier studies (Barry, 1977; Golley, 1960; Vorontsov, 1967), showed that various measures of the guts of small mammals can be used to estimate food habits. However, they did not talk about how these might change in relation to need. I present some new morphophysiological indicators below.

Data Base

Most of the data used in this chapter came from the literature; however, several studies from my lab are just now being written up, so I include some general methods we followed. Twelve pika (*Ochotona princeps*) were collected from Loveland Pass, Clear Creek County, Colorado. As samples for brown fat, four animals were collected by shotgun in January 1981, two in March 1981, two in December 1982, and four were collected alive with Tomahawk live traps in September 1982. These latter four were kept in the lab at 23°C in individual cages on natural photoperiod, given water and food ad libitum, and used in metabolic trials. Food consisted of rabbit chow and material from their hay piles. By the time the animals became cold-acclimated, the "hay" had run out, and they were given grass clippings (bluegrass) from my lawn.

Microtus ochrogaster were captured near Horsetooth Reservoir, Larimer County, Colorado. For studies of nonshivering thermogenesis (NST) the voles were trapped overnight. Measures of basal metabolism (BMR) were made the day of capture as in Wunder et al., 1977. Response to injection of norepinephrine (NE) was measured the next day (as in Wunder, 1984) in animals that had maintained body mass overnight (indicating they were eating and drinking). The dose of norepinephrine used was determined by establishing a dose-response curve for unanesthetized animals (Wunder, R. Gettinger, A. Seidel, pers. observ.). I refer to responses to injection of norepinephrine as NE responses, including both obligatory and regulatory NST (Jansky, 1973). When I refer to NST, I simply mean regulatory NST (or NE response minus BMR). These animals were then sacrificed for studies of gut morphology in the January 1984 and May 1984 samples. Thus, they had been eating rodent lab chow in the lab for 1 day. Gut measurements were made as in Gross et al., 1985. Details will be published elsewhere; but, briefly, *M. ochrogaster* used in studies of the effects of lactation on gut size were all born and raised in the lab and had not previously bred. They were held at 12°C (because it is reported that they breed best in the lab at that temperature), given a mix of horse chow and rat chow pellets and water ad libitum. They had cotton for nests. Experimental animals had litter sizes adjusted to three young at birth and were allowed to continue lactation to day 14. Controls went through pregnancy, but babies were removed at birth. Virgins had never bred. Since *M. ochrogaster* has a postpartum estrus some of the controls and experimentals were also pregnant. Fortunately, there did not appear to be any effect of pregnancy on gut measurements.

Results and Discussion

As indicated in Fig. 5.1, a small, nonhibernating mammal faced with a cold environment must modify that thermal environment to reduce cold exposure (microenvironment or insulation) or increase thermogenesis to maintain body temperature (T_B). One way to increase thermogenesis is to increase basal metabolic rate (BMR), but that is expensive because energy expenditure is increased continuously even if the animal can find a thermoneutral microhabitat for a portion of its day. Another means of increasing thermogenesis is nonshivering thermogenesis (NST). With this mode of thermogenesis the animal expends extra energy only when exposed to temperatures below its lower critical temperature and then in a most efficient manner.

Below I discuss several morphophysiological parameters that can be used to define the energy stress state of a small herbivorous mammal. These are (1) BMR, (2) NST, (3) state of brown fat, and (4) gut size parameters. At present it is difficult to quantify the relation of each of these to the energy needs of or use by a small mammal. With specific data for a particular species, however, such quantification is possible.

Basal Metabolic Rate

The BMR of small mammals frequently increases following cold exposure (Webster, 1974). Yet we (Wunder, R. Gettinger, and A. Seidel, pers. observ.) recently found that BMR does not always increase in all species of small mammals following cold acclimation. At first we thought this variation was related to diet—energy-dense (carnivore, omnivore) or energy-dilute (herbivore) food—but that relation is not clear. Wunder et al. (1977) demonstrated that the BMR of *M. ochrogaster* increased following cold exposure and may increase in winter. We now know that this change varies from year to year and seems to be correlated with weather conditions (Wunder, 1978b). In any case, if the BMR is increased, the animal has been cold-exposed, although cold exposure does not always cause an increase in the BMR. Other factors can influence the BMR (Wunder, 1979), so one must be somewhat cautious in using an increase in the BMR to indicate cold exposure; but, with proper attention to those factors one can use increases to indicate cold exposure.

Nonshivering Thermogenesis and Brown Fat

It is fairly well established that as acclimation temperature (T_{AC}) decreases, NST capacity of small mammals increases (Jansky, 1973). Thus, it seems reasonable to conclude that animals with increased NST over their "control" levels have been cold-exposed. But this relation-

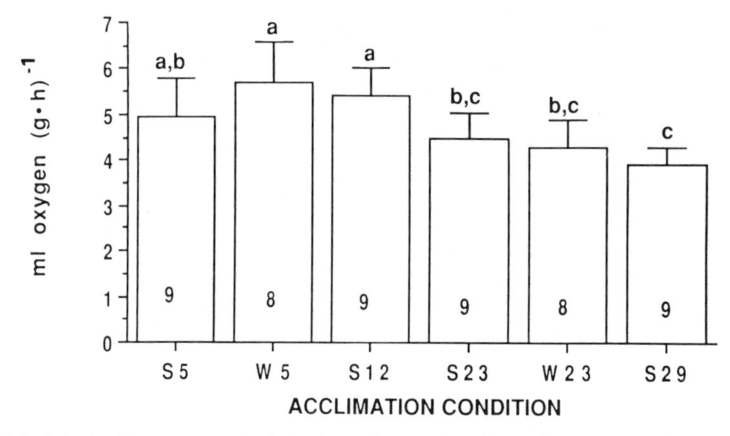

Fig. 5.2. Metabolic response to injection of norepinephrine by prairie voles (*Microtus ochrogaster*). On the acclimation condition axis, the letter refers to animals caught in summer (S) or winter (W) and the number to the temperature (°C) to which they were acclimated in the lab. Numbers in the bars are sample size. Variance is ±2 SEM. Bars sharing common letters are not statistically significantly ($P < 0.05$) different.

ship is complicated by a recently described phenomenon called diet-induced thermogenesis (Rothwell and Stock, 1979). The relationship between T_{AC} and NST has been quantitatively shown in only a few species such as rats (Jansky et al., 1967), white mice (Mejsnar and Jansky, 1971), *Peromyscus leucopus* (Lynch, 1973), golden hamsters (Vybiral et al., 1975), and *Microtus ochrogaster* (C. Simon and Wunder, pers. observ.; Fig. 5.2). These studies show that one must calibrate the NST relationship for individual species to define minimal temperatures below which further cold exposure elicits no greater NST response and to define the correlation between NST and T_{AC} for each species. When such calibrations are made, one should be able to use NST values to estimate T_{AC} and vice versa. For field biologists measurement of NST of freshly caught animals could then be used as a gauge to the field exposure temperature for the animal.

Since brown adipose tissue (BAT) is the major site of NST (Foster and Frydman, 1978), one might predict that some aspect of BAT might also be used as an indicator of thermal challenge in small mammals. Several components of BAT change following cold acclimation. The amount of BAT may be increased (Heldmaier, 1974), its histology may change from a unilocular state to a multilocular state (Tarkkonen, 1970), and its biochemistry may change (Cannon et al., 1981; Nichols and Locke, 1984). Thus, one could use BAT as in indicator of thermal challenge. BAT state may be a bit more complex than NST as we know

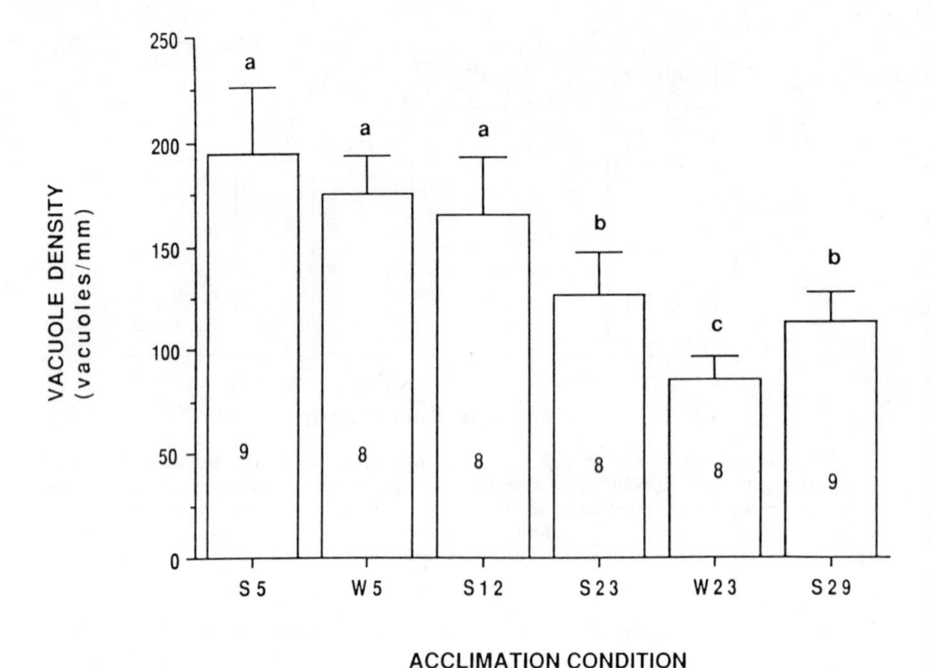

Fig. 5.3. Density index for fat vacuoles in cells of intrascapular brown adipose tissue (BAT) taken from prairie voles (*Microtus ochrogaster*). Acclimation conditions are as described for Figure 5.2. Numbers in the bars are sample sizes. Variance is ±2 SEM. Bars sharing common letters are not statistically significantly ($P < 0.05$) different.

that mass changes are not always correlated with NST capacity (Heldmaier et al., 1982). However, as seen in Fig. 5.3, vacuole density in BAT is fairly well correlated with T_{AC} of prairie voles and is not affected by season; although there is a tendency for animals tested in winter to have fewer vacuoles than those tested in summer.

Various factors other than temperature may affect the BAT and NST capacity of small herbivores. For example, photoperiod, norepinephrine (NE) or thyroxine injections, and diet may cause increases in NST in addition to T_{AC} (see Wunder, 1979, 1985). Thus, one should be cautious to sort out these effects.

Since we know that the presence of BAT and increased levels of NST indicate low temperature exposure in small mammals and that absence or reduced values indicate little or no thermal stress, what do we find in field-caught animals? Relatively few such data exist. Didow and Hayward (1969) reported that BAT mass was highest during winter in field-caught meadow voles (*M. pennsylvanicus*), and Aleksiuk (1971) found similar results for red squirrels (*Tamiasciurus hudsonicus*).

TABLE 5.1
Metabolic response (ml O_2/[g · h]) of small mammals to norepinephrine following temperature acclimation or seasonal acclimatization

Species	Acclimation[a]		Acclimatization		Reference
	Warm	Cold	Summer	Winter	
Clethrionomys rutilus	7.4 (20°C)	11.5 (5°C)	6.8	18.7	Feist and Rosenmann, 1976
Peromyscus leucopus	4 (26°C)	8 (5°C)	4	8	Lynch, 1973: values from a graph
Phodopus sungorus	6.5 (23°C)	11.5 (10°C)	—	13.3	Heldmaier et al., 1982: winter animals were held in cages outside
Microtus ochrogaster	4.1 (23°C)	5.8 (5°C)	2.6	3.5	Author

[a]Number in parentheses is acclimation temperature.

Lynch (1973) reported that in white-footed mice (*Peromyscus leucopus*) NST increased in fall and remained fairly high in winter, decreasing again in spring-summer. These studies were reasonably well correlated with effects found following cold T_{AC} in the lab. Feist and Rosenmann (1976) compared NST values for red-backed voles (*Clethrionomys rutilus*) caught in Alaska during summer and winter with values for lab-acclimated animals and found that NST was higher in winter than in summer and that field-caught animals from winter had values equal to or higher than animals acclimated to 5°C in the lab (Table 5.1). Thus, one can conclude that red-backed voles caught in Alaska are, indeed, cold-exposed in winter. Merritt (1986) also reported that NST in short-tailed shrews (*Blarina brevicauda*) captured in Pennsylvania is higher in winter than in summer, and Zegers and Merritt (1988) reported higher NST in winter than in summer for both *Peromyscus maniculatus* and *P. leucopus*. But in neither case was response to T_{AC} in the lab measured. We can now add data for prairie voles and pika (*Ochotona princeps*). We have studied NST in prairie voles (*M. ochrogaster*) in the lab and in the field for several years and have found that, in the lab, regardless of photoperiod and at a constant T_{AC} of 23°C, NST values are low in fall, increase to a peak in winter,

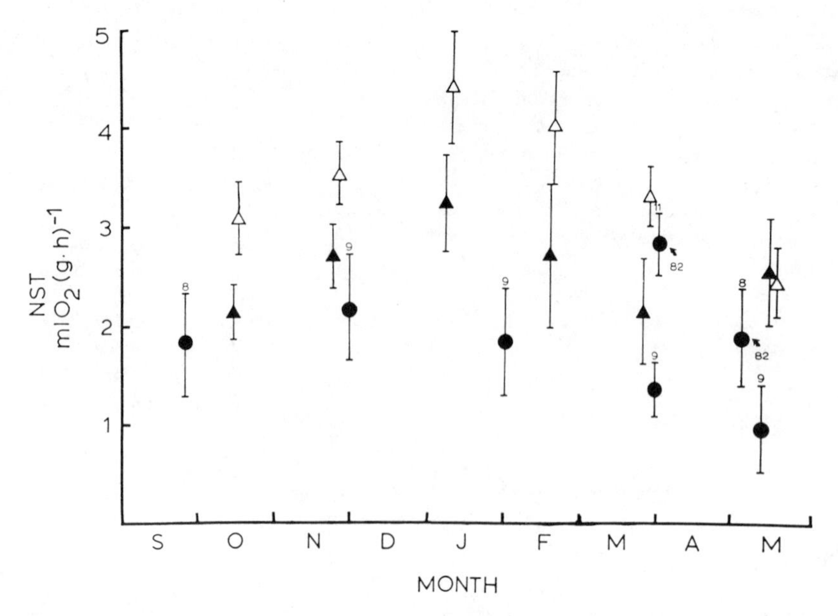

Fig. 5.4. Nonshivering thermogenic response (NE response − BMR) of prairie voles held in the lab at 23°C (shaded triangles) or held in cages outside (unshaded triangles). (From Wunder, 1984.) Shaded circles indicate values for voles captured in the field at different times of year (letters indicate months of the year). Field animals were captured in 1981 except for those marked with an 82, which were captured in 1982. Values given are means ±2 SEM with sample sizes given above the error bars.

and decrease in spring (Fig. 5.4). We originally thought this cycle was a photoperiod effect (Wunder, 1984) and now realize an endogenous rhythm exists (Wunder, 1985). When we collect *M. ochrogaster* from the field and measure NST one day later as in Wunder, 1984, we find NST increases to a peak in winter. However, there is a striking difference between values measured in lab-acclimated animals and those in field-caught animals. The NST response of field-caught animals in midwinter is lower than in animals acclimated to 23°C in the lab. This difference could be due to any of many reasons but the most likely are (1) voles may be stressed in the lab, never adjust to it, and, therefore, secrete high levels of NE, stimulating BAT and NST; (2) animals develop more BAT when given a good diet in the lab, showing what nutritionists call diet-induced thermogenesis (DIT: Rothwell and Stock, 1979); (3) voles are not exposed to low temperature during the winter in the field. I would argue against the first reason, since we have now studied several species and have looked at the allometry of their NST response. The allometry correlation is quite good, and lab animals (rats, mice, and hamsters) fit that correlation well (Wunder, R. Gettinger, and A. Seidel, pers. observ.). If much of the NST response were due to "stress", I would expect much variation between species since there should be some variation in the response of animals to the lab. Number two is possible but unknown; and, if present, must add to the T_{AC} effect producing a higher thermogenesis than T_{AC} would alone (but see the plateau effect in Fig. 5.2 with animals on chow and low T_{AC} in the lab). Rothwell and Stock (1980) found that DIT and NST are additive in rats. The last factor seems the most plausible and yet is counter to our intuitive understanding of what these animals should be experiencing in winter. One might argue that if voles live under snow in winter, that insulating effect might moderate the microclimate for them. Although near Fort Collins we do occasionally have snow cover for a few days or longer during winter, in the year when these measurements were made, we had no snow cover in the field where voles were caught. The metabolic response to norepinephrine of several species of small mammals following cold acclimation or winter acclimatization suggests an interesting hypothesis. I compared such responses for the four species for which both warm- and cold-acclimation data and summer-winter acclimatization data were available. Field data for three of the species were similar to lab data: winter responses were as high or higher than those following acclimation to cold in the lab. Only the prairie vole was different, with low field responses. The prairie vole feeds on vegetation and can gather food easily, at least for maintenance, all year. A vole, therefore, does not need to leave its nest for very long. All other species are more granivorous, must "hunt" for food, and may need to spend more time out of their nests foraging during any one bout, being exposed to ambient conditions for longer periods. This may account for the differences in response patterns. Nest usage could be monitored via radio telemetry.

Pika (O. *princeps*) also live in cold macroenvironments in Colorado. One population is reasonably easy to reach in summer and winter. These pika live at about 3660 m (12,000 feet) elevation near Loveland Pass, Clear Creek County, Colorado. We have shot animals in winter to sample for BAT. Four animals collected in January 1981 showed no BAT upon dissection. Two animals collected by shotgun in March 1981 also had no BAT. At those times animals had lots of snow cover, came out onto rocks or the snow surface for only short periods each day, and then only in low numbers. Following those dissections I concluded that pika may not develop BAT, since we had found none in midwinter. However, it was puzzling that BAT can be found in other Lagomorpha but not in these primitive forms in winter. Thus, in September 1982 I collected four live pika from this same site. Upon return to the lab I measured BMR and injected the animals with norepinephrine (0.4 mg/kg, a dose taken from our allometric relationship eliciting highest metabolism in small mammals). There was no metabolic response to norepinephrine. I then cold acclimated (5°C) the pika for 14 days. They are hard to keep alive in the lab; one died before we started cold acclimation and one died during cold acclimation. A third animal died during the second BMR determination; thus, I could test for NST in only one animal. Nonetheless, it showed a strong NST response following the NE injection (2.46 ml O_2/[g · h] for a 151-g pika). It and the animal that died during acclimation were dissected and showed high levels of BAT. I collected two more pika from the field in December 1982. They showed a lot of BAT. That year we had little snow cover at Loveland Pass by December. Thus, it appears that pika can, indeed, form and use BAT but do so only when exposed to cold (5°C). In fall, when it is cold, pika may use the tissue, but once snow arrives and insulates their habitat and dens they lose the BAT as they are living in an effectively warm environment. And they are well insulated and have increased BMR (MacArthur and Wang, 1973).

Energy Acquisition and Gut Morphology

Gut morphology of small mammals can be related to food habits (Davis and Golley, 1963; Vorontsov, 1967). Schieck and Millar (1985) recently drew conclusions about the relationships between gut morphology and food habits of small mammals, showing that herbivores tend to have longer hind guts and larger ceca than granivores or carnivores. Hansson (1985) found that *Clethrionomys glareolus* from northern Sweden eat a more energy-dilute diet than those in southern Sweden and have correspondingly larger guts. Most of these studies assume that gut size is relatively fixed or do not consider how gut size might change seasonally. However, Hansson (1985) did look at potential changes and the effects of genetic background on those changes.

Animal scientists and some ecologists have long known that gut morphology may change in relation to need or fiber content in the diet or both (Sibly, 1981), and we documented how these stressors affect gut size in prairie voles (Gross et al., 1985). We modified the animals' need for energy by placing them in the cold; and we modified their ability to get energy from food (energy dilution) by modifying the fiber content of the food. From that study we demonstrated that not only does the total size of the gut (measured as length or dry mass of tissue or both) change in response to those stressors but different portions of the gut change in response to different stressors. Generally, when stressed with decreased ambient temperature exposure (increased energy need) the size of the small intestine increases (site of chemical digestion and absorption). When stressed with fiber (energy dilution of the diet), the cecum (site of biological fermentation) increases in size. Similar changes have been reported for large ungulates (Fell and Weekes, 1975; Warner and Flatt, 1965). Loeb (1987), working with pocket gophers (*Thomomys bottae*) in the lab, found that the capacity of the stomach, small intestine, cecum, and large intestine varied with diet, increasing in size as diet quality decreased. Thus, provided one has some baseline or comparison points, one may be able to draw conclusions about the relative energy needs of animals from different sets of conditions (e.g., winter vs. summer; high vs. low food availability) by looking at relative gut size, especially the length and mass of the small intestine. Likewise by assaying the length and mass of the cecum one may be able to draw conclusions about "quality" of the diet for small herbivores.

More recently we have also found that lactation, another form of increased energy need, stimulates an increase in gut size (Wunder, J. Alldredge, and K. Hammond, pers. observ.). Interestingly, in addition to the larger small intestine found with a temperature effect, lactating females also increased stomach size over nonlactating females and nonbreeding animals (Fig. 5.5). It is unclear why lactation should affect this change in stomach size when increased need due to thermoregulatory needs did not. A possible explanation is provided by Madison's (1981) observations that activity of lactating female *M. pennsylvanicus* tends to become reduced and focused near the nest relative to nonlactating animals. Thus, lactating females may take fewer, larger meals (requiring a larger stomach for storage) and spend more time in the nest brooding young.

Given the discussion above, I suggest that changes in gut morphology could be used to indicate the level of energy stress for small herbivores from the field. Within a species of herbivore, a relatively long small intestine would indicate increased energy needs (e.g, temperature, lactation, activity). Coupled with a large stomach, lactation is indicated. An increased cecum size should indicate a change in food habits

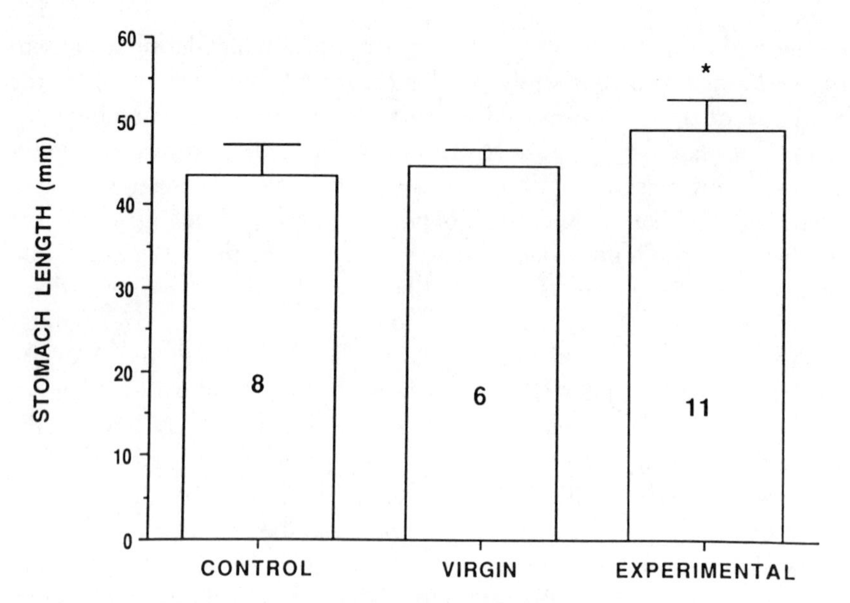

Fig. 5.5. Stomach length in prairie voles. Controls had gone through pregnancy, virgins had never bred, and experimentals had gone through 14 days of lactation with a litter size of three. Numbers in the bars are sample sizes. The asterisk indicates a value statistically significantly ($P < 0.05$) different from the others.

to foods of higher fiber content. These comparisons can be made safely only within a species and then only in a comparative sense.

I have used the logic above to predict that voles should have longer small intestines in winter because it should be cold and they should have higher thermoregulatory needs. We also predicted that the cecum should be larger in winter because plant material used as food should be dead with a higher fiber content than in summer.

Few studies have reported gut parameters of small herbivores on a seasonal basis. Only Gebczynska and Gebczynski (1971) have presented data on the effects of season and reproduction on certain parameters of the gut, studying *Microtus oeconomus* in a Polish forest. The length measures they made are the most valuable to this discussion. Unfortunately they did not measure dry mass of the tissue, only fresh wet mass. Gross et al. (1985) pointed out the problems with using wet mass for good comparisons. Gebczynska and Gebczynski (1971) noted that total intestinal length was greatest in summer, lowest in spring, and intermediate in winter. They also noted that "length of the intestines is always greatest in nursing females" (Gebczynska and Gebczynski, 1971, p. 364). These results for field-caught animals are

TABLE 5.2
Gut measurements of *Microtus ochrogaster* captured in the field

Measurement	January 1984[a] ($\overline{X}$ ±2 SEM)	May 1984[a] ($\overline{X}$ ±2 SEM)
Body mass (g)	36.7 ± 4.1	41.5 ± 4.9
Stomach length (mm)	39.9 ± 2.4*	46.8 ± 3.0*
Stomach dry mass (g)	0.115 ± 0.037	0.102 ± 0.008
Small intestine length (mm)	305 ± 13	319 ± 16
Small intestine dry mass (g)	0.1006 ± 0.008*	0.1221 ± 0.020*
Large intestine length (mm)	224 ± 12.99*	243 ± 11.38*
Large intestine dry mass (g)	0.082 ± 0.006*	0.069 ± 0.005*
Total gut length (mm)	734 ± 32*	789 ± 25*

[a]All sample sizes are 10.
*Significant ($P < 0.05$) difference between samples.

similar to our lab findings reported above. Recently, Green and Millar (1987) indicated that mass of the intestinal tract (with contents) provides a good indicator of daily food intake in deer mice (*Peromyscus maniculatus*), but concluded that gross dimensions could not be used to differentiate between energy requirements and diet quality. Since they used organ plus food contained therein, the mass of food probably overshadowed any organ mass changes and complicated their interpretation.

We also have preliminary field data (10 animals captured in January 1984 and 10 in May 1984) for *M. ochrogaster* which, surprisingly, show that our predictions given above do not hold (Table 5.2). The stomach, large intestine, and total gut length and dry mass of the small intestine were larger in May animals. Dry mass of the large intestine was greater in January. All other measures were the same in both seasons. These results allow some interesting speculation. Gut data (especially total length and dry mass of the small intestine) suggest that voles process more energy in May than in January. The larger dry mass of the large intestine in winter may be due to increased need for water or ion reabsorption. Although we did not include animals that were obviously pregnant or lactating in the May sample, it appears as though reproductive activity in spring effects a greater energy need than thermoregulation in the winter. The potential for increased energy need due to cold and decreased food availability due to high fiber in winter, may preclude gathering enough food to meet maintenance needs and breeding in voles. Thus, they may simply leave their nests for short periods to forage, spending most of their time in well-insulated nests. Voles then would have low energy need because of reduced activity and reduced exposure to cold. This conclusion is supported by our findings on NST capacity discussed above. During spring (May) voles breed and males use much energy to stay active and find females, while females would be processing extra energy for lactation. Although we have no data on the time necessary for gut size of voles to change, we do know that rats' guts increase in size within days, and once females have had a litter, 30 days are needed for the gut to return to the smaller, nonbreeding size (Fell et al., 1963). Thus, even though we eliminated females that were obviously breeding from our field sample analysis, some voles may have had enlarged guts because of prior lactation. An analysis comparing males and females, however, showed no differences due to sex, but samples of each were, obviously, small. It would be valuable to compare NST and gut sizes of voles from population cycle phases when there are many voles and they are breeding in winter. I would predict a much different picture of gut size and NST capacity.

Parotid Gland

Physiologists working with ruminants suggest that the nutritional value of tree and shrub leaves may frequently be overestimated because of a failure to account for effects of leaf tannins (Robbins et al., 1987). Condensed tannins present in many plant tissues bind and precipitate protein, making it unavailable for digestion (Haslam, 1979). In some mammals parotid salivary glands produce proline-rich secretions that bind with tannins and may prevent their binding to dietary protein. Parotid salivary glands of ruminants are three times larger in browsers than in grazers and are intermediate size in mixed feeders (Hofmann, 1973; Kay et al., 1980). Further, parotid glands increase dramatically in size in rats and mice fed diets with high tannin levels. Within 3 days on such diets the parotids of rats increased 3-fold, and secretion of proline-rich proteins was increased 12-fold (Mehansho, et al., 1983). Thus, size of this gland may be a useful gauge to the tannin level in an animal's diet. Unfortunately, I know of no such studies for wild, small mammals.

Gut Lining Histology and Physiology

Active uptake of monosaccharides (especially glucose) can be modified under various conditions but is especially sensitive to change in dietary carbohydrate level (Diamond and Karasov, 1984; Karasov et al., 1983). More recently, Ferraris and Diamond (1986) demonstrated, using a phlorizin-binding technique, that this increase in uptake of glucose is due to an increase in glucose uptake transport sites in animals consuming a higher carbohydrate diet. Specific binding to glucose transporters was 1.9 times as great in the jejunum of mice fed high-carbohydrate diets as in mice fed no-carbohydrate diets. Thus, changes in the morphological state of the gut for glucose uptake may be correlated with diet changes.

Perhaps, with proper study and calibration, this measure could be used to indicate diet quality of animals captured in the field. This line of investigation is just beginning, and more data will be necessary to make any generalizations, let alone specific predictions; however, this technique may offer a potentially powerful tool to better understand what the energy state of small mammals from the field may be.

Future Studies

This chapter emphasizes measurements that can be made easily so that mammalogists working in the field can take advantage of such measures to correlate them with population studies and investigations of other such questions where information about energetics might be helpful. By encouraging a better interaction of physiological ecologists and other field biologists, we may achieve better insight into how many small mammals adjust to varying environmental stressors. Gut size and BAT histology are two measurements that can be made by any biologist. Measurements of NST and other parameters that I have discussed are more difficult and require expensive equipment.

Throughout the text I have mentioned studies that I believe would be of value. I emphasize only a few here. From field biologists we need many more studies of gut size from animals captured at different seasons to see whether and how gut sizes change. Care must be taken to try to correlate these with reproductive condition and food habits. Measurements as simple as Van Soest analyses of fiber content of food (Goering and Van Soest, 1970) may be more valuable than the actual plant species composition of the diet. From such analyses we could gain a better understanding of how changes in diet are correlated with energy needs or food habits or both (e.g., diet fiber content). As physiologists, we need better measures of what gut size changes mean. I think we should be measuring gut volume (where and how is material entering the body) rather than simply measuring length and mass of gut tissue. These latter measures are easy to make but do not really give us insight into where and how digestion and assimilation are occurring and how they are modified by changes in gut structure and function or what compensations the animals are making and hence how these changes influence (limit?) energy or nutrient assimilation.

It would be valuable to have more studies of seasonal changes in NST from more species with different food habits or activity patterns. These would allow a better test of the hypothesis that "hunters" are more cold-exposed in winter than are vegetation feeders. Specifically, NST should be measured. As I have indicated, these measures from field animals should be correlated with responses following acclimation to various temperatures in the lab. This approach allows a comparison of what the physiology of the animals can do in the lab with what the animals actually do in the field. For field studies, BAT histology is easiest to measure. It would be valuable to see how vacuole density is correlated with NST or thermogenin (Cannon et al., 1981) content or GDP-binding capacity of BAT in various species. At present those correlations are known for very few animals. Comparisons of animals that are insulated in winter by use of the subnivean space versus those that are not would be useful.

I think this entire topic of how small mammals (especially herbivores) modify their use of energy for thermoregulation and their capacity to assimilate energy will be quite important to population biology in the future. An understanding of how herbivores handle fiber in various foods and modify protein (or other nutrient) extractions is critical to understanding what energy or nitrogen is available for reproduction. Understanding parotid function (or other detoxifying mechanisms) will be important. Such studies will necessitate that population biologists, physiologists, and biochemists work together.

ACKNOWLEDGMENTS

I thank the Chinese Academy of Science and the Northwest Plateau Institute of Biology, Xining, for inviting me to a symposium where I was forced to focus some of these thoughts. Some of the work reported here was supported by a grant from the U.S. NSF (#DCB-8303694). I thank several people for allowing me to use their or our unpublished work, among them, Cathryn Simon, Charles Ralph, Paul Arnold, Ron Gettinger, A. Seidel, Joe Merritt, Jan Alldredge, and Kim Hammond. I thank Jim Ha for help with data analysis and computer usage. Many people have contributed in discussions of these ideas over several years. I thank Dick Tracy, Jim Ha, Kim Hammond, Paul Arnold, and John Gross for patience with various ideas and Gerhard Heldmaier for helping me to understand NST. This chapter is a substantially revised manuscript from a paper previously published in "The Proceedings of the International Symposium of Alpine Meadow Ecosystem (China, October 1988).

Literature Cited

Aleksiuk, M. 1971. Seasonal dynamics of brown adipose tissue function in the red squirrel (*Tamiasciurus hudsonicus*). Comp. Biochem. Physiol., 38A:723–731.

Barry, R. E., Jr. 1977. Length and absorptive surface area apportionment of segments of the hindgut for eight species of small mammals. J. Mammal., 58:419–420.

Bartholomew, G. A. 1982. Energy metabolism. Pp. 46–93 *in* Animal Physiology: Principles and Adaptations, 4th ed. (M. S. Gordon, ed.). Macmillan, New York. 635 pp.

Bashenina, N. V. 1984. Ecophysiological characteristics of small mammals. Ekologiya, 6:40–44. (Translation.)

Cannon, B., J. Nedergaard, and U. Sundin. 1981. Thermogenesis, brown fat, and thermogenin. Pp. 99–120 *in* Survival in the Cold (X. J. Musacchia and L. Jansky, eds.). Elsevier, New York. 225 pp.

Davis, D. E. and F. B. Golley. 1963. Principles in Mammalogy. Reinhold, London. 335 pp.

Demment, M. W., and P. J. Van Soest. 1985. A nutritional explanation for body-size patterns of ruminant and nonruminant herbivores. Am. Nat., 125:641–672.

Diamond, J. M., and W. H. Karasov. 1984. Effect of dietary carbohydrate on monosaccharide uptake by mouse small intestine in vitro. J. Physiol. (Lond.), 349:419–440.

Didow, L. A., and J. S. Hayward. 1969. Seasonal variations in the mass and composition of brown adipose tissue in the meadow vole, *Microtus pennsylvanicus*. Can. J. Zool., 47:547–555.

Feist, D. D., and M. Rosenmann. 1976. Norepinephrine thermogenesis in seasonally acclimatized and cold acclimated red-backed voles in Alaska. Can. J. Physiol. Pharmacol., 54:146–153.

Fell, B. F., K. A. Smith, and R. M. Campbell. 1963. Hypertrophic and hyperplastic changes in the alimentary canal of the lactating rat. J. Pathol. Bacteriol., 85:179–188.

Fell, B. F., and T. E. C. Weeks. 1975. Food intake as a mediator of adaptation in the ruminal epithelium. Pp. 101–118 *in* Digestion and Metabolism in the Ruminant (I. W. McDonald and A. C. I. Warner, eds.). University of New England Publishing, Armidale Australia. 602 pp.

Ferraris, R. P., and J. M. Diamond. 1986. Use of phlorizin binding to demonstrate induction of intestinal glucose transporters. J. Membr. Biol., 94:77–82.

Foster, D. O., and M. L. Frydman. 1978. Non-shivering thermogenesis in the rat. II. Measurements of blood flow with microspheres point to brown adipose tissue as the dominant site of the calorigenesis induced by noradrenaline. Can. J. Physiol. Pharmacol., 56:110–122.

Gebczynska, Z., and M. Gebczynski. 1971. Length and weight of the alimentary tract of the root vole. Acta Theriol., 16:359–369.

Goering, H. K., and P. J. Van Soest. 1970. Forage fiber analysis (apparatus, reagents, procedure and some applications). Agric. Res. Ser., USDA, Washington, D.C., Agric. Handbk. No., 379:1–20.

Golley, F. B. 1960. Anatomy of the digestive tract of *Microtus*. J. Mammal., 41:89–99.

Green, D. A., and J. S. Millar. 1987. Changes in gut dimensions and capacity of *Peromyscus maniculatus* relative to diet quality and energy needs. Can. J. Zool., 65:2159–2162.

Grodzinski, W., and B. A. Wunder. 1975. Ecological energetics of small mammals. Pp. 173–204 *in* Small Mammals: Their Productivity and Population Dynamics (F. B. Golley, K. Petrusewicz, and L. Ryszkowski, eds.). Cambridge University Press, London. 451 pp.

Gross, J. E., Z. Wang, and B. A. Wunder. 1985. Effects of food quality and energy needs: changes in gut morphology and capacity of *Microtus ochrogaster*. J. Mammal., 66:661–667.

Hansson, L. 1985. Geographic differences in bank voles *Clethrionomys glareolus* in relation to ecogeographical rules and possible demographic and nutritive strategies. Ann. Zool. Fenn., 22:319–328.

Haslam, E. 1979. Vegetable tannins. Recent Adv. Phytochem., 12:475–523.

Heldmaier, G. 1974. Temperature adaptation and brown adipose tissue in hairless and albino mice. J. Comp. Physiol., 92:281–292.

Heldmaier, G., S. Steinlechner, J. Rafael, and B. Latteier. 1982. Photoperiod and ambient temperature as environmental cues for seasonal thermogenic adaptation in the Djungarian hamster, *Phodopus sungorus*. Int. J. Biometeorol., 26:339–345.

Hofmann, R. R. 1973. The ruminant stomach. Vol. 2. East African Monographs in Biology. East African Literature Bureau, Nairobi. 350 pp.

Irving, L. 1964. Terrestrial animals in cold: birds and mammals. Pp. 361–377 *in* Handbook of Physiology. Sec. 4: Adaptation to the Environment (D. B. Dill, E. F. Adolph, and C. G. Wilber, eds.). Am. Physiol. Soc., Washington, D.C. 1056 pp.

Jansky, L. 1973. Non-shivering thermogenesis and its thermoregulatory significance. Biol. Rev., 48:85–132.

Jansky, L., R. Bartunkova, and E. Zeisberger. 1967. Acclimation of the white rat to cold: noradrenaline thermogenesis. Physiol. Bohemoslov., 16:366–371.

Karasov, W. H., R. S. Pond III, D. H. Solberg, and J. M. Diamond. 1983. Regulation of proline and glucose transport in mouse intestine by dietary substrate levels. Proc. Natl. Acad. Sci. USA, 80:7674–7677.

Kay, R. N. B., W. V. Engelhardt, and R. G. White. 1980. The digestive physiology of wild ruminants. Pp. 743–761 *in* Digestive Physiology and Metabolism in Ruminants (Y. Ruckebusch and P. Thivend, eds.). MTP Press, Lancaster, England. 854 pp.

Loeb, S. C. 1987. Nutritional ecology of the Sacramento Valley pocket gopher (*Thomomys bottae navus*). Ph.D. dissertation, University of California, Davis. 139 pp.

Lynch, R. G. 1973. Seasonal changes in thermogenesis, organ weights and body composition in the white-footed mouse, *Peromyscus leucopus*. Oecologia, 13:363–376.

MacArthur, R. A., and L. C. Wang. 1973. Physiology of thermoregulation in the pika, *Ochotona princeps*. Can. J. Zool., 51:11–16.

Madison, D. M. 1981. Time patterning of nest visitation by lactating meadow voles. J. Mammal., 62:389–391.

McNab, B. K. 1981. Food habits, energetics and the population biology of mammals. Am. Nat., 116:106–124.

——. 1986. The influence of food habits on the energetics of eutherian mammals. Ecol. Monogr., 56:1–19.

Mehansho, H., A. Hagerman, S. Clements, L. Butler, J. Rogler, and D. M. Carlson. 1983. Modulation of proline-rich protein biosynthesis in rat parotid glands by sorghums with high tanin levels. Proc. Natl. Acad. Sci. USA, 80:3948–3952.

Mejsnar, J., and L. Jansky. 1971. Non-shivering thermogenesis and calorigenic action of catecholamines in white mouse. Physiol. Bohemoslov., 20:157–162.

Merritt, J. F. 1986. Winter survival adaptations of the short-tailed shrew (*Blarina brevicauda*) in an Appalachian montane forest. J. Mammal., 67:450–464.

Nagy, K. A. 1987. Field metabolic rate and food requirement scaling in mammals and birds. Ecol. Monogr., 57:111–128.

Nagy, K. A., and R. W. Martin. 1985. Field metabolic rate, water flux, food consumption and time budget of koalas, *Phascolarctos cinereus* (Marsupialia: Phascolarctidae) in Victoria. Aust. J. Zool., 33:655–665.

Nagy, K. A., and G. C. Suckling. 1985. Field energetics and water balance of sugar gliders, *Petaurus breviceps* (Marsupialia: Petauridae). Aust. J. Zool., 33:683–691.

Nicholls, D. G., and R. M. Locke. 1984. Thermogenic mechanisms in brown fat. Physiol. Rev., 64:1–64.

Robbins, C. T., S. Mole, A. E. Hagerman, and T. A. Hanley. 1987. Role of tannins in defending plants against ruminants: reduction in dry matter digestion? Ecology, 68:1606–1615.

Rothwell, N. J., and M. J. Stock. 1979. A role for brown adipose tissue in diet induced thermogenesis. Nature, 281:31–34.

——. 1980. Similarities between cold and diet-induced thermogenesis in the rat. Can. J. Physiol. Pharmacol., 58:842–848.

Rowsemitt, C. T., T. H. Kunz, and R. H. Tamarin. 1975. The timing and patterns of molt in *Microtus breweri*. Occas. Pap. Mus. Nat. Hist. Univ. Kansas, 34:1–11.

Schieck, J. O., and J. S. Millar. 1985. Alimentary tract measurements as indicators of diets of small mammals. Mammalia, 49:93–104.

Shvarts, S. S. 1975. Morpho-physiological characteristics as indices of population processes. Pp. 129–152 *in* Small Mammals: Their Productivity and Population Dynamics (F. B. Golley, K. Petrusewicz, and L. Ryszkowski, eds.). Cambridge University Press, London. 451 pp.

Sibly, R. M. 1981. Strategies in digestion and defecation. Pp. 109–139 *in* Physiological Ecology (C. R. Townsend and P. Calow, eds.). Sinauer, Sunderland, Mass. 393 pp.

Stenseth, N. C., E. Framstad, P. Migula, P. Trojan, and B. Wojciechowska-Trojan. 1980. Energy models for the common vole *Microtus arvalis*: energy as a limiting resource for reproductive output. Oikos, 34:1–22.

Tarkkonen, H. 1970. The interscapular BAT during cold acclimatization in mice and its role in heat production. Ann. Med. Exp. Biol. Fenn., 48:168–171.

Van Soest, P. J., T. Foose, and J. B. Robertson. 1983. Comparative digestive capacities of herbivorous animals. Pp. 51–59 *in* Proc. Cornell Nutr. Conf.

Vorontsov, N. 1967. Evolution of the alimentary system of myomorph rodents. Indian Nat. Sci. Doc. Centre, New Delhi. 346 pp. (Translation.)

Vybiral, S. L., L. Jansky, and J. Mejsnar. 1975. Non-shivering thermogenesis in awake hibernators. Pp. 31–36 *in* Depressed Metabolism and Cold Thermogenesis (L. Jansky. ed.). Charles University Press, Prague. 286 pp.

Warner, A. C. I. 1981. Rate of passage of digesta through the gut of mammals and birds. Nutr. Abstr. Rev. Ser. B, 51:789–820.

Warner, A. C. I., and W. P. Flatt. 1965. Anatomical development of the ruminant stomach. Pp. 24–38 *in* Physiology of Digestion in the Ruminant (R. W. Dougherty, ed.). Butterworths, London. 480 pp.

Webster, A. J. F. 1974. Adaptation to cold. Pp. 71–106 *in* Environmental Physiology (D. Robertshaw, ed.). University Park Press, Baltimore. 326 pp.

White, T. C. R. 1978. The importance of relative food shortage in animal ecology. Oecologia, 33:71–86.

Wunder, B. A. 1978a. Implications of a conceptual model for the allocation of energy resources by small mammals. Pp. 68–75 *in* Populations of Small Mammals under Natural Conditions (D. Snyder, ed.). Pymatuning Lab. Ecol., Univ. Pitt. Spec. Publ. Ser., 5:1–237.

——. 1978b. Yearly differences in seasonal thermogenic shifts of prairie voles (*Microtus ochrogaster*). J. Therm. Biol., 3:98.

——. 1979. Hormonal mechanisms. Pp. 143–158 *in* Comparative Mechanisms of Cold Adaptation (L. S. Underwood, L. L. Tieszan, A. B. Callahan, and G. E. Folk, eds.). Academic Press, New York. 379 pp.

——. 1984. Strategies for, and environmental cueing mechanisms of, seasonal changes in thermoregulatory parameters of small mammals. Pp. 165–172 *in* Winter Ecology of Small Mammals (J. Merritt, ed.). Spec. Publ. Carnegie Mus. Nat. Hist., 10:1–392.

——. 1985. Energetics and thermoregulation. Pp. 812–844 *in* Biology of New World *Microtus* (R. H. Tamarin, ed.). Spec. Publ. Am. Soc. Mammal., 8:1–893.

Wunder, B. A., D. S. Dobkin, and R. D. Gettinger. 1977. Shifts of thermogenesis in the prairie vole (*Microtus ochrogaster*): strategies for survival in a seasonal environment. Oecologia, 29:11–26.

Zegers, D. A., and J. F. Merritt. 1988. Adaptations of *Peromyscus* for winter survival in an Appalachian montane forest. J. Mammal., 69:516–523.

Alan R. French

6. Mammalian Dormancy

Most of the food that an adult mammal eats is used for the production of heat to maintain a relatively high and constant body temperature. Such endothermy confers a large measure of independence from environmental conditions, but the energetic costs of this heat production can become prohibitive, especially in habitats that are seasonally cold and unproductive. In response to immediate or anticipated shortfalls in energy, many small mammals are able to reduce the level at which they regulate body temperature and, as a result, the energy needed for thermoregulation. Slight reductions in temperature and energy use occur during sleep in most species, although traditionally these are considered to be fluctuations within the normeothermic range. More-profound reductions are referred to as either shallow or deep torpor, depending on whether body temperature is above or below some arbitrary level, usually around 15°C (Hudson, 1967).

The use of torpor is quite variable (Lyman et al., 1982), but two basic strategies are evident. Many species that forage year round use brief episodes of torpor only during short-term energetic emergencies. Because such hypothermia usually occurs during the inactive portion of the daily cycle and thus is less than a day in duration, it is commonly referred to as daily torpor. Torpor bouts lasting two or three days, however, have been observed in a few species that presumably forage throughout the year (Brower and Cade, 1971; Brown and Bartholomew, 1969). At the opposite extreme are true hibernators that use

Department of Biological Sciences, State University of New York, Binghamton, New York 13902-6000 torpor to make stored energy supplies last during a period of seasonal dormancy. Multiday bouts of torpor, sometimes over a month in duration (Menaker, 1959), are common in hibernators in cold environments. Hibernating species of bats are noted for their use of daily torpor on cold days during the active season, which they then extend into multiday bouts of torpor during winter dormancy (Davis, 1970). But this pattern is not widespread among other mammals.

A mammal's capacity for torpor appears to represent a compromise between the energetic benefits and the ecological costs of low body temperatures (French, 1986, 1988). Any drop in body temperature followed immediately by rewarming (an arousal) requires less energy than continuous euthermy (Tucker, 1965). The lower the body temperature and the longer the duration of torpor, the greater the energetic savings. However, sensory function and mobility also decrease as body temperature declines, and torpid mammals are not only unable to defend territories or to reproduce, but are much less aware of their environment and more susceptible to the theft of their resources and to predation than are euthermic individuals. From these observations, I have argued that mammals should remain torpid no more than necessary to ensure survival (French, 1986). The wide variation in the use of torpor among mammals thus may have an adaptive explanation when analyzed within an energetics framework. This chapter examines the energetics of seasonal dormancy, but many of the principles hold for daily torpor as well.

Energy expended during a season of dormancy depends primarily on time spent torpid and euthermic, and the metabolic costs of those thermoregulatory states. This expenditure can be separated into six major components: cost of a dormant season = ([total hours of euthermy] [cost/*h*]) + ([total hours of torpor] [cost/*h*]) + ([total number of arousals] [cost/arousal]). Other variables, such as the costs of activity while euthermic within the hibernaculum and the costs of cooling during entrance into torpor, are relatively minor in importance and have been ignored for the sake of simplicity. None of the components of this equation is a constant. All vary with ambient temperature and also vary among individuals as a function of their body size, length of the dormant season, and energy supply (Fig. 6.1). The interactions among these influences on energy use are complex, and each will be analyzed separately using data from experiments where, for the most part, the other factors were held constant.

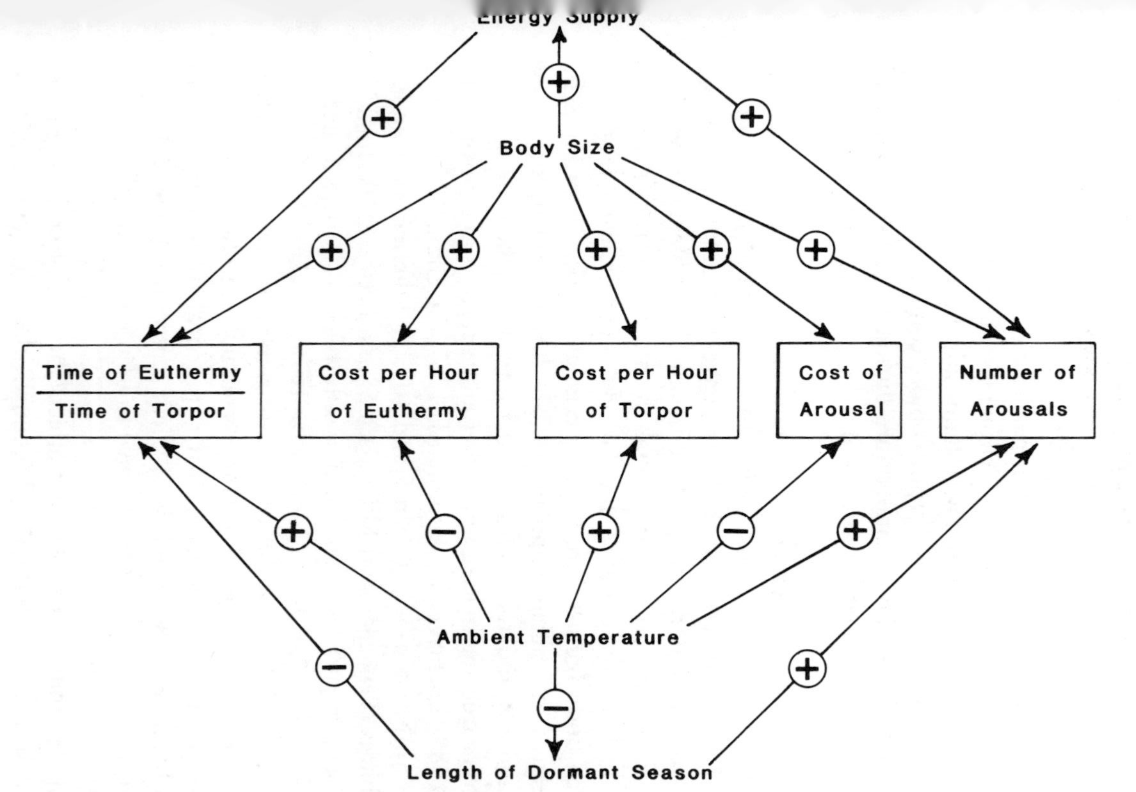

Fig. 6.1. The influence of various parameters on energy costs of a season of dormancy. Positive arrows indicate a direct relationship, and negative arrows an inverse relationship. The gradient between body and ambient temperatures during torpor is assumed to remain relatively constant. The effect of length of the dormant season on number of arousals is complicated. The number of arousals is directly related to the length of dormancy of individuals that alter the dates of entry and emergence from year to year. Species with long hibernation seasons, however, tend to arouse less frequently than do similar-sized species with short seasons, but do not necessarily have fewer total number of arousals because their seasons of dormancy last longer. (Modified from French, 1976.)

Ambient Temperature

Metabolism during both euthermia and torpor is directly proportional to the gradient maintained between body and ambient temperatures. This gradient increases in a euthermic mammal as ambient temperature falls, and consequently cost per hour of euthermy and cost of arousal are inversely related to ambient temperature below the limits of thermal neutrality (Fig. 6.1). The relationship between ambient temperature and cost per hour of torpor is more complicated. If the hypothalamic set-point for body temperature during torpor is below ambient temperature, body temperature will fall to near-ambient levels. The thermal gradient will be minimal, and it will be related to body size and heat capacity. Under such conditions, body temperature, and hence cost per hour of torpor, is directly related to ambient temperature (Fig. 6.1). All hibernators, however, have a set-point below which they will not allow body temperature to fall. As ambient temperature declines below this critical point, metabolism during torpor is elevated to maintain body temperature at relatively constant levels (Fig. 6.2). Thus when body temperature is being regulated, cost of torpor is inversely related to ambient temperature. The lowest temperature below which mammals undergo such regulated torpor may be only a degree or two above freezing in long-term hibernators from cold environments (Geiser, 1988; Lachiver and Boulouard, 1967; Meehan, 1976; Soivio et al., 1968), but can be between 5 and 10°C in hibernators from less harsh habitats (Pengelley and Kelley, 1966) and between 10 and 15°C in many species that use daily torpor only during short-term energy emergencies (Geiser, 1986a; Morhardt, 1970; Tucker, 1965; Wang and Hudson, 1970). However, temperatures to which long-term hibernators are exposed during dormancy generally are above those that elicit regulated torpor, and as such it is practical to view the cost of torpor as being directly related to ambient temperature (Fig. 6.1). Unless stated otherwise, torpor is assumed to be nonregulated throughout the rest of this chapter.

A change in ambient temperature affects the cost of euthermy more than the cost of torpor (Fig. 6.2). Mammals, therefore, use more total energy at low ambient temperatures than at high ambient temperatures unless a large amount of time is spent in torpor (French, 1976). Time spent in torpor, however, is also influenced by ambient temperature. Duration of bouts of torpor, and hence the frequency of arousals, changes seasonally in hibernators that are kept under constant environ-

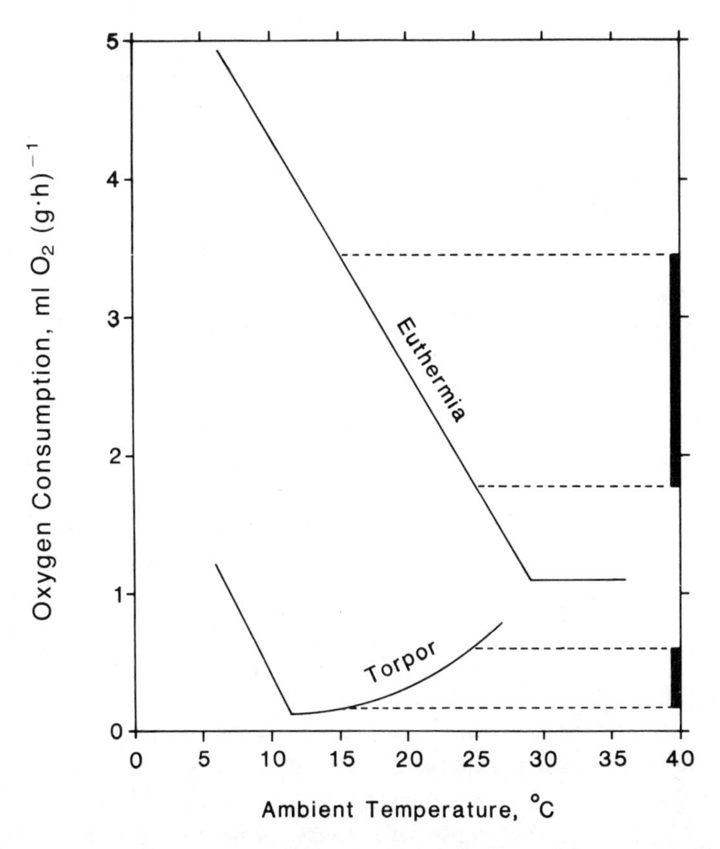

Fig. 6.2. The relationship between ambient temperature and rates of oxygen consumption derived from the dasyurid marsupial *Antechinomys laniger* when at rest during euthermia and during torpor. As ambient temperature falls from about 27 to 12°C, body temperature during torpor approximates ambient temperature, and metabolic rate declines. Body temperature during torpor is regulated above about 12°C in the face of further reductions in ambient temperature, and metabolic rates increase as a result. Note that as ambient temperature falls, metabolic costs of euthermia increase much more than costs of nonregulated torpor decline (solid bars projected to the right represent the effects of a decline from 25 to 15°C). (Data from Geiser, 1986b.)

mental conditions (Lyman et al., 1982). Torpor bouts of most species are short at the beginning and end of dormancy, but are long and relatively constant in duration throughout most of the winter. These are the autumn, winter, and spring phases of hibernation according to the terminology proposed by Twente and Twente (1967). The magnitude of these seasonal changes in torpor duration is profoundly affected by changes in ambient temperature. Maximum torpor duration is inversely related to temperature (Geiser, 1988; Twente and Twente, 1965; Twente et al., 1977) such that number of arousals and relative amount of time spent euthermic increase as air temperature rises (Fig. 6.1). Arousal frequency in midwinter and metabolic rates during deep (nonregulated) torpor scale similarly with respect to body temperature (Hammel et al., 1968; Twente and Twente, 1965), suggesting that some metabolic factor may influence the timing of arousals during the winter phase of hibernation. When body temperatures are regulated during torpor, torpor duration tends to be shorter than expected from an extrapolation of data gathered at warmer temperatures. This again suggests a link between metabolism and timing of arousals, but the correlation is imprecise (Geiser, 1988).

There is no universal answer to the question of what ambient temperature is energetically optimum for hibernation, because the temperature-dependent ratio of time euthermic to time torpid varies greatly among species. For hibernators such as badgers, which enter shallow, regulated torpor infrequently during a 2–3 month dormant season (Harlow, 1981), total energy expenditures clearly are inversely related to burrow temperatures. For hibernators that spend most of dormancy in deep torpor, however, just the opposite is true. For example, the jumping mouse *Zapus princeps* depletes its fat deposits twice as rapidly at 15°C as at 4.6°C (Cranford, 1978). As ambient temperature declines, the energetic benefits of the increase in torpor duration more than compensate for the increased cost of arousal and cost per hour of euthermy. The optimum temperature from an energetic perspective appears to be the one that permits the longest bouts of torpor; namely the lowest ambient temperature above the point where the hibernator begins to regulate body temperature at a gradient above ambient.

Most hibernators that overwinter in burrows probably do not have many options as to the thermal environment in which they reside. Successful individuals generally choose hibernacula deep enough to avoid freezing. But the depth at which the ground freezes often varies from year to year, and animals must be conservative in deciding where to build their winter nests. Depth of the hibernaculum may reflect a balance between safety and need to maximize torpor duration. Because the least amount of energy apparently is used at the coolest temperature above which regulated torpor occurs, hibernating in comparatively shallow nests may be energetically advantageous for species that allow body temperatures to fall to near 0°C. Residing close to the surface, however, may increase chances of freezing during torpor or being discovered by predators. The above balance could change over the course of the hibernation season, but this possibility has not been investigated systematically and little is known about the thermal preferences of most burrowing hibernators during dormancy.

Cave-dwelling bats, on the other hand, are known to select temperatures with remarkable precision. Many observations indicate that bats hibernate only in the cooler portions of suitable caverns (Beer and Richards, 1956; Goehring, 1972) and that these bats make intracavern movements as the microclimates change during the dormant season (Daan, 1973; Twente, 1955). Species are often segregated by temperature in the same cave. Energetic consequences of temperature selection probably are best demonstrated by the behavior of horseshoe bats (*Rhinolophus ferrumequinum*) (Ransome, 1968, 1971). This species feeds during the winter when temperatures are mild. Insects fly when outside temperatures rise above 15°C, and if upon arousal the bats find food, they return to their caves and select a relatively warm temperature at which to resume hibernation. As a consequence, they arouse and have the opportunity to feed again within a couple of days. Conversely, if they fail to find food, the bats select a colder temperature and remain torpid for nearly a week. The frequency of arousal at any air temperature changes seasonally, but horseshoe bats compensate for this seasonal change by simultaneously changing their thermal preferences such that the above durations of torpor remain relatively constant.

Body Size

At any ambient temperature, large mammals have higher absolute rates of metabolism than do small mammals, although metabolic rates during euthermy and deep torpor appear to scale differently with respect to body size. Thermal conductance, and hence the rate at which energy is used to maintain euthermia, is proportional to $mass^{0.5}$ at cool temperatures where resting metabolism is elevated above basal levels (Herreid and Kessel, 1967). Limited data suggest that metabolic rates during deep torpor scale close to $mass^{1.0}$ (French, 1985; Kayser, 1961), but confidence in the precision of that exponent is not high. Because of high rates of metabolism while euthermic, large hibernators save absolutely more energy upon entering deep torpor than do small hibernators, but because of the dissimilar allometries of metabolic rates, proportional savings are greatest for the smallest species (Fig. 6.3). This disproportionate reduction of metabolism has two sources:

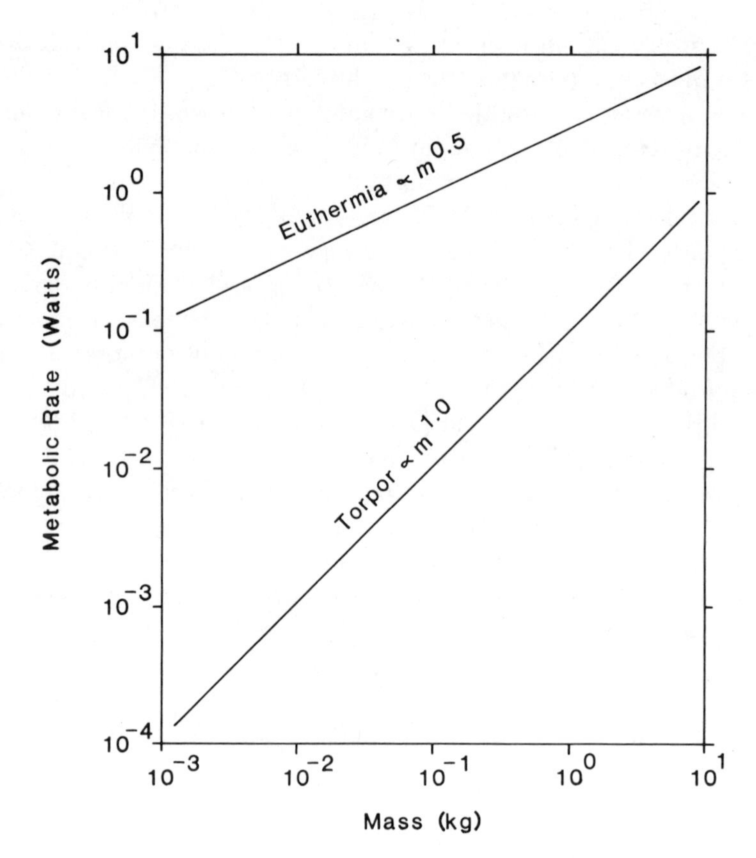

Fig. 6.3. A comparison of the allometries of metabolic rates during euthermia and deep (nonregulated) torpor at 5°C. m = mass. The regression for euthermia was calculated from the equation for thermal conductance derived by Herreid and Kessel (1967), and the regression for torpor was calculated from the average values reported by French (1985).

(1) the minimum gradient between body and ambient temperatures during torpor is a positive function of mass (Kayser, 1964; Morrison, 1960); and (2) the effect of body temperature on rates of metabolism (i.e., Q_{10}) during entry into torpor is inversely related to mass (French, 1985; Geiser and Kenagy, 1988). In other words, at any low ambient temperature, small species appear to undergo a disproportionately greater decrease in metabolic rate per degree reduction in body temperature and also to attain lower body temperatures during deep torpor than do large species.

Large hibernators more than compensate for their high rates of metabolism by storing more energy than do small hibernators. Hence the direct relationship between body size and energy supplies during dormancy in Figure 6.1. The proportion of body mass that is fat at the start of dormancy does not vary systematically with size (French, 1986; Morrison, 1960), indicating that fat content scales approximately with mass$^{1.0}$. In contrast, metabolic rate during euthermia at any cold ambient temperature is proportional to mass$^{0.5}$. The ability to store energy in the form of body fat, therefore, increases with increasing body size faster than does energy use during euthermia. A large mammal, consequently, can fast at euthermic body temperatures for a longer period of time than can a small mammal if both have the same percentage of body fat. Small mammals will starve within a few days unless they become torpid, but large mammals have the potential to remain warm for many months without eating. In fact, although many large mammals such as bears (Hock, 1960) have long dormant seasons, no mammal above about 5 kg in mass has been shown to have the capacity for deep torpor. Furthermore, the smaller the hibernator, the more time it must spend at low body temperatures during dormancy, all other things being equal.

Large hibernators appear to have some energetic advantages over small ones. They do not consistently have longer hibernation seasons or deposit proportionately less fat, but do spend more time at high body temperatures during dormancy than small species. The frequency at which long-term hibernators kept at 5°C arouse in midwinter is proportional to mass$^{0.1}$, and, more important energetically, duration of those euthermic episodes scales with mass$^{0.38}$ (French, 1985). This trend in time spent at high body temperatures tends to equalize rates at which fat stores are depleted among long-term hibernators of different sizes. The allometry of energy use during dormancy is a function of allometries of the costs of euthermic intervals, arousals, and bouts of torpor, all of which can be roughly estimated from available data. Each of these three parameters represents the product of metabolic rate and time spent in that thermoregulatory state. At temperatures below thermal neutrality, metabolism during euthermia scales with mass$^{0.5}$ (Herreid and Kessel, 1967); maximal sustained metabolism, as would occur at all body temperatures during the arousal process, appears to be a constant multiple of standard metabolic rate (Schmidt-Nielsen, 1984) and thus should scale approximately with mass$^{0.72}$; and metabolism during deep torpor scales approximately with mass$^{1.0}$ (French, 1985). Duration of euthermic intervals, arousals, and bouts of torpor scale with mass$^{0.38}$ (French, 1985), mass$^{0.40}$ (Heinrich and Bartholomew, 1971), and mass$^{-0.10}$ (French, 1985), respectively. Therefore, cost of a euthermic interval is proportional to $mass^{0.88}$ ($mass^{0.5} \times mass^{0.38}$), cost of an arousal is proportional to $mass^{1.12}$ ($mass^{0.72} \times mass^{0.40}$), and cost of a bout of torpor is proportional to $mass^{0.90}$ ($mass^{1.0} \times mass^{-0.10}$). Although these allometries of energy use are just approximations, they all are close to the presumed allometry for energy storage as fat ($mass^{1.0}$). Consequently, large and small hibernators starting with the same percentage of body fat can have hibernating seasons of about the same length (French, 1986, 1988).

The ratio of time of euthermy to time of torpor increases in the spring phase of dormancy in most hibernators, but the magnitude of that increase is a positive function of body size. A rise in time spent at high body temperatures increases a hibernator's ability to assess environmental conditions and thereby more accurately time emergence from dormancy. A complete cessation of torpor permits early breeding. Large hibernators tend to resume activity and begin breeding earlier in the spring than ecologically similar hibernators that are small. Species with long dormant seasons have relatively little time for reproduction and preparation for the next hibernation season. These time constraints are greater for large species than small, because times for gestation and juvenile growth increase with size (Western, 1979). An early emergence, however, is perilous, especially if food is not yet continuously available. Large hibernators emerge early because they need more time for reproduction, but they can afford to take this energetic gamble because they use their fat reserves at a relatively slower rate when euthermic than do small species (French, 1986, 1988).

Intraspecific differences in rate of energy use in the spring occur when different sexes and age classes within a population have different energetic constraints or different reproductive strategies or both. Large adult yellow-bellied marmots (*Marmota flaviventris*) terminate hibernation spontaneously in the spring (French, 1990) and begin breeding before food becomes available, whereas small yearlings, which do not breed, are more conservative than adults in that they emerge about a month later when food is plentiful (Carey, 1985). Small youngsters hedge their bets by increasing the frequency at which they arouse in the spring without ceasing torpor until they feed (French, 1990). Similarly, in the case of the ground squirrels *Spermophilus beldingi* and *S. lateralis*, only the largest males terminate hibernation spontaneously (Barnes, 1984; French, 1982). Small males, which do not breed following dormancy, and females continue to hibernate until food is available. Large male ground squirrels do not produce sperm immediately upon cessation of torpor and often remain underground at high body temperatures until they attain reproductive competency. I have observed this accelerated use of energy near the end of dormancy in captive *S. beldingi* kept in artificial burrow systems, and this behavior was deduced also for *S. lateralis* by Barnes et al. (1986). Even males of the tiny jumping mouse *Zapus princeps* emerge before females and also gamble more with their fat reserves by arousing more frequently in the spring phase of dormancy (French, 1986).

Length of the Dormant Season

The date at which hibernators emerge from dormancy often differs slightly from year to year in response to variations in climate (Michener, 1984), and individuals must be flexible and sufficiently conservative in their hibernation responses to survive during the worst-case conditions. In any one year, however, the duration of dormancy can vary greatly among species, as reflected in large variations in the percentage of time hibernators spend torpid. Badgers (*Taxidea taxus*), for example, are dormant for only 2–3 months, during which time they undergo infrequent shallow torpors of about a day in duration (Harlow, 1981). This pattern contrasts greatly with the frequent, long, and deep torpors of yellow-bellied marmots, which are nearly the same size as badgers but are inactive about 8 months each year. Furthermore, individuals that have short dormant seasons are more likely to gamble with their fat reserves and terminate hibernation spontaneously than are members of similar-sized species that have long hibernation seasons (French, 1988). These data again suggest that mammals spend only as much time in torpor as necessary for survival.

Energy Supply

Although related to energetics, interspecific differences in patterns of thermoregulation during dormancy are undoubtedly fixed genetically. Some individual flexibility does exist, however, and certain hibernators are able to partially compensate for an accumulation of a relatively small store of energy by increasing the time they spend in torpor during dormancy (French, 1989). Spontaneous termination of hibernation in fat but not lean male Belding's and golden-mantled squirrels is one example of this energetic feedback. More subtle correlations between fat content and rate of fat utilization have been documented best in the spring phase of dormancy in those *S. beldingi* that do not spontaneously terminate hibernation. Squirrels that enter the spring phase with small fat reserves do not shorten the duration of their torpors as much as heavier squirrels, and the leanest animals may not show a distinct spring phase at all (French, 1982). As these squirrels begin to run out of fat, episodes of torpor begin to lengthen once again. Timing of this starvation period differs among individuals as a function of the amount of fat deposited prior to dormancy. Furthermore, the likelihood that females at 5°C will have euthermic episodes longer than 36 h at the transition between winter and spring phases of dormancy increases with mass (French, 1989), and at 15°C where long intervals of euthermia occur throughout the spring phase, mean duration of euthermic episodes shows a positive correlation with mass (French, 1982).

Constraints of body size on energy availability during dormancy have been circumvented by species that store food rather than body fat. In general, species that store food have a relative abundance of energy, and, as a result, arouse much more frequently throughout dormancy than do similar-sized hibernators that rely on fat (French, 1986). For example, many pocket mice in the genus *Perognathus* are seasonal hibernators, and some remain underground 10 months of each year (O'Farrell et al., 1975). These mice may experience over a hundred arousals, each lasting several hours, whereas a similar-sized bat may arouse for only an hour or two on less than two dozen occasions during dormancy (French, 1977; Twente et al., 1985). Maximum torpor duration of pocket mice is less than a week (French, 1977), but that of some bats can be over a month (Menaker, 1959; Twente et al., 1985). Short torpors lasting less than a week are characteristic of hibernators that store food, such as chipmunks (Jameson, 1964; Panuska, 1959; Pivorun, 1976) and hamsters (Hall and Goldman, 1980).

Hibernators that store food appear to be more sensitive to alterations in their energy supply than are species that deposit fat (French, 1989). I found that at 5°C the pocket mouse *P. inornatus* and the chipmunk *Tamias striatus* both undergo seasonal changes in the durations of bouts of torpor and euthermia similar to those described for hibernators that store fat. In other words, both species gamble more with their energy supplies as spring approaches by increasing the amount of time spent at high body temperatures. The magnitude of those seasonal changes, however, is strongly dependent on the amount of seeds ani-

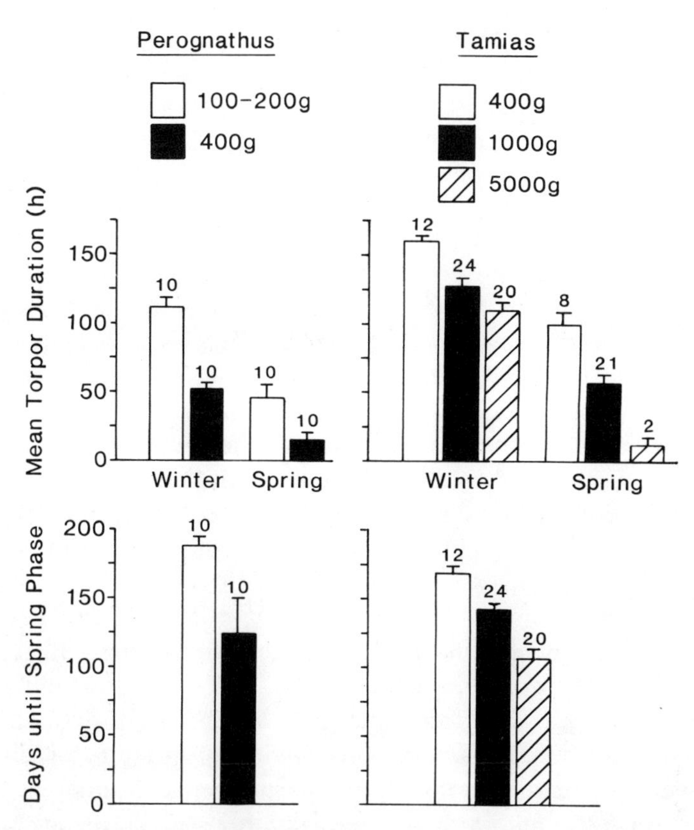

Fig. 6.4. Effects of amount of food stored prior to dormancy (top) on mean duration of episodes of torpor during both winter and spring phases of hibernation and (bottom) on time spent in hibernation before energy use was accelerated in the spring phase. Means ±SE are for (left) the pocket mouse *Perognathus inornatus* and (right) the chipmunk *Tamias striatus*. Numbers are sample sizes. Data from pocket mice given 100 g and 200 g of seeds were similar and so were combined.

mals are given at the start of dormancy. Pocket mice given 400 g of seeds had episodes of torpor, both in the winter and spring phases of hibernation, that averaged less than half as long as those of individuals that started with either 100 g or 200 g of seeds (Fig. 6.4). Chipmunks also showed an inverse relationship between average torpor duration and energy availability. As food supply increased, however, chipmunks increased their propensity for terminating hibernation spontaneously in the spring, whereas all the tiny pocket mice continued to hibernate until they ran out of food. Furthermore, individuals of both species began the springtime increase in energy consumption sooner when they had abundant supplies of food than when their energy supplies were more limited (Fig. 6.4). Again, hibernating mammals appear to reduce the time they spend at low body temperatures whenever possible.

Conclusions

Mammalian dormancy, defined as a period of inactivity, spans a continuum of physiological states ranging from sleep to seasonal hibernation, and the energetic savings of each strongly depends on the level at which body temperature is regulated. There is considerable variability in the patterns of dormancy, but two fundamental strategies exist: some mammals employ torpor as a facultative response to an immediate energetic emergency, whereas others enter a seasonal dormancy and use torpor to slow the rate of depletion of their energy stores and thereby avoid a shortfall of energy in the future.

The need to lower body temperatures to conserve energy is countered by the advantages of euthermy inherent to all mammals, and as a consequence hibernators use torpor sparingly. These conflicting pressures change seasonally and lead to concomitant changes in the time spent torpid during the dormant season. The relative magnitudes of these pressures also differ among species and between sexes and age-classes within species. Comparisons among hibernators of different sizes, with different methods of energy storage, with different reproductive strategies, and with different durations of dormancy suggest that time spent in torpor has been matched evolutionarily to the animals' energy deficits. Furthermore, individuals have some flexibility in their torpor responses. When they have been successful at foraging and thereby have accumulated a relatively large store of energy, hibernators spend less time torpid during the dormant season than they would otherwise. Species that deposit fat have relatively small stores of energy and can afford to gamble less with those stores than similar-sized species that cache food. Consequently, hibernators that store food respond to an abundance of energy by increasing the frequency at which they arouse throughout the hibernation season, whereas individuals that have comparatively large fat deposits can afford to increase time spent euthermic only in the spring when the value of high body temperatures is greatest.

Literature Cited

Barnes, B. M. 1984. The influence of energy stores on the activation of reproductive function in the golden-mantled ground squirrel. J. Comp. Physiol. B, 154:421–425.

Barnes, B. M., M. Kretzmann, P. Licht, and I. Zucker. 1986. The influence of hibernation on testis growth and spermatogenesis in the golden-mantled ground squirrel, *Spermophilus lateralis*. Biol. Reprod., 35:1289–1297.

Beer, J. R., and A. G. Richards. 1956. Hibernation in the big brown bat. J. Mammal., 37:31–41.

Brower, J. E., and T. J. Cade. 1971. Bicircadian torpor in pocket mice. Bioscience, 21:181–182.

Brown, J. H., and G. A. Bartholomew. 1969. Periodicity and energetics of torpor in the kangaroo mouse, *Microdipodops pallidus*. Ecology, 50:705–709.

Carey, H. V. 1985. The use of foraging areas by yellow-bellied marmots. Oikos, 44:273–279.

Cranford, J. A. 1978. Hibernation in the western jumping mouse (*Zapus princeps*). J. Mammal., 59:496–509.

Daan, S. 1973. Activity during natural hibernation in three species of vespertilionid bats. Neth. J. Zool., 23:1–71.

Davis, W. H. 1970. Hibernation: ecology and physiological ecology. Pp. 266-300 *in* Biology of Bats, vol. 1 (W. A. Wimsatt, ed.). Academic Press, New York. 406 pp.

French, A. R. 1976. Selection of high temperatures for hibernation by the pocket mouse, *Perognathus longimembris*: ecological advantages and energetic consequences. Ecology, 57:185–191.

——. 1977. Periodicity of recurrent hypothermia during hibernation in the pocket mouse, *Perognathus longimembris*. J. Comp. Physiol., 115:87–100.

——. 1982. Intraspecific differences in the pattern of hibernation in the ground squirrel *Spermophilus beldingi*. J. Comp. Physiol., 148:83–91.

——. 1985. Allometries of the durations of torpid and euthermic intervals during mammalian hibernation: a test of the theory of metabolic control of the timing of changes in body temperature. J. Comp. Physiol. B, 156:13–19.

——. 1986. Patterns of thermoregulation during hibernation. Pp. 393–402 *in* Living in the Cold: Physiological and Biochemical Adaptations (H. C. Heller, X. J. Musacchia, and L. C. H. Wang, eds.). Elsevier, New York. 587 pp.

——. 1988. The patterns of mammalian hibernation. Am. Sci., 76:568–575.

——. 1989. The impact of variations in energy availability on the time spent torpid during the hibernation season. Pp. 129–136 *in* Living in the Cold, II (A. Malan and B. Canguilhem, eds.). Colloque INSERM/John Libby Eurotext, London. 525 pp.

——. 1990. Age-class differences in the pattern of hibernation in yellow-bellied marmots, *Marmota flaviventris*. Oecologia, 82:93–96.

Geiser, F. 1986a. Temperature regulation in heterothermic marsupials. Pp. 207–212 *in* Living in the Cold: Physiological and Biochemical Adaptations (H. C. Heller, X. J. Musacchia, and L. C. H. Wang, eds.). Elsevier, New York. 587 pp.

——. 1986b. Thermoregulation and torpor in the kultarr, *Antechinomys laniger* (Marsupialia: Dasyuridae). J. Comp. Physiol. B, 156:751–757.

——. 1988. Reduction of metabolism during hibernation and daily torpor in mammals and birds: temperature effect or physiological inhibition? J. Comp. Physiol. B, 158:25–37.

Geiser, F., and G. J. Kenagy. 1988. Torpor duration in relation to temperature and metabolism in hibernating ground squirrels. Physiol. Zool., 61:442–449.

Goehring, H. H. 1972. Twenty-year study of *Eptesicus fuscus* in Minnesota. J. Mammal., 53:201–207.

Hall, V., and B. Goldman. 1980. Effects of gonadal steroid hormones on hibernation in the Turkish hamster (*Mesocricetus brandti*). J. Comp. Physiol., 135:107–114.

Hammel, H. T., T. J. Dawson, R. M. Abrams, and H. J. Anderson. 1968. Total calorimetric measurements on *Citellus lateralis* in hibernation. Physiol. Zool., 41: 341–357.

Harlow, H. J. 1981. Torpor and other physiological adaptations of the badger (*Taxidea taxus*) to cold environments. Physiol. Zool., 54:267–275.

Heinrich, B., and G. A. Bartholomew. 1971. An analysis of pre-flight warm-up in the sphinx moth, *Manduca sexta*. J. Exp. Biol., 55:223–239.

Herreid, C. F., II, and B. Kessel. 1967. Thermal conductance in birds and mammals. Comp. Biochem. Physiol., 21:405–414.

Hock, R. J. 1960. Seasonal variations in physiologic functions of arctic ground squirrels and black bears. Bull. Mus. Comp. Zool., 124:155–169.

Hudson, J. W. 1967. Variations in the patterns of torpidity of small homeotherms. Pp. 30–46 *in* Mammalian Hibernation, III (K. C. Fisher, A. R. Dawe, C. P. Lyman, E. Schonbaum, and F. E. South, Jr., eds.). Oliver and Boyd, Edinburgh. 535 pp.

Jameson, E. W. Jr. 1964. Patterns of hibernation of captive *Citellus lateralis* and *Eutamias speciosus*. J. Mammal., 45:455–460.

Kayser, C. 1961. The physiology of natural hibernation. Pergamon Press, Oxford.

——. 1964. La dépense d'énergie des Mammifères en hibernation. Arch. Sci. Physiol., 18:137–150.

Lachiver, F., and R. Boulouard. 1967. Evolution de la périodicité des phases de sommeil et de réveil chez le Lérot (*Eliomys quercinus*) au cours du sommeil hivernal et de la léthargie induite en été. J. Physiol. (Paris), 59:250–251.

Lyman, C. P., J. S. Willis, A. Malan, and L. C. H. Wang. 1982. Hibernation and torpor in mammals and birds. Academic Press, New York. 317 pp.

Meehan, T. E. 1976. The occurrence, energetic significance and initiation of spontaneous torpor in the Great Basin pocket mouse, *Perognathus parvus*. Ph.D. dissertation, University of California, Irvine. 111 pp.

Menaker, M. 1959. The frequency of spontaneous arousal from hibernation in bats. Nature, 203:540–541.

Michener, G. R. 1984. Age, sex, and species differences in the annual cycles of ground-dwelling sciurids: implications for sociality. Pp. 81–107 *in* Biology of Ground-dwelling Sciurids (J. O. Murie, and G. R. Michener, eds.). University of Nebraska Press, Lincoln. 459 pp.

Morhardt, J. E. 1970. Body temperatures of white-footed mice *Peromyscus* during daily torpor. Comp. Biochem. Physiol., 33A:423–439.

Morrison, P. 1960. Some interrelations between weight and hibernation function. Bull. Mus. Comp. Zool., 124:75–91.

O'Farrell, T. P., R. J. Olson, R. O. Gilbert, and J. D. Hedlund. 1975. A population of Great Basin pocket mice, *Perognathus parvus*, in the shrub-steppe of south-central Washington. Ecol. Monogr., 45:1–28.

Panuska, J. A. 1959. Weight patterns and hibernation in *Tamias striatus*. J. Mammal., 40:554–556.

Pengelley, E. T., and K. H. Kelley. 1966. A "circannian" rhythm in hibernating species of the genus *Citellus* with observations on their physiological evolution. Comp. Biochem. Physiol., 19:603–617.

Pivorun, E. B. 1976. A biotelemetry study of the thermoregulatory patterns of *Tamias striatus* and *Eutamias minimus* during hibernation. Comp. Biochem. Physiol., 53A:265–271.

Ransome, R. D. 1968. The distribution of the greater horse-shoe bat, *Rhinolophus ferrumequinum*, during hibernation, in relation to environmental factors. J. Zool., 154:77–112.

——. 1971. The effect of ambient temperature on the arousal frequency of the hibernating greater horseshoe bat, *Rhinolophus ferrumequinum*, in relation to site selection and the hibernation state. J. Zool., 164:353–371.

Schmidt-Nielsen, K. 1984. Scaling: Why Is Animal Size So Important? Cambridge University Press, Cambridge. 241 pp.

Soivio, A., H. Tahti, and R. Kristoffersson. 1968. Studies on the periodicity of hibernation in the hedgehog (*Erinaceus europaeus L.*). III. Hibernation in a constant ambient temperature of −5°C. Ann. Zool. Fenn., 5:224–226.

Tucker, V. A. 1965. Oxygen consumption, thermal conductance, and torpor in the California pocket mouse *Perognathus californicus*. J. Cell. Comp. Physiol., 65:393–403.

Twente, J. W. 1955. Some aspects of habitat selection and other behavior of cavern-dwelling bats. Ecology, 36:706–732.

Twente, J. W., and J. A. Twente. 1965. Effects of core temperature upon duration of hibernation of *Citellus lateralis*. J. Appl. Physiol., 20:411–416.

——. 1967. Seasonal variation in the hibernating behavior of *Citellus lateralis*. Pp. 47–63 *in* Mammalian Hibernation, III (K. C. Fisher, A. R. Dawe, C. P. Lyman, E. Schonbaum, and F. E. South, Jr., eds.). Oliver and Boyd, Edinburgh. 535 pp.

Twente, J. H., J. Twente, and V. Brack Jr. 1985. The duration of the period of hibernation of three species of vespertilionid bats. II. Laboratory studies. Can. J. Zool., 63:2955–2961.

Twente, J. H., J. A. Twente, and R. M. Moy. 1977. Regulation of arousal from hibernation by temperature in three species of *Citellus*. J. Appl. Physiol. Resp. Environ. Exercise Physiol., 42:191–195.

Wang, L. C. H., and J. W. Hudson. 1970. Some physiological aspects of temperature regulation in the normothermic and torpid hispid pocket mouse, *Perognathus hispidus*. Comp. Biochem. Physiol., 32:275–293.

Western, D. 1979. Size, life history and ecology in mammals. Afr. J. Ecol., 17:185–204.

Richard W. Hill

7. The Altricial/Precocial Contrast in the Thermal Relations and Energetics of Small Mammals

The physiological ecology of nestling and weanling small mammals has been relatively neglected. This is no doubt because nestlings and weanlings are less obvious than adults and can be more difficult to study, especially if one wants to keep conditions as true to nature as possible. Nonetheless, the nestling and weanling stages are so important ecologically and evolutionarily that a strong effort must be made to bolster our knowledge of them.

This chapter is largely an exploration of the contrasts between altricial and precocial development, emphasizing arguments based on energetics for the adaptive evolution of thermoregulatory altriciality. Thermal relations are so important to energetics and growth that they feature prominently in the chapter. Unfortunately, much of what we know about the significance and implications of altricial and precocial development is hypothetical. Some of my major goals are (1) to explore not only the hypotheses but the assumptions behind them, (2) to disentangle certain major lines of argument that heretofore have been needlessly comingled, (3) to correct a major misconception that has permeated much of the previous writing on this topic, (4) to show how this correction must substantially alter thinking about the altricial/precocial contrast, and (5) to identify questions that are particularly in need of answers.

Before turning to these major themes, I want to examine mortality during the nestling and weanling stages of certain small mammals for

The Museum and Department of Zoology, Michigan State University, East Lansing, Michigan 48824.

which extensive demographic data are available. To my knowledge, the mortality rates of nestlings and weanlings have not been extensively summarized elsewhere. Certainly, they have gone almost unmentioned in the physiological literature. Yet, as will be seen, the data on mortality have enormous implications.

Death in the Weeks following Birth

Large sets of data are available on the death rates of nestlings and weanlings in natural populations for certain *Peromyscus* species and voles. I have attempted a full review of the literature on *Peromyscus*. As might be guessed, the death rates of interest are difficult to assess. The usual approach is first to estimate the number of animals born in an area by estimating the number of parturitions and multiplying by average litter size; each instance that a female is observed pregnant and later lactating is counted as a parturition. The number of juveniles caught is then compared with the estimated number of births to assess mortality between birth and trappable age. Since *Peromyscus* do not enter traps until 4–6 weeks old, in effect the mortality rates estimated for *Peromyscus* are for the period from birth to 4–6 weeks of age. A variety of obvious methodological problems (some mentioned later) can arise in such studies. Nonetheless, these types of estimates of early mortality rates are in general the best available. Sometimes, novel approaches are taken to estimating death rates. For example, instead of using traps to follow his population of *P. maniculatus*, Howard (1949) used nest boxes. He was able to enumerate directly and mark newborn animals and then estimate their mortality on a continuous basis. In this present summary, I have paid close attention to the methodological details of each study, and in certain cases I have excluded mortality-rate estimates because of methodological considerations (see note to Table 7.1).

Table 7.1 summarizes the estimates of mortality rate in *Peromyscus*. The mortality rate within a species can vary with factors such as season (e.g., Hansen and Batzli, 1978). However, the summary statistics deserve stress here: 71% of *P. leucopus* are estimated to die before they are 4–6 weeks old, and 51% of *P. maniculatus* die before that age.

Harland et al. (1979) noted that all extant juveniles are unlikely to be trapped and that incomplete trapping of juveniles will tend to cause overestimation of the early mortality rate. They presented an estimate

TABLE 7.1

Percent survival between birth and 4–6 weeks of age in natural populations of two species of *Peromyscus*

	Percent survival	Number of births	Number of survivors	Reference
			Peromyscus leucopus	
	4	140	6	Bendell, 1959[a]
	59	103	61	Hansen & Batzli, 1978: spring[b]
	15	110	16	Hansen & Batzli, 1978: summer[b]
	20	179	35	Hansen & Batzli, 1978: fall[b]
	69	91	63	Harland et al., 1979[c]
TOTAL	29	623	181	
			Peromyscus maniculatus	
	71	217	154	Fairbairn, 1977
	25	169	42	Howard, 1949: before mid-Sept.[d]
	70	153	107	Howard, 1949: after mid-Sept.[d]
	59	217	129	Mihok, 1979[e]
	31	373	117	Millar & Innes, 1983[f]
	5	59	3	Sullivan, 1977: mainland
	72	61	44	Sullivan, 1977: large island
	94	98	92	Sullivan, 1977: small island
	54	154	83	Sullivan, 1979: forest
	57	190	109	Sullivan, 1979: logged/burned area
	46	54	25	Sullivan, 1979: logged area
	35	83	29	Van Horne, 1981: logged 2–5 yr before study
	59	39	23	Van Horne, 1981: logged 7–9 yr before study
	29	137	40	Van Horne, 1981: logged 23–25 yr before study
	37	59	22	Van Horne, 1981: not logged
TOTAL	49	2063	1019	

Note: In some instances, subcategories of data presented in original papers were combined. I excluded a number of studies from this summary because of methodological considerations. Specifically, I excluded trapping studies in which the trap interval was longer than 3 weeks because longer trapping intervals could entirely miss numerous pregnancies. I also excluded studies (except those on islands) where the study area was in one type of habitat (e.g., field) closely surrounded by another type (e.g., forest); in such settings, emigration of juveniles seems unlikely to be balanced by immigration, thus heightening concerns about the confounding of emigration and death. I also excluded data for study plots that were food-supplemented or subjected to deliberate removal of animals; and I excluded studies where there was a strong likelihood of confounding prenatal and postnatal mortality.

[a]See Loeb and Schwab, 1987, for potential criticisms, which in any case will not alter greatly the percent survival estimate.

[b]Data listed are from standard trapping study, sexes combined. This report also lists data from another, more questionable methodology.

[c]The data listed here are those most methodologically comparable to other studies (see text for further discussion). It is noteworthy that the estimate of number of survivors is for 4-week-olds. Additional, sizable mortality was known to occur by 6 weeks of age but was not fully quantified.

[d]Five-week survivorship.

[e]Two years combined, using the average estimate of number of litters in each year.

[f]All grids and years combined. Mortality was subdivided into that attributable to disappearance of the mother and that not. About 30% of mortality was caused by disappearance of the mother.

of juvenile trappability and concluded that, if that trappability were taken into account, the mortality rate up to 4 weeks of age in their study might have been as low as 12% (rather than the 31% given in Table 7.1). Most investigators provide no estimate of juvenile trappability. I have not applied Harland et al.'s correction to all the studies summarized for three reasons: (1) the size of the correction is uncertain; (2) application to other studies would be an extrapolation; and (3) to get closest to the truth, such a correction should not be applied unless corrections are also applied for sources of opposite error such as incomplete maternal trappability (incomplete trapping of mothers will tend to cause *under*estimation of early mortality rates: see Anderson and Boonstra, 1979, for noteworthy comments on this). In general (as already suggested), the straightforward estimates of early mortality rate in Table 7.1 are as accurate as can be had at present, even though we know it would be desirable to correct them for a number of sources of potential error.

The study by Goundie and Vessey (1986) on *P. leucopus* is excluded from Table 7.1 because the woodlot they worked in was closely surrounded by field (see note to Table 7.1). However, their estimates of the death rate prior to the age of possible emigration should be reliable. By following animals born in nest boxes, they estimated the death rate between birth and 3 weeks of age to be 60% in the spring and 88% in fall.

Voles are another group for which extensive demographic data are available. Gliwicz (1983) summarized estimates of early mortality of *Clethrionomys glareolus* and found an overall death rate of 60–70% between birth and 6 weeks of age. Early mortality rates (birth to 4–6 weeks old) in New World *Microtus* seem generally to be as high or higher. Boonstra and Krebs (1978) reported 70% in *M. townsendii*. For *M. pennsylvanicus*, mortality rates of 88% (Getz, 1960), 68% (Getz et al., 1979: weighted average for two habitats), 92% (Golley, 1961: 34 days of age), and 85% (Krebs et al., 1969) have been found. For *M. ochrogaster*, rates of 78–90% (Cole and Batzli, 1979), 89% (Getz et al., 1979: weighted average for three habitats), and 80% (Krebs et al., 1969) were reported.

The Significance of Early Death Rates

A chapter on physiology may at first seem a strange place to summarize death rates. However, they tell an important tale.

We learn from the death rates that for an estimated 50–80% (or more) of small mammals born the life of a nestling or weanling is the only life ever led. Thus, if we want to know about the physiology of an "average" *Peromyscus* or *Microtus*, we must understand the physiology of the nestling and weanling stages. Adult physiology provides only a truncated perspective.

We also learn from early death rates that the nestling and weanling stages are potentially times of enormous natural selection. We may well be unable to comprehend the evolutionary biology of small mammals until we have a far better understanding than at present of the processes at work during their early life stages.

Altricial and Precocial Development

Some species of small mammals are born in a relatively naked, uncoordinated, and helpless state and pass through a relatively long postnatal period of dependency on their parents. These are said to undergo *altricial development*. Other species are more developmentally advanced when born and are said to show *precocial development*. They are typically furred at birth or soon after; their eyes typically open soon; and they mature comparatively rapidly in locomotor ability and other respects, achieving independence of their parents relatively early. McClure (1987), for example, provided a summary of the differences between two species of rodents that are similar in adult size but differ in whether their mode of development is altricial (*Neotoma floridana*) or precocial (*Sigmodon hispidus*). Altricial and precocial development are not monolithic categories; intergrades exist between the extremes, and a species may be relatively altricial in certain respects and relatively precocial in others (e.g., Kurta and Kunz, 1987). Nonetheless, the distinction between altricial and precocial development is a useful organizing principle in the analysis of mammalian postnatal ontogeny.

To the degree that parallels can be drawn between the concept of precociality in reptiles and that in mammals, all extant reptiles must be classed as precocial (Case, 1978a). At the time of hatching, they have open eyes, are capable of adultlike locomotion, and must be self-sufficient in obtaining food. This condition of extant reptiles suggests that in the evolution of mammals (and birds), precociality was the preexisting form of development.

Advantages of precocial development come easily to mind. For example, among small mammals, precocial young are typically quicker than altricial ones to become mature enough to survive death of their parents. Newborn precocial young have larger brains relative to body size than newborn altricial young (Harvey and Bennett, 1983; Pagel and Harvey, 1988); precociality may represent a heterochronic means of advancing postnatal brain size. Precocial young are more able than altricial ones to evade predators (Pagel and Harvey, 1988). Why, then, has altricial development evolved?

Various theories of the evolution of mammalian altriciality have been proposed (Case, 1978a). One creditable notion is that selective forces favoring small adult size have been a primary impetus for the evolution of altricial development (e.g., Hopson, 1973). According to this argument, adults of small body size must bear young of such small size that the young must of necessity be highly immature at birth. Hopson (1973), indeed, concluded that small adult size led the first mammals to be universally altricial, meaning that within the mammals, precociality would then be consistently a derived condition. The validity of his argument is debatable, however, because the adult size of Triassic mammals—often 20–30 g (Crompton and Jenkins, 1979)—was by no means so small as truly to necessitate altricial development of the young (see later). This question aside, if we consider modern species, it certainly would be difficult to believe that animals the size of adult masked shrews (*Sorex cinereus*) could bear precocial young (Pearson, 1948), and they do not. According to the small-size argument in its pure form, altriciality is simply a necessary and possibly disadvantageous (nonadaptive) byproduct of the evolution of adult body dimensions.

Although this size argument probably applies in some instances, it is, as already suggested, easily shown not to apply in all. Four-day-old Eastern woodrats (*N. floridana*), for example, are altricial by all standards; they are very sparsely furred, have closed eyes and little locomotor competence, and are utterly dependent on their parents for essentials such as food; placed individually at a moderate ambient temperature of 20°C, they cool to 25°C colonic temperature in just 30 min (McClure and Randolph, 1980). These young, however, weigh about 20 g and thus are similar in size to adult white-footed mice (*P. leucopus*), which are fully furred and independent creatures that can keep their body temperature near 37°C even when the air temperature is −20°C (Wickler, 1980). Four-day-old white-footed mice weigh 3.3 g (Hill, 1976) and thus are similar in weight to adult masked shrews (*S.

cinereus), and yet unlike the shrews, the young mice are naked and helpless. The body size at which size alone becomes sufficient to account for altricial features can be debated from a variety of perspectives (e.g., Pearson, 1948; Turner and Schroter, 1985). Case (1978a) concluded from an empirical survey of the modern birds and mammals that it is at an adult size of 10 g and under that body size becomes sufficient to dictate altricial development.

In cases where altricial young are large enough in principle to function in the manner of adult mammals, it seems both theoretically and empirically (Case, 1978a) true that the species has faced real evolutionary options in respect to mode of development: the mode is not constrained by extrinsic morphological or physiological factors to be altricial or precocial, but could in principle be either. In such cases, it is reasonable to entertain the possibility that altricial development pertains because it has intrinsic virtues; that is, altricial development possibly has been directly selected for, rather than being merely a byproduct of other evolutionary events. What might be the intrinsic virtues of altricial development? How could altricial development be adaptive?

As Case (1978a) and Pagel and Harvey (1988) stressed, life history features of a species have probably often influenced whether the evolution of altricial (or precocial) development has been favored. For instance, if a species uses vigorous pursuit to gather prey, birth in an immature state might have been favored because it would have freed the mother from the encumbrance of retaining young within her for a protracted prenatal development (Case, 1978a). Nest security could also be an influential factor (Case, 1978a); in the example just mentioned, if the species is able to find or create safe nesting sites, the evolution of altricial development would seem more likely than otherwise. Altricial development might also be favored as a means of increasing peak reproductive rate (Martin and MacLarnon, 1985; Pagel and Harvey, 1988).

As noted, altriciality involves many features of the animal and is not monolithic. Ultimately, an understanding of the evolution of altriciality may demand a somewhat separate understanding of the evolution of each of its major individual elements. The remainder of this chapter focuses on one specific element: *thermoregulatory altriciality*. The principal concern will be whether thermoregulatory altriciality is adaptive and, if so, how. Implicitly, I shall be concerned with species that are of great enough adult body size for their altriciality to be not merely dictated by size. I explore adaptive arguments—using the condition of modern species as my basis—while acknowledging criticisms of the approach (Hafner and Hafner, 1988) because I believe such explorations provide insight if carried out with discipline. My goal is to bring increased discipline to this area of argumentation.

Introduction to Thermal Physiology

Up to now I have said almost nothing about the thermal physiology of nestling small mammals. In particular, I have deliberately avoided the common practice of using thermoregulatory status as one of the defining features for altriciality and precociality. My reason for following this course is not that thermoregulatory status is irrelevant or uncorrelated with altriciality and precociality. Indeed, altricial and precocial species differ dramatically in their pace of thermoregulatory maturation after birth. Nonetheless, the way the distinctions are usually phrased is inaccurate, and I have put off introducing thermoregulatory considerations until they could be dealt with carefully. Case (1978a) and others categorize altricial young as devoid of physiological thermoregulatory ability (*ectothermic* in Case's terms). In fact, as we shall see, this is a misconception that has led to serious inaccuracies in prior argumentation concerning the evolution of altricial development.

To highlight the problems of interpretation that have existed and their causes, I take a historical approach. First, I describe the traditional types of data that have been the mainstay for analysis of nestling thermal physiology. Then, I elaborate a prominent theory of adaptation that has been based on those data. Finally, I focus on the shortcomings of the traditional data and on revisions to theory that have been stimulated by recent insights into nestling thermal physiology.

Thermal Phenomenology in Isolated Altricial and Precocial Young

A review of the literature reveals that there are numerous papers on the ontogeny of homeothermy in small mammals, encompassing many species. A closer look at these papers reveals that almost all follow a single basic experimental design: nestlings were isolated from their par-

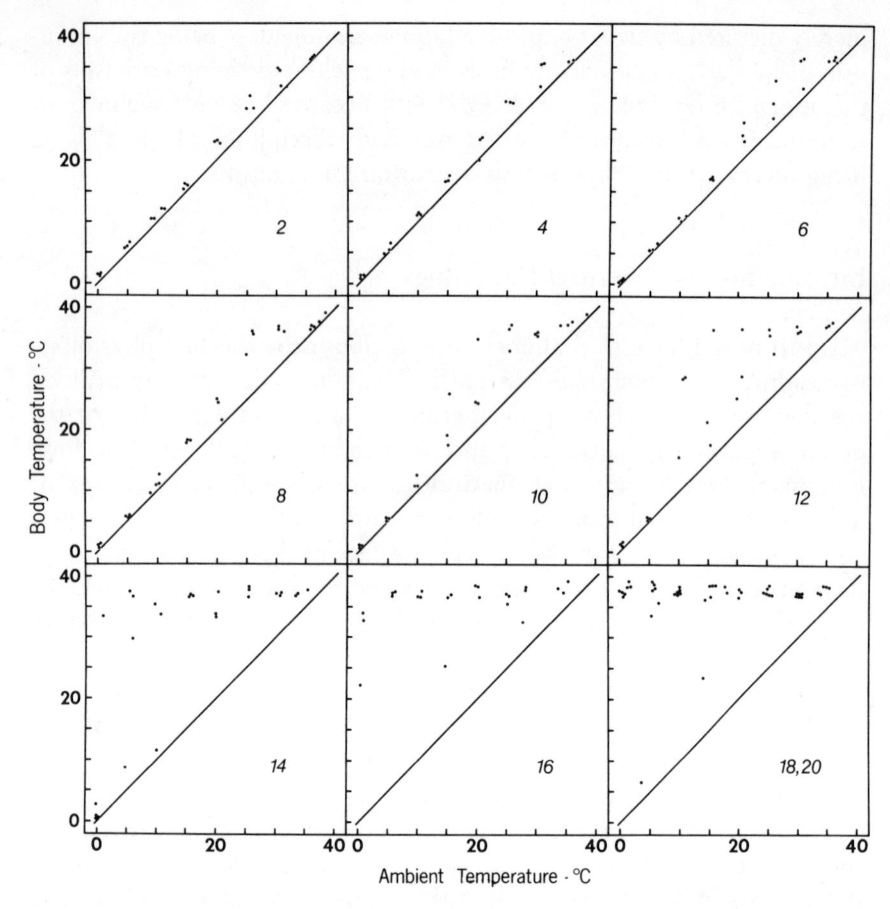

Fig. 7.1. Body temperature as a function of ambient temperature in isolated *Peromyscus leucopus* of various ages. (From Hill, 1976, courtesy of *Physiological Zoology*. © 1976 by the University of Chicago.) Age in days is indicated by the number at the lower right of each block. Temperatures were measured after 2.5–3.0 h of exposure. Diagonals are lines of equality between body temperature and ambient temperature.

ents, siblings, and nest for study. What do we learn from this traditional experimental design?

Figures 7.1 and 7.2 illustrate the typical responses of individual altricial nestlings—in this case, *P. leucopus*—studied in isolation (Hill, 1976). In the days after birth, the isolated nestlings resembled poikilotherms in the responses of their body temperature and metabolic rate to changes in ambient temperature. The overall pattern was that as the ambient temperature was lowered, their body temperature fell in

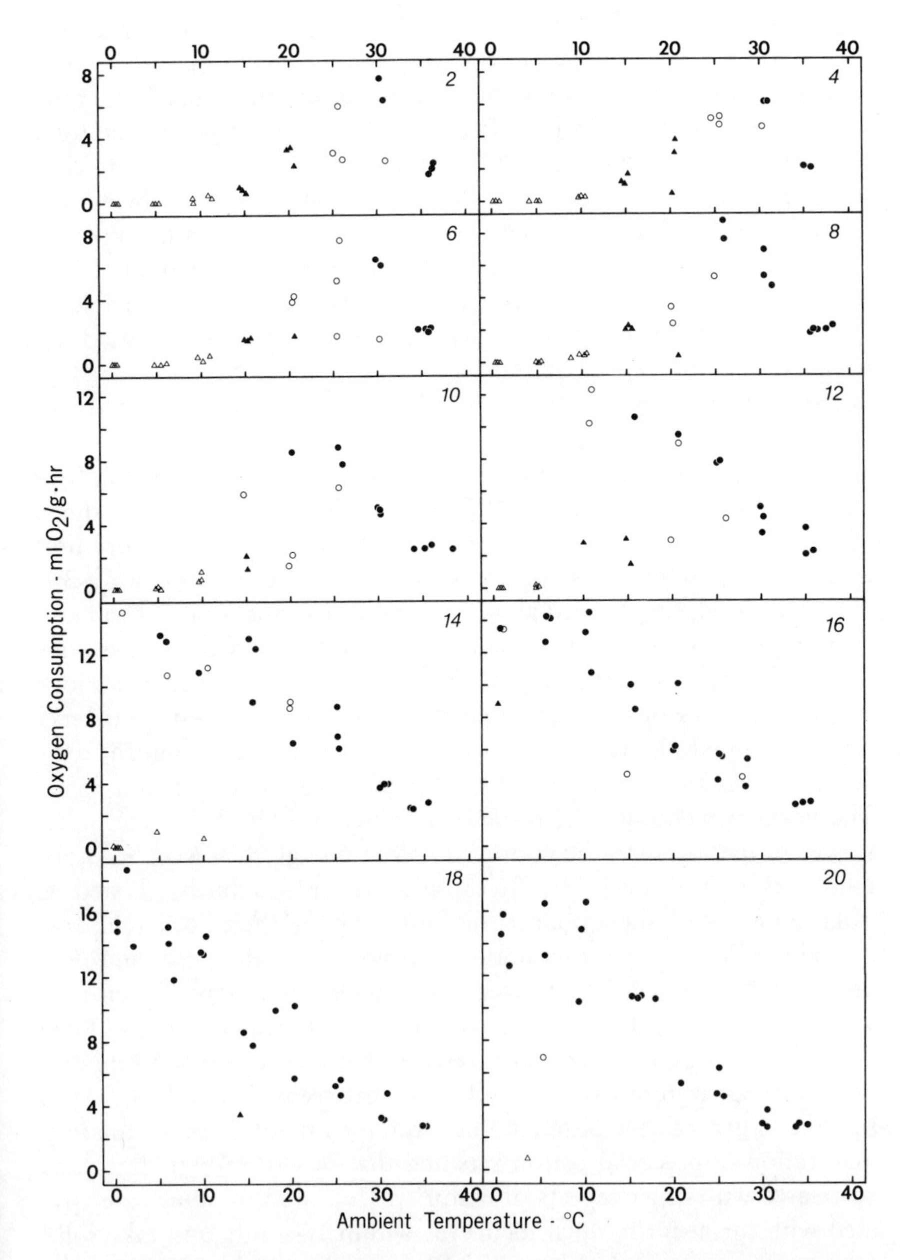

Fig. 7.2. Rate of oxygen consumption as a function of ambient temperature in isolated *Peromyscus leucopus* of various ages. (From Hill, 1976, courtesy of *Physiological Zoology*. © 1976 by the University of Chicago.) Age in days is indicated by the number at the upper right of each block. Rates were measured after 2.0–2.8 h of exposure. Symbols indicate body temperatures at the ends of experiments. Closed circles indicate 34° or higher; open circles, 24.0–33.9°C; closed triangles, 14.0–23.9°C; open triangles, 13.9°C or lower.

parallel, remaining close to ambient temperature; their metabolic rate, except for a rise from 35 to 30°C, also fell with ambient temperature. By the time the nestlings reached weaning age (about 21 days), in contrast, their responses were those typical of homeotherms. Their body temperature remained relatively constant regardless of ambient temperature, and their metabolic rate rose as ambient temperature fell. The transition from the response pattern of newborns to that of weanlings was gradual. Thermoregulatory maturation was particularly rapid between 10 and 14 days of age.

The type of contrast that can exist between nestlings of altricial and precocial species was nicely illustrated by the studies of P. A. McClure and her colleagues on woodrats and cotton rats (McClure, 1987; McClure and Randolph, 1980; Webb and McClure, 1989). The nestlings of the cotton rats (*S. hispidus*) displayed relative thermoregulatory precociality. Exposed in isolation to an air temperature of 20°C for a 30-min test period, they cooled almost to ambient temperature when newborn, but they matured rapidly and maintained their body temperature at adult levels by about 10–12 days of age, when they weighed about 17 g. By 5 days of age, the isolated cotton rats exhibited an inverse (i.e., homeothermic) relation between metabolic rate and ambient temperature over the entire range of temperatures between 35 and 15°C. The nestling woodrats (*N. floridana*), which are altricial, were much slower to mature than the cotton rats even though they were substantially larger at all ages (e.g., over twice as large at birth). Tested in isolation at 20°C, the woodrats did not maintain their body temperature above 25°C for 30 min until they were over 7 days old, and they did not maintain it at adult levels until about 20–22 days, when they weighed over 50 g. The isolated woodrats had to reach 12 days of age before they exhibited an inverse relation between metabolic rate and ambient temperature over the full range between 35 and 15°C. As is the rule in general, the difference between the rate of thermoregulatory maturation of precocial cotton rats and that of altricial woodrats was correlated with other aspects of maturity. Cotton rat nestlings are covered with fur at birth, open their eyes within 18–36 h, and take solid food and are weaned by about 10–12 days. Woodrat nestlings, on the other hand, do not become covered with fur until about 10 days, do not open their eyes until about 14 days, and are 20–22 days old before they take solid food and are weaned. Waldschmidt and Müller (1988) reported similar contrasts between precocial *Acomys* and altricial *Gerbillus*.

The Adaptiveness of Thermoregulatory Altriciality

How might the thermoregulatory altriciality seen in animals such as *P. leucopus* and *N. floridana* be adaptive? Because physiological thermoregulation is energetically costly, many investigators have concluded that the adaptive advantages of thermoregulatory altriciality lie in its implications for energy use.

The energy-demanding processes of nurturing altricial nestlings to weaning can be placed into three categories (Fig. 7.3):

1. *Energy-demanding processes that can be carried out only by the young using their own chemical-energy resources.* Notable among these processes are growth, development, and such maintenance functions as circulation, respiration, and digestion. The young must circulate their own blood, and only they can effect their own growth. They must carry out these functions using the energy available from their milk.
2. *Energy-demanding processes that can be carried out only by the parent(s) using parental energy resources.* Food gathering is one example. When nurturing young, the mother must gather food above and beyond what she would otherwise require. The energy cost of gathering this extra food is properly included among the costs of nurturing the young, and it is a cost that only the parent can meet (altricial young cannot gather their own food during most or all of nestling life).
3. *Energy-demanding processes that can be carried out by either the young or the parent(s).* The largest cost in this category is that of keeping the young warm. The young can carry out their own thermoregulation, in which case heat production to keep them warm will draw on their own chemical-energy resources. Or the parent(s) can keep the young warm (through heat flow from the parent(s) to the young), in which case the heat production to warm the young will draw directly on parental chemical-energy resources.

One way to consider the energy implications of thermoregulation is to focus on how the young use the chemical energy provided them. Certain energy-requiring processes, such as growth, *must* be carried out using this energy. On the other hand, certain other processes—notably thermoregulation—can be carried out either with the chemical-energy resources given the young or with other parental energy sources. It will

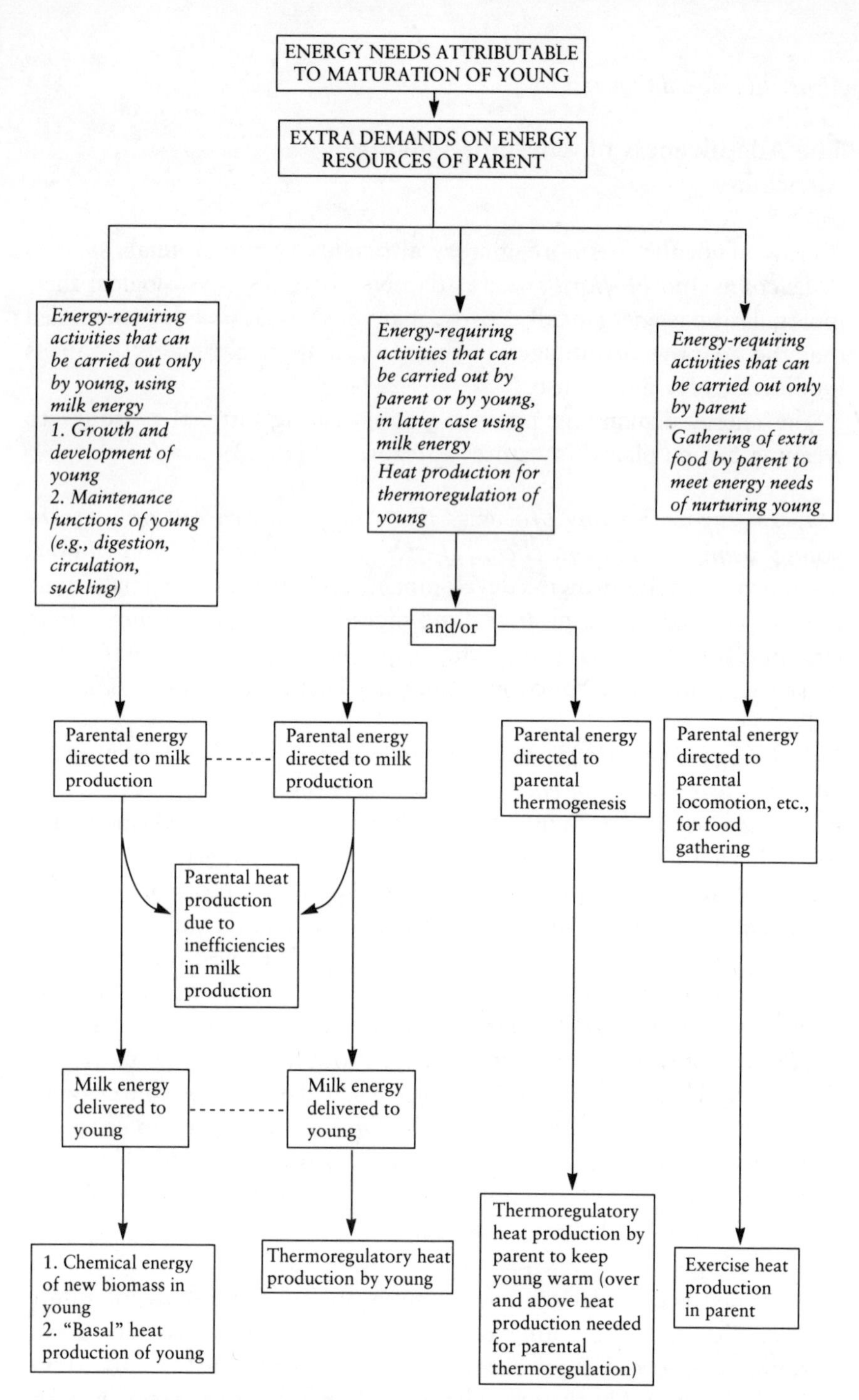

Fig. 7.3. A classification of the energy requirements of growth and maturation of altricial young from birth to weaning. The only energy needs considered are those attributable to growth and maturation of the young, above and beyond other energy demands of the parent. The second box from the top emphasizes that these needs must all be met by the parent. Three categories of energy-requiring activity are identified in the italic boxes near the top (lists of examples are not comprehensive). Boxes below indicate (1) how the parent provides energy for each category of energy requirement and (2) the fate of energy directed along each path. Dashed lines connect boxes of identical meaning.

be evident that avoiding use of the youngs' chemical-energy resources for the latter processes could increase the energy available for the former processes.

From this view of the energetics of development has come the most common argument for the adaptiveness of thermoregulatory altriciality. Dawson (1962) and Dawson and Evans (1957), in papers on birds, were among the first to voice the argument. Since then, it has been applied equally to mammals and reiterated many times (e.g., Case, 1978a; Chew and Spencer, 1967; McClure and Randolph, 1980; Waldschmidt and Müller, 1988). Altricial development is posited to mean that early-age nestlings are unable to thermoregulate physiologically. This inability of the young to warm themselves is in turn argued to be advantageous because it promotes rapid growth and development by denying the use of the youngs' chemical-energy resources for processes that do not require use of those resources. It is important for the young to be warm, but the warming is done by the parent(s) using energy above and beyond that provided as milk.

This argument involves several assumptions that are rarely stated or evaluated:

Assumption 1. The argument just presented is always voiced with the implicit presumption that the young will receive a constant chemical-energy supply regardless of whether it is they or their parents that keep them warm, and thus that any of the youngs' chemical energy that can be saved from use in thermoregulation will in fact be available for other uses. There is an assumption, in other words, that warming of the nestlings by the parents will not entail much (if any) cutback in the amount of chemical energy the parents can provide the young. There are at least two mechanisms by which this assumption could be violated. First, the parents could be energy-limited, so that energy used to make extra heat for warming the young would subtract from the energy that could be given the young in chemical form. Second, the parents could be time-limited, so that the time spent warming the young would detract sufficiently from foraging time to reduce the amount of food passed to the young. We know little about either possibility, and there is a considerable need to know more. The actual state of affairs probably varies with species and environmental conditions.

To illustrate the types of specific concern raised here, consider the common assumption that parents keep their young warm with but a trifling increase in heat production above what they would be expending anyway to keep just themselves warm (e.g., Bartholomew, 1982), suggesting that keeping the young warm would not have an impact of the energy-limitation type on feeding of the young. Although never recognized, the assumption of a trifling increase in heat production is in fact based on a further assumption that the parents would be fully homeothermic even if they had no young. The adults of many species of small mammals undergo daily torpor when without young, *especially when energy-limited* (Hill, 1983; Hudson, 1978). If rearing young causes torpor to cease (as it does in certain bat species, for example: Tuttle, 1975), and the cessation of torpor permits the parents to warm the young, then the warming of the young would be achieved at a considerable increase in parental energy expenditure. In times of energy shortage, this could encroach on energy resources available for feeding the young.

In a perceptive article, Bronson (1985) stressed that parental mammals the size of house mice can readily become energy-limited in cool environments when food is so distributed that they must spend a lot of time foraging.

All of this said, it *does* seem that, for many species under many (realistic) conditions, the parents probably have a great deal of latitude in the extent to which they can warm their early-age young without significantly impairing feeding of the young.

Assumption 2. There is also an assumption that if chemical-energy resources of the young can be spared from use in heat production, then those resources can be used to accelerate growth or maturation. Restated, this assumption is that growth or maturation or both are energy-limited: the processes could occur faster if more energy were directed to them. There is quite a bit of support for this assumption. Perhaps the most fascinating comes from an early experiment in which mouse pups were provided with rat dams as foster mothers. Able to obtain much more milk than usual, these mouse pups grew into giants (Parkes, 1929). More-recent and sophisticated studies on laboratory rats and mice have shown that when the amount of milk received by nestlings is varied, there is a direct relation between amount of milk received and growth (e.g., Dickerson and Widdowson, 1960; Fuchs, 1982; Heggeness et al., 1961; Widdowson and McCance, 1960). Kaufman and Kaufman (1987) reported that the removal of young from litters of *Peromyscus polionotus* led to increased growth rates in the remaining individuals.

In general, providing more milk energy to nestling mammals increases their rate of weight gain and their weight at weaning, but effects on other aspects of growth and maturation can be complex. The previously cited studies by Widdowson and colleagues on laboratory rats, for example, found that the growth of the skeleton and many internal organs was accelerated by increased provision of milk, but some aspects of skeletal growth and tissue maturation showed less acceleration than did weight gain. In general, the timing of discrete maturational events such as incisor eruption, appearance of hair, and eye opening is *not* altered much, if at all, by increasing the energy resources of nestlings (Heggeness et al., 1961; Parkes, 1929; Widdowson and McCance, 1960).

Assumption 3. A third assumption is that it is important to keep the young fully warm as much of the time as possible. Among other things, high body temperatures are assumed to be important in facilitating growth and maturation.

Although there is strong evidence in support of this assumption, the amount of good evidence is much less than typically thought. The usual experimental design in research on thermal effects on growth and maturation has been simply to house families (parents plus young) in cold and warm rooms (e.g., Biggers et al., 1958). With this design, the body temperatures of the "cold-reared" and "warm-reared" young are not assuredly different because the parents can keep the young warm regardless of outside air temperature, and there are other interpretive difficulties (e.g., apparent effects of air temperature on the young may in part be indirect manifestations of effects on the parents, for the parents as well as the young are exposed to the temperature differences; see Barnett, 1973).

I have removed litters of *P. leucopus* from their parents daily between the ages of 2 and 15–16 days for 4.5 h/d and exposed half of the nestlings to air at 10°C while keeping the others at 35°C; the parental cage and parents always remained at room temperature (ca. 23°C), and the young were in the parental cage except when undergoing the thermal treatments. With this experimental design, the cold-treated nestlings were known to experience lower body temperatures than the warm-treated ones, and thermal effects on the young were not confounded with ones on the parents. The cold treatments markedly slowed both growth and maturation (Hill, 1983). For example, the cold-treated young opened their eyes 2 days later than the warm-treated ones and were 10 percent lighter at 15–16 days. Thermoregulatory maturation was also slowed, such that at 15–16 days, the cold-treated young (tested in isolation) were much less able than the warm-treated to maintain a high body temperature during cold exposure. Spiers and Adair (1987) reported evidence from a similar experimental design that cold treatments (20°C) slowed the growth of rats (*Rattus*).

Assumption 4. A final assumption worthy of note is that high rates of growth and maturation are an advantage. There are at least two ways in which an advantage could be realized. First, high rates of growth and maturation could hasten weaning. Second, they could lead to relatively large size or advanced maturation at weaning.

I am aware of only one study on a free-living population of small mammals that provides data on whether size at weaning is related to probability of survival within a single population. Beacham (1979) studied the weights of dispersing voles (*Microtus townsendii*) that settled on grids and found that the animals that remained more than 2 weeks were heavier than those that remained less, suggesting that large size aids establishment in a population. Besides this field study, there are a number of laboratory studies that deal with the implications of size at weaning. The laboratory studies have the advantage of providing extensive and relatively straightforward data, but they do not obviate the need for further work in ecologically realistic settings; among other things, differential mortality due to ecologically realistic stresses is not operative in most laboratory studies. In an analysis of a large breeding colony of *P. maniculatus*, Myers and Master (1983) found that animals relatively large at weaning grew up to be relatively large at reproductive age. In turn, relatively large mothers tended to produce more young per litter and to produce young of larger size at weaning than small mothers. The overall pattern would clearly support the hypothesis that large size at weaning is an advantage. Earle and Lavigne (1990) also found in *P. maniculatus* that relatively large mothers tended to produce relatively large numbers of young per litter; the total weight of litters at birth was positively correlated with maternal size, but at weaning neither the weight of individual young nor that of litters was clearly correlated. Kaufman and Kaufman (1987) examined *P. polionotus* and found that size of the mother was positively correlated with size of the young at weaning and the number of young per litter. Derrickson (1988) found that *P. leucopus* that were relatively heavy at weaning tended to be relatively heavy at reproductive age, and relatively heavy mothers tended to produce more young per unit time than lighter ones. Fuchs (1982) reported that female laboratory mice weaned at relatively small size tended to be relatively old when they first gave birth and to have relatively few young in their first litter. In *Spermophilus lateralis* reared in the laboratory, young that were relatively large at weaning reached adequate weight for successful hibernation sooner than young small at weaning; for young of the year in nature, relatively early entry into hibernation is of survival advantage (Phillips, 1981). A major interpretive difficulty with many of the sorts of laboratory studies described is that the correlations reported are likely to include a genetic component, and yet the question of interest here is whether a phenotypically plastic increase in weaning size has advantages. The one laboratory test of tolerance to stress in weanlings is that of Fleming and Rauscher (1978). Studying weanling *P. leucopus*, they found a strong positive correlation between body size and endurance time in cold water.

Although the weight of evidence is that large weaning size is of advantage, the extreme paucity of field data means that there remains a critical need for more research on the ecological implications of size and maturity at weaning. One aspect of this need that deserves emphasis is that body size is one of the few characteristics of young mammals that can be readily measured during studies of ontogeny in the wild. Thus, there is a temptation to use body size as an index of responses to conditions, predictor of probability of success, and so forth. Such use of body size as a proxie for relatively unmeasurable properties urgently needs to be placed on a firmer foundation.

The Actual Thermoregulatory Status of Altricial Nestlings

Returning now to the data that exist on the thermoregulatory physiology of altricial nestlings, I want to stress once more that the vast majority of studies have been done on isolated animals, and these kinds of studies have led to the conclusion that early-age nestlings are incapable of keeping themselves warm. This conclusion, in turn, has been pivotal in framing the dominant theory for the adaptive evolution of thermoregulatory altriciality. In fact, the conclusion is probably wrong for many small mammals, and the theory must be reconsidered in that light.

Data on isolated altricial nestlings have the serious limitation that altricial nestlings do not in fact develop in isolation. Their normal environment is within an insulating nest, huddled with nestmates. This natural environment is much more thermally protective than the artificial environment of an isolated animal in a metabolic chamber, and we must expect substantial differences in performance in the two. Clearly, for ecological and evolutionary analysis, we need to anchor our thoughts to data gathered in natural or seminatural environments.

Reconsider the data (Figs. 7.1 and 7.2) on 4-day-old white-footed mice studied in isolation. As earlier noted, their body temperatures approximated ambient temperature, and except for a rise from 35 to 30°C, their metabolic rates fell as ambient temperature fell. The data in Figure 7.4 on 4-day-olds studied in huddled groups in nests could

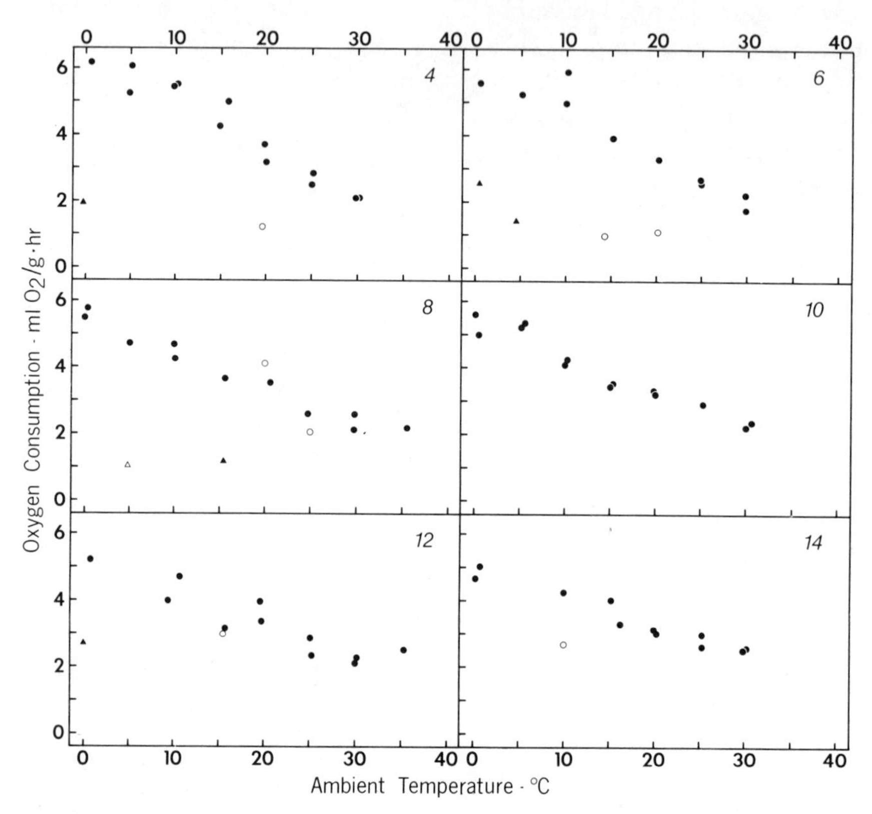

Fig. 7.4. Rate of oxygen consumption and body temperature as functions of ambient temperature for litters of four nestling *Peromyscus leucopus* of various ages studied in cotton nests. (From Hill, 1976, courtesy of *Physiological Zoology*. © 1976 by the University of Chicago.) Age in days is indicated by the number at the upper right of each block. Litters were exposed for 2.5–3.0 h, and measures were taken toward the end of the period. For each experiment, the symbol indicates the average body temperature of two selected young from the litter according to the key specified in the legend to Figure 7.2.

hardly present a more-different picture. Under these circumstances, the young were homeothermic. Their body temperatures remained above 34°C at most ambient temperatures, and their metabolic rates increased steadily as ambient temperature fell. At an early age, nestling white-footed mice exhibit a physiological thermoregulatory response to cold, and evidently they have evolved sufficient thermogenic abilities that they typically can keep themselves warm for several hours even at the lowest air temperatures they are likely to encounter (Millar and Gyug, 1981) when they are in the normal environment of nestlings (Hill, 1976).

Hints of the thermoregulatory nature of the nestlings are in fact evident in the data for isolated young. Even 2-day-old isolated *P. leucopus* increase their metabolic rate in the manner of homeotherms when the ambient temperature is lowered from 35 to 30°C (Fig. 7.2), and some stabilization of their body temperature is evident in this range (Fig. 7.1). For the most part, however, the thermoregulatory ability of the young is obscured in the isolated environment because it is overwhelmed. The metabolic rates shown in Figure 7.2 are those prevailing after 2.0–2.8 h. In fact, when isolated early-age nestlings are first placed at a cold ambient temperature, they often exhibit a marked, transient increase in metabolic rate (Hill, 1970), but without the insulation of nest and siblings, even their peak rate of thermogenesis is inadequate to keep their body temperature from falling. Then, as the temperature of their tissues decreases, the rate at which the tissues can produce heat is depressed, leading to a still greater fall in body temperature, and so on until body temperature approximates ambient temperature.

Unfortunately, very few studies are available on the thermoregulatory physiology of small-mammal nestlings in their natural setting, huddled in a nest. The results available, however, generally parallel those for the white-footed mouse. One of the most thorough studies is that by Sheck (1982) on nestling *S. hispidus*. Although *S. hispidus* undergoes relatively precocial development, isolated nestlings exhibit only meager thermoregulatory abilities at the early age of 3 days. Sheck compared the abilities of isolated 3-day-olds with those of groups of 3-day-olds huddled in nests, using 3.5-h exposures to ambient temperatures. The isolates held their body temperatures stable as the ambient temperature was lowered from 35 to 30°C, but then they allowed their temperatures to fall, and at 0°C ambient temperature, body temperatures were 0°C and metabolic rates were near zero. Litters in nests, in sharp contrast, were homeothermic at 0°C; they maintained body temperatures near 35°C for the full 3.5-h test period by elevating their metabolic rates to well above basal levels.

Gebczynski (1975) showed that *Clethrionomys glareolus* exhibits an increase in metabolic rate as ambient temperature falls even when just 1–3 days old. He also found that three litters aged 2, 3, and 5 days—which were being reared by their mothers in outdoor nest boxes with nests—had average body temperatures of 31, 33, and 33°C, respectively, after a half-hour of maternal absence even though the outside air temperature was 6–8°C. The latter data indicate that the litters were able to maintain their temperatures at relatively constant levels inasmuch as the temperatures of litters *with* their mothers were 33–34°C, and it seems inevitable that this maintenance of body temperature was attributable at least in part to the young animals' demonstrated homeothermic response in thermogenesis. In laboratory tests at an ambient temperature of 5°C, litters aged 1–6 days in artificial nests showed a drop in body temperature of 0.2–0.3°C/min over an unspecified period of time. The reason for this greater rate of cooling in the laboratory is unknown. Possibly, 5°C was so cold as to overwhelm thermoregulatory abilities at these early ages, or the artificial nests may have been inadequate compared with the field nests.

Gelineo and Gelineo (1952) studied laboratory rats in nests during 30-min absences of their mother at air temperatures of 13–19°C. Two of three litters in which subcutaneous temperatures were measured kept their temperatures above 30°C by 5–6 days of age. A fourth litter (subcutaneous temperature unmeasured) kept the temperature of its nest at 32°C when just 3 days old. Schmidt et al. (1987) reported that litters of laboratory rats aged 5–8 days, studied in nests at an air temperature of 25°C, maintained average colonic temperatures of 34.5–36.3°C during 30–60 min absences of their mothers. The young exhibited a daily rhythm in the core temperatures they maintained, highest temperatures occurring near the time of lights off and lowest occurring near lights on. Manifestation of this rhythm was accentuated by lengthening the duration of maternal absences, but the temperatures maintained by the litters remained high nonetheless. After 4 h of maternal absence, litters had colonic temperatures averaging 35.9°C near the time of lights off and 32.4°C near lights on. Taylor (1960) showed that laboratory rats increase their metabolic rates in response to declining ambient temperatures within a day after birth.

Casey (1981) reported cooling curves for lemming (*Lemmus tri-*

mucronatus) litters in nests at an outside air temperature of 0°C. By 4 days of age, there was essentially no drop in pharyngeal temperature over an hour of absence of the mother (although, for unexplained reasons, the measured temperature was just 24°C even at the start of the tests).

Bryant and Hails (1975) studied effects of huddling and nests in young laboratory mice exposed to 25°C. When the mice were 8–12 days old, they could maintain adultlike body temperatures in isolation; huddling reduced their metabolic rates (their energy costs of thermoregulation), and a nest reduced them still further. When the mice were just 3–4 days old, on the other hand, huddling increased their metabolic rates. The authors found this increase to be paradoxical and presented evidence that it was due to a rise in locomotor activity with huddling. However, body temperatures of the huddled young were not measured, and thus an additional explanation was not recognized. Consider that isolated young of these earlier ages became hypothermic. Furthermore, laboratory mice are known to exhibit physiological thermoregulatory responses within 1–2 days of birth, as documented later. Accordingly, the rise in metabolic rate observed in the huddled groups of 3- to 4-day-olds could have been explained in part by their thermoregulating more effectively than isolates. Provision of a nest to the huddled 3- to 4-day-olds lowered their metabolic rates, as expected for thermoregulating animals. In contrast to Bryant and Hails's results on 3- to 4-day-olds, Vinter et al. (1982) found that, in 2-day-olds at 30°C (a temperature below thermoneutrality), huddled groups maintained higher body temperatures with lower metabolic rates than isolates.

Weigold (1973) monitored body temperatures of young *Myotis myotis* in natural colonies during extended periods (up to 1 h or more) of parental absence. Isolated young had temperatures near ambient. Young in the centers of clusters, by contrast, had temperatures of 32–36°C (8–17°C above ambient) and showed little decline in temperature while the parents were away.

I have already mentioned some studies in which the effects of huddling without a nest were evaluated. Others include the work of Alberts (1978), Cosnier et al. (1965), and Kornienko and Son'kin (1974) on *Rattus*; Stanier (1975), on *Mus*; and Gebczynski (1970), on *Glis*. In each of these instances, thermoregulatory abilities were enhanced in the nestlings that were grouped when compared with isolates. Nagel (1989a), in contrast, recently reported that groups of nestling white-toothed shrews (*Crocidura russula*) without nests maintained lower body temperatures and metabolic rates than isolates; this unusual result was attributed to a greater proclivity to enter torpor in the groups.

Although there have been few studies in which nestlings have actually been studied as huddled groups in nests, extrapolations of relevance can be ventured from the abundant literature on isolates. As already stressed, the fundamental homeothermic nature of early-age *P. leucopus* and *S. hispidus* is decipherable from the data available on isolated individuals of the species: the early-age isolates show a negative-feedback relation between metabolic rate and ambient temperature within a narrow range of high (but below thermoneutral) ambient temperatures (Fig. 7.2; Sheck, 1982). Placing the *P. leucopus* or *S. hispidus* in litters in nests permits their homeostatic thermogenic responses to have increased effect, resulting in a broad range of thermoregulation. From examples such as these, we can conclude that any clear evidence of a homeothermic metabolic response in early-age isolates can be taken as likely evidence for significant thermoregulatory ability at an early age in the natural nestling habitat (see also Hull, 1973).

A review of the large literature on metabolic responses of isolates leaves no doubt that the majority of altricial placental small mammals that have been studied show signs of activating thermogenesis in response to declining ambient temperature at an early age. To add just a few examples to those already mentioned, isolated *Neotoma floridana* (Webb and McClure, 1989), *Sylvilagus floridanus* (Gates, 1974), *Gerbillus perpallidus* (Waldschmidt and Müller, 1988), and *Microtus arvalis* (Gelineo, 1962) show increases in metabolism in response to lowered ambient temperature within a day of birth. Isolated laboratory rats (Taylor, 1960) and domestic rabbits (Edson and Hull, 1977) do also. Laboratory mice show increases within 1–2 days (Cassin, 1963; Hill, 1970; Pichotka, 1964; Stanier, 1975; Vinter et al., 1982). *P. maniculatus* (Chew and Spencer, 1967) and *Lemmus lemmus* (Hissa, 1968) exhibit such responses within 2 days.

The question of thermoregulation by litters in nests could hardly be more important for the analysis of nestling physiology, and yet it will be evident from this review of the pertinent evidence that we sorely need more data to reach a full understanding. We have seen that in certain cases, when studies in nests have been carried out, the body temperatures of the nestlings have fallen. In some of these studies, however, the ambient temperature has been near freezing; responses to near-freezing temperatures are not irrelevant, but often microclimatic conditions in nature will be more favorable, and nestlings that fail to thermoregulate near freezing may succeed at higher ambient temperatures. Another point to recognize is that the studies done have generally not permitted mothers to build nests of self-selected size and composition (usually, nest size and composition have been fixed by the investigator). In nature, females may build better-insulating nests as ambient temperatures fall (e.g., Barnett, 1956; Lynch, 1974; Shump, 1978; Thorne, 1958), thus assisting thermoregulation by their young. The potential facilitating effects of communal nesting on thermoregulation have also generally not been taken into account except for bats. Finally, note that it is not necessary for nestlings to maintain body temperatures at the full adult level for them to experience the twofold consequences of thermoregulatory effort: (1) increased energy use and (2) maintenance of higher body temperatures than would otherwise pertain.

Overall, it can be concluded, on the basis of available evidence, that nestlings of altricial placental small mammals generally carry out some of their own physiological thermoregulation starting at an early age. There are exceptions. Hamsters (*Mesocricetus auratus*), for example, seem to show no evidence of endogenous thermoregulatory responses to cold until 7–10 days old, as judged from studies of isolates (Hissa, 1968; Hissa and Lagerspetz, 1964; Rink, 1969; Vishinesku et al., 1965); this trait is possibly related to their exceptionally short gestation time compared with other eutherians. Furthermore, marsupials do not carry out endogenous thermoregulation for a long time after birth (e.g., Morrison and Petajan, 1962). Nonetheless, the majority of the altricial placental small mammals probably start to thermoregulate at an early age, and we must now explore why they would do so and what the implications are for theory.

Why Do Early-Age Nestlings Thermoregulate?

Mothers can spend many hours of each night away from their young. *P. leucopus* mothers in outdoor enclosures, for example, average 5.1–6.3 h/d away from their young, and continuous absences of over 3 h are common (Harland and Millar, 1980; Hill, 1972). *C. glareolus* mothers in outdoor pens average about 6 h/d away, with some absences lasting 30–45 min (Gebczynski, 1975). We have already reviewed the evidence that, if the body temperatures of nestlings fall, their growth and development can be substantially retarded (e.g., Hill,

1983). Combining these perspectives, we see that nestlings in principle face two options during maternal absences, each option composed of internally contradictory elements. One option is not to thermoregulate. Given that maternal absences can be long, nestlings following this option would confront a high risk of becoming cold. The consequences of the option would thus be twofold: the nestlings would *conserve* their energy resources by not using them for heat production, but they would likely be unable to *use* the conserved resources effectively for growth and development during maternal absences because of the fall of their body temperatures. They would thus be likely to lose time in completing their development. The other option is for the nestlings to thermoregulate. Young following that option would use some of their energy resources for heat production rather than growth. However, they would stay warm during the absence of their mother, and thus they would be able to use their remaining energy resources to grow and mature at a pace relatively unimpeded by hypothermia.

The evidence available indicates that natural selection has favored the latter option in the majority of small placental mammals, even when their development is altricial.

Theory Reconsidered

As stated earlier, the traditional and most commonly cited theory for the adaptive advantage of thermoregulatory altriciality has been that altriciality facilitates growth and development by transferring the energy costs of thermoregulation from the young to the parents. This traditional argument in fact confounds two distinct ideas. One is an idea of *adaptive advantage*, and the other is an idea of the *means* of gaining that advantage:

Idea 1. There is an adaptive advantage to having the parents keep the young warm. Parental warming frees chemical-energy resources of the young for use in growth and maturation. Growth and maturation are thus accelerated, to the presumed advantage of the young.

Idea 2. Thermoregulatory altriciality is the mechanism of having the parents keep the young warm. The notion here is that during a substantial period following birth, a characteristic of altriciality is that the young are functionally *unable* to use their own chemical-energy resources to produce heat for thermoregulation. Thus, altriciality has been argued to serve as a *means* of assuring that the young are warmed by their parents, not themselves.

The evidence is that the second of these ideas is in fact false for most placental small mammals. This does not necessarily mean, however, that there are no grains of truth in the traditional argument. A moment's reflection reveals that the two ideas within the traditional argument have been conflated needlessly and are in truth independent. Nestlings could partake of the advantages described in the first idea despite not adhering to the second idea.

A rigid inability to increase thermogenesis to keep warm is not the only mechanism by which nestlings could be made to defer to their parents for warming. Instead, nestlings that possess an ability to carry out their own physiological thermoregulation could suspend or curtail their thermoregulation facultatively. Thus, they could defer to their mother for warming when she is present and yet keep themselves warm when she is absent, enjoying the best of two possible worlds.

Facultative suspension of endogenous thermoregulation in the presence of the mother is an option available to precocial as well as altricial nestlings. Thus, in considering this option, we are no longer dealing with a theory for the adaptivness of altriciality. I shall return to the specific questions surrounding altriciality shortly.

There are two possible ways in which endogenous physiological thermoregulation by nestlings could be curtailed or suspended, *passive* and *active*:

1. *Passive.* In the presence of their mother, young that are capable of thermoregulation could be led to throttle back their own rate of thermoregulatory thermogenesis simply because warming by the mother would place them in a less-demanding thermoregulatory environment. I term such a curtailment of thermogenesis *passive* because it would occur via the normal graded operation of the feedback controls intrinsic to the thermoregulatory system.
2. *Active.* Alternatively, thermoregulatory thermogenesis by the young could be curtailed or suspended by mechanisms that interrupt the normal feedback controls of the thermoregulatory system. Mechanisms homologous or analogous to adult torpor would be one possibility. For example, when the parents are present, tactile or other stimuli received from them could lead to a wholesale suspension of the operation of the youngs' endogenous thermoregulatory apparatus.

Passive curtailment of endogenous thermoregulation in the presence of the mother seems inevitable but might not completely abolish the nestlings' thermoregulatory thermogenesis. Active mechanisms could have the advantage of fully suspending such thermogenesis.

We know that the adults of many species of small mammals are capable of torpor (e.g., Hill, 1983; Hudson, 1978). Although this would suggest that the young would also be widely capable of a torporlike suspension of thermoregulation, I know of little firm evidence—positive or negative—for the existence of such a phenomenon in the young, especially preweanling young. There is a pressing need to explore the possibility further. Nagel (1977, 1989b) argued that episodes of torpor occur in preweanling white-toothed shrews (*Crocidura russula*) separated from their parents, and Weigold (1973) obtained evidence that isolated *Myotis myotis* undergo nocturnal exit from and entry into torpor as early as 6–10 days of age. Nestlings of laboratory rats aged 5–10 days are known to suppress their thermogenic response to cold when starved, and the suppression is abolished by decerebration (Bignall et al., 1975, 1977). The effect of decerebration shows that, in young rats, the thermoregulatory stimulation of metabolism in response to cold is under active, central nervous system control; but, like the data on shrews and bats, the results on rats do not relate to whether parental presence would be sufficient to cause active suppression of thermoregulatory thermogenesis. In my own studies of metabolism in young *P. leucopus*, I have gathered numerous records of oxygen consumption over time, and some of those strongly suggest that the young are capable of rapid activation and deactivation of thermoregulation (Hill, 1970), although again there is no information on responses to parental presence. Figure 7.5 shows a record for an isolated 12-day-old exposed to 21°C. The resemblance of this record to that of an adult exiting torpor is striking. This animal had a body temperature of about 37°C at the end. Data on it and others of its age indicate that its body temperature had to be much lower—near 28°C—at 1 h into the experiment. Evidently, this animal activated thermoregulatory thermogenesis during the experiment and raised its body temperature accordingly. Figure 7.6 shows a metabolic record for a 12-day-old at 15°C that closely resembles records for adults entering torpor. Calculations indicate that the animal had a body temperature of over 34°C for the first 90 min. Its final temperature was 21°C. The decline in metabolism occurred much more suddenly than is typical for animals that are

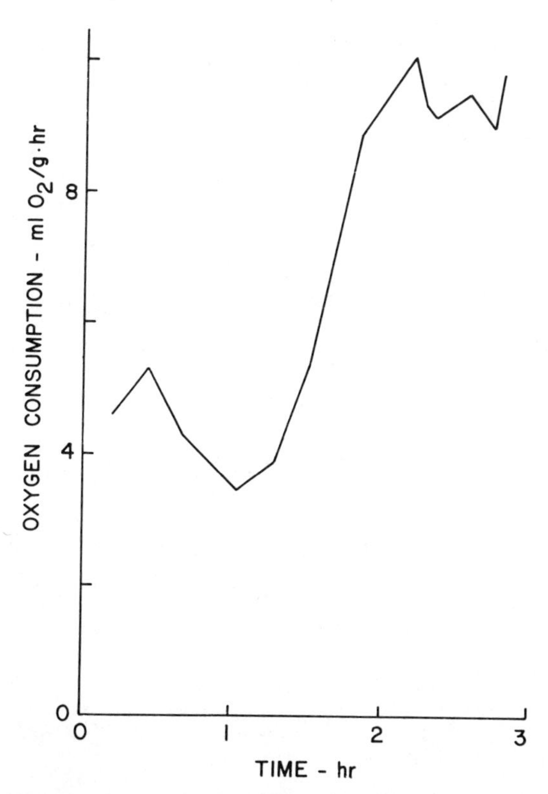

Fig. 7.5. Rate of oxygen consumption as a function of time in an isolated 12-day-old *Peromyscus leucopus* exposed to an ambient temperature of 21°C. Animals were not observed during these experiments, and thus changes in locomotor activity and their effect on metabolic rates are unknown. This animal's body temperature was 36.7°C at the end. Body temperatures were not measured between the start and end of these experiments, but when animals of similar age at similar ambient temperatures completed experiments with metabolic rates of 3–4 ml O_2/(g · h) (rates like those seen for this animal at about 1 h), their body temperatures were always below 30°C.

simply gradually cooling, and it suggests an abrupt deactivation of thermoregulation.

Bryant and Hails (1975) presented a method for estimating the metabolic rates of nestling small mammals in the presence of their mother. They reported that 4- and 8-day-old litters of *Mus* reduced their metabolic rates when their mothers joined them, suggesting that their thermoregulatory effort was reduced either passively or actively.

Geiser and his colleagues tested postweanlings of several species for the ability to undergo torpor. Although the postweanlings of several

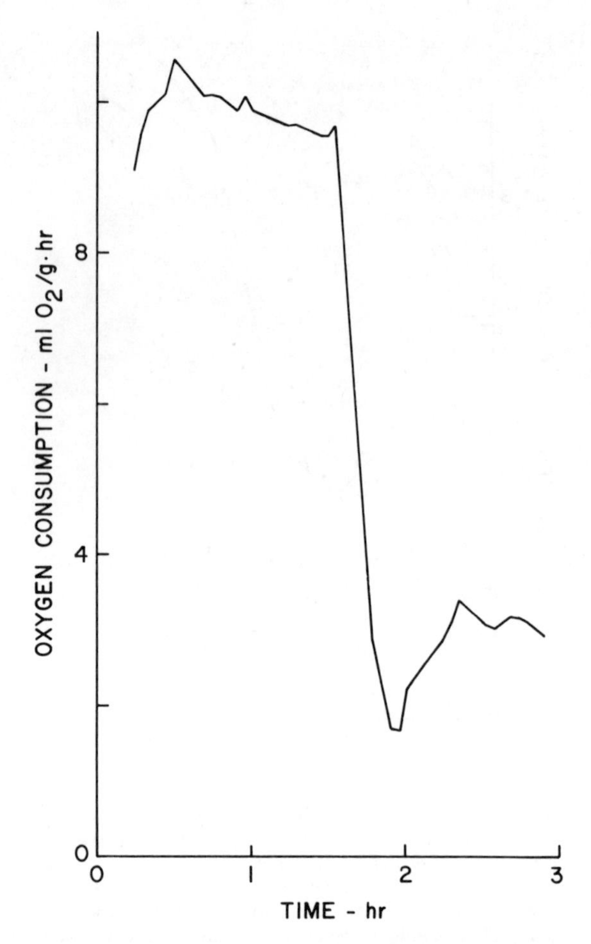

Fig. 7.6. Rate of oxygen consumption as a function of time in an isolated 12-day-old *Peromyscus leucopus* exposed to an ambient temperature of 15°C. Individual was different from that shown in Figure 7.5. The animal's body temperature at the end was 21.4°C. If body temperature, ambient temperature, and metabolic rate are symbolized T_b, T_a, and M, respectively, and if $(T_b - T_a)/M$ is assumed constant (Hill and Wyse, 1989), then the body temperature during the first 90 min would have been over 34°C.

dasyurid marsupials displayed episodes of torpor, those of *Spermophilus saturatus* did not (Geiser and Kenagy, 1990).

Empirical Evidence on the Energetics of Altricial Nestlings

Recognizing that altricial placental nestlings typically exhibit powers of physiological thermoregulation at an early age, we have seen that facultative curtailment or suspension of thermoregulation is the option available to them for deferring to their parents for warming. That option is presumably available to precocial nestlings as well as to altricial ones. The traditional theory for the adaptiveness of thermoregulatory altriciality does not apply or has a much more limited application than was heretofore thought. In particular, from the perspective of the traditional theory, there can be no simple, energetics-based prediction as to whether altricial or precocial young would grow faster.

In fact, altricial and precocial mammals do not differ systematically in postnatal growth rate (Case, 1978b). For a more exacting analysis, however, we need to recognize that growth rates depend on—and conflate—two major and quite different factors: (1) the rates that the young are provisioned with chemical energy by their mothers and (2) the efficiencies of the young in using their available chemical energy for growth (Case, 1978b). Each of these factors ultimately will need to be examined in its own right to see if it differs systematically between altricial and precocial young.

The question I pursue briefly here is whether the properties of altricial young lead them to be different from precocial ones in the efficiency with which they can direct their available milk energy into growth. This is the question that remains as an unanswered legacy of the traditional theory for the adaptiveness of thermoregulatory altriciality.

To my knowledge, only two studies have explicitly compared altricial and precocial nestlings in the same laboratory, using the same techniques, to evaluate their relative allocation of energy into growth. As shown in Table 7.2, the altricial nestlings in both cases were estimated to direct more of their milk energy into growth. However, only limited confidence can be placed in this trend at present. The sample size is small, the differences between the altricial and precocial species are small, neither study took account of the effects of 24-h activity patterns, uncertainties surround the results for *Sigmodon* and *Neotoma* (see note to Table 7.2), and neither study adequately took account of the concern stressed in this chapter that the metabolic expenditures of even resting nestlings may be very different in natural environments from those in artificial ones. Clearly, an acute need exists for more data. If, as data accumulate, the trend in Table 7.2 proves to be real and general, we will need to understand why it pertains.

With the traditional theory discounted, there remain several possible mechanisms by which altriciality might enable young to direct more of

TABLE 7.2

Estimated percentage of milk energy devoted to growth between birth and weaning in two pairs of rodent species, each pair consisting of an altricial and a precocial form of similar adult body size

Species	Percentage of milk energy directed to growth	Reference
Neotoma floridana (altricial)	37[a]	McClure & Randolph, 1980, fig. 9
Sigmodon hispidus (precocial)	24[a]	McClure & Randolph, 1980, fig. 9
Gerbillus perpallidus (altricial)	37[b]	Waldschmidt & Müller, 1988, fig. 13
Acomys cahirinus (precocial)	31[b]	Waldschmidt & Müller, 1988, fig. 13

[a]McClure (1987) stated that these values are to be revised, with the revised figures apparently bringing the species closer together. McClure (1987) also stated, however, that the revised value for *Neotoma* will still exceed that for *Sigmodon*. Neither the revised values nor their documentation is yet available. Oswald and McClure (1990) recently published new data that are not focused on the issue at hand but from which the percentage of milk energy directed to growth can be calculated. The metabolic rates of *S. hispidus* and *N. floridana* in the new data were estimated at an ambient temperature of 35°C, rather than the 20°C used by McClure and Randolph (1980). The change in ambient temperature greatly altered the picture for *S. hispidus*, leading to a percentage of milk energy directed to growth (46%) that was greater than for *N. floridana* (39%). Oswald and McClure's data are not presented as a test of the altricial/precocial contrast in growth efficiency; indeed, an ambient temperature of 35°C might mask differences between young with different thermoregulatory modes by removing any stimulus for thermoregulatory thermogenesis. Nonetheless, the difference between McClure, 1987 and Oswald and McClure, 1990 illustrates the tenuousness of our knowledge of these matters.

[b]Weaning age, 25 days for *Gerbillus* and 15 days for *Acomys*.

their energy resources into growth. Unfortunately, we know almost nothing about any of them. (1) Altricial nestlings might be incubated by their mother for a greater fraction of all their nestling hours than precocial ones; (2) the altricial nestlings might have greater capacities for facultative suspension of their endogenous thermoregulation; (3) they may use these capacities more frequently; and (4) altricial nestlings may tend to have fundamentally lower overall metabolic rates than precocial ones (Webb and McClure, 1989).

If altricial nestlings prove to have especially high growth efficiencies, this would not necessarily mean that they would grow faster than precocial ones. However, it would mean that they would grow faster than *they themselves* would grow without the elevation of their efficiencies. These two ideas need to be distinguished, and the distinction hinges on the point already made, that growth rates depend not only on growth efficiencies but also on rates of energy provisioning. In her writings, McClure (McClure, 1987; McClure and Randolph, 1980) raises the interesting possibility that, in *Neotoma*, altricial development may represent a compensation for an energy-limited ecological situation. *Neotoma* mothers give their nestlings less milk energy per unit time than *Sigmodon* mothers, but the growth rates of the *Neotoma* are argued to be more similar to those of the *Sigmodon* than this fact would suggest because the *Neotoma* nestlings direct more of their milk energy into growth.

Mammals versus Birds

Altricial bird nestlings seem typically to be utterly devoid of thermoregulatory thermogenic responses for a substantial part of their nestling life (Hill and Beaver, 1982). The traditional argument for the advantages of thermoregulatory altriciality may thus apply to birds in its original form, as proposed by Dawson (1962) and Dawson and Evans (1957), with only modest quantitative amendments (Hill and Beaver, 1982). Altricial bird nestlings also show a clear tendency to grow faster than precocial ones (Case, 1978b; Ricklefs, 1973). It may be that the heightened growth rates of altricial young result straightforwardly—at least in part—from the inability of early-age nestlings to use their energy resources for thermoregulation.

We need to recognize that, for many reasons that are beyond the scope of this chapter, thermoregulatory altriciality probably has different ecological implications for birds than for mammals (one reason, for example, is that there are large differences in patterns of energy provisioning of the young in the two groups). Thus, we should not be surprised to find altriciality take different forms in the two groups. The extrapolation of the traditional theory from birds to mammals as if there were no difference seems to have been an error.

The Future

To understand the altricial/precocial contrast in mammals, we will need to integrate analyses of pre- and postnatal development. This will be no easy task (Gittleman and Thompson, 1988); just the study of nestling development has enormous complexity. Nonetheless, the task has been started (e.g., McClure, 1987) and must be pursued. Another important need will be to gather sufficiently extensive data to take account of diversity. Because of variations in context among taxa, we should not expect altricial development to have the same features and meaning in all small mammals, and key insights may remain hidden until contextual patterns are elucidated. For example, mothers and nestlings of species with relatively unreliable food sources (e.g., some insectivores) may have evolved different options for responding to food shortage than those of species with reliable food sources. Gregariousness of parents during nesting could also be a factor of note because numerous parameters important to growth, energetics, and thermoregulation may be altered by group nesting (Mennella et al., 1990; Sayler and Salmon, 1969); in communal nests, for instance, thermoregulation by the young may be facilitated by the increased numbers of young in a group, and individual mothers may enjoy enhanced freedom to engage in energy-saving hypothermia without subjecting their offspring to excessive cooling. Throughout future efforts to understand the altricial/precocial contrast, mammalogists will need to break the habit of thinking of early-age altricial young as necessarily ectothermic or poikilothermic. Anyone who has read the literature in this area will recognize that this misconception has had a pervasive influence. A fresh look is needed.

In the 1950s and 1960s most studies of energetics and thermoregulation during nestling development were concerned principally with measuring growth curves and the thermoregulatory responses of isolated young. Since then, this research plan has been continued with disturbing monotony. We need to be very clear about what we learn from studies of isolated nestlings. The chief value of such studies to an understanding of natural history is that they reveal the pace at which young mammals develop the thermoregulatory ability to function as *adults*. From such studies, we learn that precocial species are prepared to function as adults faster than altricial ones. The studies of isolates, however, may reveal very little about how the young themselves function in nature. Thus, whereas the traditional research plan deserves to be continued for certain purposes, other research plans are required to answer many of the central questions concerning the physiological ecology and evolutionary biology of nestling development.

ACKNOWLEDGMENTS

This chapter was written in part during my tenure in the Frank A. Brown, Jr. Memorial Readership at the Marine Biological Laboratory, Woods Hole. I thank Thomas Tomasi and three anonymous reviewers for helpful comments. Thomas Kunz aided my understanding of the literature on bats. Joanna Gliwicz assisted with access to the literature on bank voles. Margaret Eppstein helped with the literature search.

Literature Cited

Alberts, J. R. 1978. Huddling by rat pups: Group behavioral mechanisms of temperature regulation and energy conservation. J. Comp. Physiol. Psychol., 92:231–245.

Anderson, J. L., and R. Boonstra. 1979. Some aspects of reproduction in the vole *Microtus townsendii*. Can. J. Zool., 57:18–24.

Barnett, S. A. 1956. Endothermy and ectothermy in mice at −3°C. J. Exp. Biol., 33:124–133.

——. 1973. Maternal processes in the cold-adaptation of mice. Biol. Rev., 48:477–508.

Bartholomew, G. A. 1982. Body temperature and energy metabolism. Pp. 333–406 *in* Animal Physiology (M. S. Gordon, ed.). Macmillan, New York. 635 pp.

Beacham, T. D. 1979. Size and growth characteristics of dispersing voles, *Microtus townsendii*. Oecologia, 42:1–10.

Bendell, J. F. 1959. Food as a control of a population of white-footed mice, *Peromyscus leucopus noveboracensis* (Fischer). Can. J. Zool., 37:173–209.

Biggers, J. D., M. R. Ashoub, A. McLaren, and D. Michie. 1958. The growth and development of mice in three climatic environments. J. Exp. Biol., 35:144–155.

Bignall, K. E., F. W. Heggeness, and J. E. Palmer. 1975. Effect of neonatal decerebration on thermogenesis during starvation and cold exposure in the rat. Exp. Neurol., 49:174–188.

——. 1977. Sympathetic inhibition of thermogenesis in the infant rat: possible glucostatic control. Am. J. Physiol., 233:R23–R29.

Boonstra, R., and C. J. Krebs. 1978. Pitfall trapping of *Microtus townsendii*. J. Mammal. 59:136–148.

Bronson, F. H. 1985. Mammalian reproduction: an ecological perspective. Biol. Reprod., 32:1–26.

Bryant, D. M., and C. J. Hails. 1975. Mechanisms of heat conservation in the litters of mice (*Mus musculus* L.). Comp. Biochem. Physiol., 50A:99–104.

Case, T. J. 1978a. Endothermy and parental care in the terrestrial vertebrates. Am. Nat., 112:861–874.

——. 1978b. On the evolution and adaptive significance of postnatal growth rates in the terrestrial vertebrates. Q. Rev. Biol., 53:243–282.

Casey, T. M. 1981. Nest insulation: energy savings to brown lemmings using a winter nest. Oecologia, 50:199–204.

Cassin, S. 1963. Critical oxygen tensions in newborn, young, and adult mice. Am. J. Physiol., 205:325–330.

Chew, R. M., and E. Spencer. 1967. Development of metabolic response to cold in young mice of four species. Comp. Biochem. Physiol., 22:873–888.

Cole, F. R., and G. O. Batzli. 1979. Nutrition and population dynamics of the prairie vole, *Microtus ochrogaster*, in central Illinois. J. Anim. Ecol., 48:455–470.

Cosnier, J., A. Duveau, and J. Chanel. 1965. Consommation d'oxygène et grégarisme du rat nouveau-né. C. R. Soc. Biol., 159:1579–1581.

Crompton, A. W., and F. A. Jenkins, Jr. 1979. Origin of mammals. Pp. 59–73 *in* Mesozoic Mammals (J. A. Lillegraven, Z. Kielan-Jaworowska, and W. A. Clemens, eds.). University of California Press, Berkeley. 311 pp.

Dawson, W. R. 1962. Evolution of temperature regulation in birds. Pp. 45–71 *in* Comparative Physiology of Temperature Regulation, part 1 (J. P. Hannon and E. Viereck, eds.). Arctic Aeromedical Laboratory, Fort Wainwright, Alaska. 131 pp.

Dawson, W. R., and F. C. Evans. 1957. Relation of growth and development to temperature regulation in nestling field and chipping sparrows. Physiol. Zool., 30:315–327.

Derrickson, E. M. 1988. Patterns of postnatal growth in a laboratory colony of *Peromyscus leucopus*. J. Mammal., 69:57–66.

Dickerson, J. W. T., and E. M. Widdowson. 1960. Some effects of accelerating growth. II. Skeletal development. Proc. R. Soc. Lond. B Biol. Sci., 152:207–217.

Earle, M., and D. M. Lavigne. 1990. Intraspecific variation in body size, metabolic rate, and reproduction of deer mice (*Peromyscus maniculatus*). Can. J. Zool., 68:381–388.

Edson, J. L., and D. Hull. 1977. The effect of maternal starvation on the metabolic response to cold of the newborn rabbit. Pediatr. Res., 11:793–795.

Fairbairn, D. J. 1977. Why breed early? A study of reproductive tactics in *Peromyscus*. Can. J. Zool., 55:862–871.

Fleming, T. H., and R. J. Rauscher. 1978. On the evolution of litter size in *Peromyscus leucopus*. Evolution, 32:45–55.

Fuchs, S. 1982. Optimality of parental investment: the influence of nursing on reproductive success of mother and female young house mice. Behav. Ecol. Sociobiol., 10:39–51.

Gates, D. R. 1974. The ontogeny of thermoregulation in the cottontail rabbit (*Sylvilagus floridanus*). M.S. thesis, Western Illinois University, Dekalb. 70 pp.

Gebczynski, M. 1970. Development of temperature regulation in the fat dormouse. Acta Theriol., 15:357–372.

——. 1975. Heat economy and energy cost of growth in the bank vole during the first month of postnatal life. Acta Theriol., 20:379–434.

Geiser, F., and G. J. Kenagy. 1990. Development of thermoregulation and torpor in the golden-mantled ground squirrel *Spermophilus saturatus*. J. Mammal., 71:286–290.

Gelineo, S. 1962. Razvitak hemijske termoregulacije u poljske voluharice *Microtus arvalis* Pall. Rad Jugosl. Akad. Znan. Umjet., 329:41–64.

Gelineo, S., and A. Gelineo. 1952. La température du nid du rat et sa signification biologique. Bull. Acad. Serbe Sci., 4:197–210.

Getz, L. L. 1960. A population study of the vole, *Microtus pennsylvanicus*. Am. Midl. Nat., 64:392–405.

Getz, L. L., L. Verner, F. R. Cole, J. E. Hofmann, and D. E. Avalos. 1979. Comparisons of population demography of *Microtus ochrogaster* and *M. pennsylvanicus*. Acta Theriol., 24:319–349.

Gittleman, J. L., and S. D. Thompson. 1988. Energy allocation in mammalian reproduction. Am. Zool., 28:863–875.

Gliwicz, J. 1983. Survival and life span. Pp. 161–172 *in* Ecology of the Bank Vole (K. Petrusewicz, ed.). Acta Theriol., 28 (Suppl. 1).

Golley, F. B. 1961. Interaction of natality, mortality and movement during one annual cycle in a *Microtus* population. Am. Midl. Nat., 66:152–159.

Goundie, T. R., and S. H. Vessey. 1986. Survival and dispersal of young white-footed mice born in nest boxes. J. Mammal., 67:53–60.
Hafner, J. C., and M. S. Hafner. 1988. Heterochrony in rodents. Pp. 217–235 *in* Heterochrony in Evolution. A Multidisciplinary Approach (M. L. McKinney, ed.). Plenum Press, New York. 348 pp.
Hansen, L., and G. O. Batzli. 1978. The influence of food availability on the white-footed mouse: populations in isolated woodlots. Can. J. Zool., 56:2530–2541.
Harland, R. M., P. J. Blancher, and J. S. Millar. 1979. Demography of a population of *Peromyscus leucopus*. Can. J. Zool., 57:323–328.
Harland, R. M., and J. S. Millar. 1980. Activity of breeding *Peromyscus leucopus*. Can. J. Zool., 58:313–316.
Harvey, P. H., and P. M. Bennett. 1983. Brain size, energetics, ecology and life history patterns. Nature, 306:314–315.
Heggeness, F. W., D. Bindschadler, J. Chadwick, P. Conklin, S. Hulnick, and M. Oaks. 1961. Weight gains of overnourished and undernourished preweanling rats. J. Nutr., 75:39–44.
Hill, R. W. 1970. The ontogeny of homeothermy in neonatal *Peromyscus leucopus noveboracensis*. Ph.D. dissertation, University of Michigan, Ann Arbor. 203 pp.
——. 1972. The amount of maternal care in *Peromyscus leucopus* and its thermal significance for the young. J. Mammal., 53:774–790.
——. 1976. The ontogeny of homeothermy in neonatal *Peromyscus leucopus*. Physiol. Zool., 49:292–306.
——. 1983. Thermal physiology and energetics of *Peromyscus*; ontogeny, body temperature, metabolism, insulation, and microclimatology. J. Mammal., 64:19–37.
Hill, R. W., and D. L. Beaver. 1982. Inertial thermostability and thermoregulation in broods of redwing blackbirds. Physiol. Zool., 55:250–266.
Hill, R. W., and G. A. Wyse. 1989. Animal Physiology, 2d ed. Harper & Row, New York. 656 pp.
Hissa, R. 1968. Postnatal development of thermoregulation in the Norwegian lemming and the golden hamster. Ann. Zool. Fenn., 5:345–383.
Hissa, R., and K. Lagerspetz. 1964. The postnatal development of homoiothermy in the golden hamster. Ann. Med. Exp. Biol. Fenn., 42:43–45.
Hopson, J. A. 1973. Endothermy, small size, and the origin of mammalian reproduction. Am. Nat., 107:446–452.
Howard, W. E. 1949. Dispersal, amount of inbreeding, and longevity in a local population of prairie deermice on the George Reserve, southern Michigan. Contrib. Lab. Vertebr. Biol. Univ. Mich., 43:1–50.
Hudson, J. W. 1978. Shallow, daily torpor: a thermoregulatory adaptation. Pp. 67–108 *in* Strategies in Cold: Natural Torpidity and Thermogenesis (L. C. H. Wang and J. W. Hudson, eds). Academic Press, New York. 715 pp.
Hull, D. 1973. Thermoregulation in young mammals. Pp. 167–200 *in* Comparative Physiology of Thermoregulation. Vol. 3 (G. C. Whittow, ed.). Academic Press, New York. 278 pp.
Kaufman, D. W., and G. A. Kaufman. 1987. Reproduction by *Peromyscus polionotus*: number, size, and survival of offspring. J. Mammal., 68:275–280.
Kornienko, I. A., and V. D. Son'kin. 1974. The group effect in the thermoregulation reactions of young baby rats. Ekologiya, 5(2):46–51. (Translation published as Sov. J. Ecol., 5:130–134.)
Krebs, C. J., B. L. Keller, and R. H. Tamarin. 1969. *Microtus* population biology:

demographic changes in fluctuating populations of *M. ochrogaster* and *M. pennsylvanicus* in southern Indiana. Ecology, 50:587–607.

Kurta, A., and T. H. Kunz. 1987. Size of bats at birth and maternal investment during pregnancy. Symp. Zool. Soc. Lond., 57:79–106.

Loeb, S. C., and R. G. Schwab. 1987. Estimation of litter size in small mammals: bias due to chronology of embryo resorption. J. Mammal., 68:671–675.

Lynch, C. B. 1974. Environmental modification of nest-building in the white-footed mouse, *Peromyscus leucopus*. Anim. Behav., 22:405–409.

Martin, R. D., and A. M. MacLarnon. 1985. Gestation period, neonatal size and maternal investment in placental mammals. Nature, 313:220–223.

McClure, P. A. 1987. The energetics of reproduction and life histories of cricetine rodents (*Neotoma floridana* and *Sigmodon hispidus*). Symp. Zool. Soc. Lond., 57:241–258.

McClure, P. A., and J. C. Randolph. 1980. Relative allocation of energy to growth and development of homeothermy in the eastern wood rat (*Neotoma floridana*) and hispid cotton rat (*Sigmodon hispidus*). Ecol. Monogr., 50:199–219.

Mennella, J. A., M. S. Blumberg, M. K. McClintock, and H. Moltz. 1990. Inter-litter competition and communal nursing among Norway rats: advantages of birth synchrony. Behav. Ecol. Sociobiol., 27:183–190.

Mihok, S. 1979. Behavioral structure and demography of subarctic *Clethrionomys gapperi* and *Peromyscus maniculatus*. Can. J. Zool., 57:1520–1535.

Millar, J. S., and L. W. Gyug. 1981. Initiation of breeding by northern *Peromyscus* in relation to temperature. Can. J. Zool., 59:1094–1098.

Millar, J. S., and D. G. L. Innes. 1983. Demographic and life cycle characteristics of montane deer mice. Can. J. Zool., 61:574–585.

Morrison, P., and J. H. Petajan. 1962. The development of temperature regulation in the opossum, *Didelphis marsupialis virginiana*. Physiol. Zool., 35:52–65.

Myers, P., and L. L. Master. 1983. Reproduction by *Peromyscus maniculatus*: size and compromise. J. Mammal., 64:1–18.

Nagel, A. 1977. Torpor in the European white-toothed shrews. Experientia, 33:1455–1456.

——. 1989a. Paradoxical social effects on response to cold during ontogeny of the common white-toothed shrew, *Crocidura russula*. J. Comp. Physiol., 159B:301–304.

——. 1989b. Development of temperature regulation in the common white-toothed shrew, *Crocidura russula*. Comp. Biochem. Physiol., 92A:409–413.

Oswald, C., and P. A. McClure. 1990. Energetics of concurrent pregnancy and lactation in cotton rats and woodrats. J. Mammal., 71:500–509.

Pagel, M. D., and P. H. Harvey. 1988. How mammals produce large-brained offspring. Evolution, 42:948–957.

Parkes, A. S. 1929. Note on the growth of young mice suckled by rats. Ann. Appl. Biol., 16:171–173.

Pearson, O. P. 1948. Metabolism of small mammals, with remarks on the lower limit of mammalian size. Science, 108:44.

Phillips, J. A. 1981. Growth and its relationship to the initial annual cycle of the golden-mantled ground squirrel, *Spermophilus lateralis*. Can. J. Zool., 59:865–871.

Pichotka, J. 1964. Chemische Temperaturregulation bei neugeborenen Maeusen. Helgol. Wiss. Meeresunters., 9:274–284.

Ricklefs, R. E. 1973. Patterns of growth in birds. II. Growth rate and mode of development. Ibis, 115:177–210.

Rink, R. D. 1969. Oxygen consumption, body temperature, and brown adipose tissue in the postnatal golden hamster (*Mesocricetus auratus*). J. Exp. Zool., 170:117–123.

Sayler, A., and M. Salmon. 1969. Communal nursing in mice: influence of multiple mothers on the growth of the young. Science, 164:1309–1310.

Schmidt, I., R. Kaul, and G. Heldmaier. 1987. Thermoregulation and diurnal rhythms in 1-week-old rat pups. Can. J. Physiol. Pharmacol., 65:1355–1364.

Sheck, S. H. 1982. Development of thermoregulatory abilities in the neonatal hispid cotton rat, *Sigmodon hispidus texianus*, from northern Kansas and south-central Texas. Physiol. Zool., 55:91–104.

Shump, K. A., Jr. 1978. Ecological importance of nest construction in the hispid cotton rat (*Sigmodon hispidus*). Am. Midl. Nat., 100:103–115.

Spiers, D. E., and E. R. Adair. 1987. Thermoregulatory responses of the immature rat following repeated postnatal exposures to 2,450-MHz microwaves. Bioelectromagnetics, 8:283–294.

Stanier, M. W. 1975. Effect of body weight, ambient temperature and huddling on oxygen consumption and body temperature of young mice. Comp. Biochem. Physiol., 51A:79–82.

Sullivan, T. P. 1977. Demography and dispersal in island and mainland populations of the deer mouse, *Peromyscus maniculatus*. Ecology, 58:964–978.

——. 1979. Demography of populations of deer mice in coastal forest and clear-cut (logged) habitats. Can. J. Zool., 57:1636–1648.

Taylor, P. M. 1960. Oxygen consumption in new-born rats. J. Physiol. Lond., 154: 153–168.

Thorne, O. 1958. Shredding behavior of the white-footed mouse, *Peromyscus maniculatus osgoodi*, with special reference to nest building, temperature, and light. Thorne Ecol. Res. Stn. Boulder Colo. Bull., 6:1–75.

Turner, J. S., and R. C. Schroter. 1985. Why are small homeotherms born naked? Insulation and the critical radius concept. J. Therm. Biol., 10:233–238.

Tuttle, M. D. 1975. Population ecology of the gray bat (*Myotis grisescens*): factors influencing early growth and development. Occas. Pap. Mus. Nat. Hist. Univ. Kansas, 36:1–24.

Van Horne, B. 1981. Demography of *Peromyscus maniculatus* populations in seral stages of coastal coniferous forest in southeast Alaska. Can. J. Zool., 59:1045–1061.

Vinter, J., D. Hull, and M. C. Elphick. 1982. Onset of thermogenesis in response to cold in newborn mice. Biol. Neonate, 42:145–151.

Vishinesku, N., K. Nersesian-Vazilinu, and D. Radu. 1965. Nablyudeniya nad termoregulyatsiei v ontogenezise u *Mesocricetus auratus*. Rev. Roum. Biol. Ser. Zool., 10:249–256.

Waldschmidt, A., and E. F. Müller. 1988. A comparison of postnatal thermal physiology and energetics in an altricial (*Gerbillus perpallidus*) and a precocial (*Acomys cahirinus*) rodent species. Comp. Biochem. Physiol., 90A:169–181.

Webb, D. R., and P. A. McClure. 1989. Development of heat production in altricial and precocial rodents: implications for the energy allocation hypothesis. Physiol. Zool., 62:1293–1315.

Weigold, H. 1973. Jugendentwicklung der Temperaturregulation bei der Mausohrfledermaus, *Myotis myotis* (Borkhausen, 1797). J. Comp. Physiol., 85:169–212.

Wickler, S. J. 1980. Maximal thermogenic capacity and body temperatures of white-footed mice (*Peromyscus*) in summer and winter. Physiol. Zool., 53:338–346.

Widdowson, E. M., and R. A. McCance. 1960. Some effects of accelerating growth. I. General somatic development. Proc. R. Soc. Lond. B Biol. Sci., 152:188–206.

Teresa H. Horton and Carol N. Rowsemitt

8. Natural Selection and Variation in Reproductive Physiology

Physiologists, ecologists, and evolutionary biologists have all conducted numerous studies on rodent species to determine which factors in the environment can regulate life history patterns. While the organizational level at which these different researchers approach the topic may influence their view of what is an answer, it is clear that the common goal is the understanding of basic principles regulating life history. To attain this common goal, and occasionally to be able to recognize it, we need to step back from our own disciplines at times and incorporate ideas from other fields of study. This practice can be particularly valuable when trying to understand something as complex as the life history of rodents. For example, the life history of a rodent describes its development, maturation, reproduction, and senescence. Ecologists and evolutionary biologists may find that a source of important information is the comparison of life history patterns among individuals to determine which patterns enhance reproductive output (Boyce, 1988; Gadgil and Bossert, 1970; Partridge and Harvey, 1988; Smith-Gill, 1983; Tuomi et al., 1983). Central to the description of a life history pattern is the timing of transitions between life stages (Caswell, 1983; Cole, 1954; Partridge and Harvey, 1988). These transitions are controlled by physiological mechanisms. For physiologists the important information may be the specific response patterns of neuroendocrine systems that must be stimulated by an environmental signal in order to induce a transition

Teresa H. Horton: Department of Biological Sciences, Kent State University, Kent, Ohio 44242. Carol N. Rowsemitt: Department of Biology, University of Utah, Salt Lake City, Utah 84112.

between life stages. The observation of variation among species, or among individuals of the same species, provides valuable insight into alternative designs for mechanisms regulating the functions of organisms.

In this chapter we, as physiologists, discuss the existence of individual variation in responsiveness to environmental cues by rodents. Any variation in responsiveness of the neuroendocrine system to environmental cues can alter transitions of individuals between life stages, inducing variation in life history patterns. To understand this variation we need to know what the rodent is physiologically capable of doing, in a mechanistic view, and what this variability contributes to the reproductive success of the individual. Thus, we believe as do many others, that an understanding of the concepts of natural selection and the importance of individual variation within populations is essential to the successful interpretation of physiological processes in general and reproductive responses to environmental cues specifically (Bennett, 1987; Crews, 1987; McClintock, 1987).

Our aims in this chapter are several: (1) to present some basic concepts of natural selection to provide a common background for all readers; (2) to identify the types of environmental information rodents are known to utilize to schedule reproductive efforts; (3) to emphasize the importance of individual variation in responses to environmental information, and; (4) utilizing the results of the first three aims, to give examples using the concepts of natural selection to provide insight into and alternative interpretations of phenomena observed in both laboratory and field studies of rodents.

Principles of Evolution, Natural Selection, and Variation

The study of evolution incorporates more than the description of phylogenetic relationships; it seeks to understand the conditions and mechanisms conducive to change. Three commonly discussed components that contribute to evolution by natural selection are (1) variation among individuals, (2) heritability of the differences that cause variation, and (3) differential reproductive success. Because of the central role of variation in natural selection, variability in physiological responses should be viewed as fundamental to the biology of organisms. Extensive research in evolutionary biology utilizing both experimental approaches and mathematical models has produced a body of theory that can facilitate the interpretation of individual variation in physiological responses to environmental cues. In the following sections we introduce several concepts from the field of evolutionary biology. These concepts are commonly discussed by evolutionary biologists; however, as our goal is to facilitate discussion among physiologists, ecologists, and evolutionary biologists, the principles are presented here to establish a common ground.

Recognition of Individual Variation

Rodent species use diverse physical, biotic, and social factors to regulate sexual maturation and reproduction. Moreover, individuals of the same species may respond differently to the same factors. The information contained in interindividual variation in physiological processes has often gone unrecognized (Bennett, 1987). In this chapter we view the ability to respond to environmental factors as a species characteristic, thus the phenotype (i.e., observed response to the environment) may reflect the genetic potential of the individual. Like other genetically determined characters, phenotypes may vary because of genetic polymorphisms or modification by the environment. When viewed in this manner, variation in responsiveness to environmental cues can be subjected to the same quantitative analyses as other phenotypic characters (Bennett, 1987; Bull, 1987; Falconer, 1981; Via and Lande 1985; Wallace, 1981). In the following sections we will discuss several sources of individual variability, genetic and environmental. In addition, Blank (this volume) discusses genetic polymorphisms in responses to environmental signals.

Researchers' evaluations of variation change when they begin to view species or populations as collections of individuals rather than as homogeneous units. As described by Mayr (1976), an often overlooked contribution of Darwin's *On the Origin of Species* (1860) was the introduction into the scientific literature of "population thinking" which focuses on the genetic uniqueness of the individuals that make up a population. This view permits the recognition that survival and reproduction are dependent on the combination of many traits of the individual and the environment. Corollaries to this view are that (1) individuals possessing different combinations of traits may be successful in one environment, but not in another, and (2) a trait, which by itself may not be advantageous, may be maintained because in combination with other traits reproductive success of the individual is enhanced. By focusing on the uniqueness of individuals, one begins to evaluate traits by what they contribute to the reproductive success of individuals under specific environmental conditions and to recognize that traits do not exist "for the good of the species."

The goal of laboratory studies of physiological processes is often to describe the specific biochemical or cellular mechanisms that contribute to reproduction. It is often assumed that the biochemical and cellular events will be identical for individuals of all species. To study the specifics of a mechanism, one needs to isolate the system and study it under controlled and usually simple conditions. The need to study simple systems can obscure recognition of the intrinsic biological importance of variation to the organism (Bennett, 1987). The integrative processes required for the function of an organism may produce individual variation in reproductive responses to environmental cues that are fundamentally important rather than evidence of an imprecise regulatory process.

Recognition of the importance of variation may lead physiologists to discover genetically or developmentally determined differences in sensitivity to environmental factors. These discoveries may yield a better understanding of the integrative mechanisms regulating reproductive responses to the environment (Crews, 1987). Population biologists may gain by recognizing the extent to which variation in responsiveness to the environment may account for differences among populations (Negus and Berger, 1987).

Fitness and Models

Central to the theory of evolution is the idea that phenotypic variation leads to differential reproductive success: individuals displaying a particular trait will contribute more to future gene pools than individuals without that trait. The hypothesis of individual selection proposes that natural selection results from the competition between individuals to make the largest genetic contribution to future generations (Brandon, 1978, reprinted in Sober, 1984; Grafen, 1988). The reproductive fitness of an individual can be defined as that animal's "propensity to survive and reproduce in a particularly specified environment and population" (Mills and Beatty, 1979, p. 270). Many aspects of life history contribute to the lifetime reproductive success of individuals: survival to breeding age, reproductive life span, fecundity or mating success, and offspring survival (Clutton-Brock, 1988, p. 1). Theories of life history evolution address how the distribution of these different life stages (i.e., the life history pattern) influences fitness.

Mathematical models have been used to describe the costs and benefits of different life history patterns on reproductive success; lifetime reproductive success of individuals is often assumed to indicate the Darwinian fitness of the life history pattern (see Grafen, 1988 for a discussion of this assumption). The development of these models can be seen in nearly 40 years of publications (Bell, 1980; Charlesworth and Leon, 1976; Cole, 1954; Fisher, 1958; Partridge and Harvey, 1988; Reznick, 1985; Stearns, 1976). Changes in behavior or physiology are assumed to alter the cost and benefit relationships; thus, the mathematical model can be used to predict the value of changes in life history to the reproductive success of an individual, and by inference, its Darwinian fitness.

Current theories of life history evolution assume that resources are wasted and fitness is reduced if an attempt at reproduction fails and that each attempted reproduction decreases the potential for future reproduction. These arguments are based on the assumption that reproduction is energetically expensive and may impair the survival of the reproducing individual. See Reznick, 1985, and Partridge and Harvey, 1988, for reviews and Kenagy, 1987, and Tuomi et al., 1983 for alternative views. Laboratory and field measurements of energetic requirements of mating and reproduction (i.e., territorial defense, courtship behavior, gestation, lactation, and so on) have been made for few species. To fully understand the variability in reproductive activity and life history patterns will require studies of the energetics of a wider variety of species and life stages (Clutton-Brock, 1988, p. 2; Gittleman and Thompson, 1988; Thompson, this volume). Despite data from a limited number of species, it is assumed for rodents that the periods of late gestation and lactation are the energetically most demanding; the timing of reproduction is assumed to reflect selection of individuals that reproduce when the environment provides sufficient resources to meet the needs of gestation and lactation (Bronson, 1989; Randolph et al., 1977; Sadleir, 1969). The use of environmental cues to regulate reproduction is hypothesized to result from the value of those cues in indicating the availability of resources necessary for successful reproduction (Bronson, 1989; Kenagy and Barnes, 1984).

The cost-benefit functions presented in the mathematical models can be weighted by the probability that the individual will reproduce successfully at a certain time, the probability that it could do better if it delayed reproduction, and the probability of dying before achieving another opportunity to reproduce. These probability functions may be dependent on the environmental conditions and the animal's own physiological condition as determined by its age and energy reserves. The models predict that the reproductive success of individuals capable of responding to environmental factors may be improved relative to other members of the population through adjustments in either the timing of reproduction or the number of young produced (Bell, 1980; Cole, 1954; Millar, 1977). If the ability to respond to environmental cues enhances fitness, then individuals expressing the appropriate phenotype will produce more young. Over time, the proportion of animals responding to the cue will increase in the population. The extensive literature on life history theory suggests four major reasons why a single response to each cue does not become fixed in the population. (1) Conditions change; a pattern of responses favored in one generation or one habitat may not be favored in another. (2) Successful reproduction rarely relies on only one environmental characteristic or physiological trait; thus, animals may be able to modify their response to a cue when presented with conflicting information. (3) Individual animals are not all in the same physical condition when they encounter a given cue. (4) In some cases, the frequency with which one response is favored depends on the responses of other individuals in the population.

A class of model that has been valuable in understanding alternative strategies is the model of an evolutionarily stable strategy (ESS) (Hines, 1987; Maynard Smith, 1982). Although ESS theory was primarily developed for studies of frequency-dependent selection, we believe it also has heuristic value for studies of the environmental regulation of physiology and behavior. Derived from the game theories used in political science and economics, an ESS is defined by Maynard Smith (1982, p. 204) as, "a strategy such that, if all the members of a population adopt it, no mutant strategy could invade the population under the influence of natural selection." In this case, the term *strategy* is defined as "a specification of what an individual will do in any situation in which it may find itself" (Maynard Smith, 1982, p. 204). As applied to ESS theory, strategies are the physiological and behavioral phenotypes prevalent among individuals of the population; strategy does not imply that conscious decisions are made by individuals. When applied to the environmental regulation of physiology or behavior, the concept of strategy represents adjustments based on the probability that the physical or biotic environment will present a specific set of conditions in the future. By viewing responses to environmental cues as strategies, researchers can propose hypotheses regarding the conditions under which a genetic mutation producing an alternative strategy might invade the population. Often the ESS proposes a conditional response, or mixed strategy: under one set of conditions the individual should do A, under a different set of conditions it should do B.

ESS theory provides a framework for studying the conditions favoring the evolution of variation, but it does not examine the genetic or physiological sources of variation. Quantitative and population geneticists recognize that phenotypic variation of a trait typically results from many genetic and environmental components (Bennett, 1987; Blank, this volume; Falconer, 1981; Wallace, 1981).

Sexual Differences in Parental Investment

One source of phenotypic variation that is genetic, and which fits the classical concept of polymorphisms at specific gene loci, is the gender of an animal. Williams (1975a, p. 129) stated: "The principle that one sex [male] tries to achieve fertilization to a maximum degree and the other [female] tries to optimize it in a qualitative sense is a key to understanding many aspects of courtship and sexual behavior." See Darwin (1874) for further discussion. Although energy costs are incurred by both sexes, differences exist in the timing of their reproductive costs and investment in each offspring (Clutton-Brock, 1988; Clutton-Brock et al., 1982). Except in purely monogamous species, fitness of males usually is increased by inseminating multiple females. Females can produce only a limited number of offspring, and they invest a relatively large amount of energy in those few offspring. These conditions typically result in females being more selective about the quality of their mate than males under most circumstances, an attribute termed "coyness" by Darwin (1874). This is not to suggest that selection results in males that mate indiscriminately (Berger, 1989). The implication that females should be more selective in mate quality than males has led to many studies on the phenomenon of sexual selection (Williams, 1975b).

While data are available for only a few species, it is assumed that most of the female rodent's energetic expenses for reproduction occur after mating, during gestation and lactation (Bronson, 1989; Kenagy, 1987; Randolph et al., 1977). The male always invests energy before mating; energy is committed to development of reproductive structures as well as to male-male competition in the form of aggression and territorial defense. In cases where males are involved in parental care of the young, sires also have some postfertilization costs. Thus, several aspects of reproductive effort differ between the sexes: the cost per individual offspring, the factor being optimized (quality vs. quantity), and the timing of the costs of reproduction. In general, gender differences in patterns and magnitude of energetic investments may result in the sexes responding to different stimuli or processing the same stimuli differently. Our assessments of specific gender differences may change as additional measures of reproductive costs become available (Clutton-Brock, 1988; Clutton-Brock et al., 1982; Gittleman and Thompson, 1988; Thompson, this volume).

Components of Variance

The observed, or phenotypic variation, in a trait reflects the sum of the genetic and environmentally induced variation in a trait (Falconer, 1981). Genetic variation is the proportion of variance due to the presence of different alleles for genes (also called genetic polymorphisms) (Bennett, 1987; Falconer 1981). Evidence for genetic polymorphisms in responses to environmental cues are more thoroughly discussed by Blank (this volume).

The environmental component of variance results from the nongenetic circumstances that influence the phenotype of an individual (Falconer, 1981). Because physiological processes are ultimately regulated by the activation of genes, the observed phenotypic variance reflects the susceptibility of the genetic program to environmental perturbation. Thus, if the ability to respond to a particular environmental cue is the genetic trait under consideration, then responses to that cue may vary because of other factors in the environment.

Environmentally induced variation can result from the lack of a tightly regulated pattern of gene expression that permits random variation in a variable environment or from the environmental triggering of alternative developmental pathways (Bull, 1987; Falconer, 1981; Smith-Gill, 1983; Via and Lande, 1985). While specific, environmentally induced phenotypes resulting from either process are not heritable, the susceptibility of the genome to environmental perturbations is heritable. Thus the potential for phenotypic variation resulting from environmental factors is subject to natural selection (Bull, 1987; Via and Lande, 1985).

Environmental Cues

Environmental factors that may be used as cues are derived from physical, biotic, and social aspects of the environment. In the following discussion *factor* is used to indicate any component found in the environment that may potentially influence an animal's physiology, *cue* is reserved to describe those factors for which specific and predictable effects on reproduction have been demonstrated. Environmental cues influence reproduction by activating physiological processes within the body of an organism; thus the ability to respond to cues can be influenced by the bodily condition of the animal (i.e., age and energy reserves) as well as other factors in the habitat or external environment (i.e., photoperiod, food, conspecifics, and temperature: Table 8.1). The ability of the neuroendocrine system to integrate external environmental cues and bodily condition may result in variation in the responses elicited by any single cue. If genetically based, then the ability to integrate information is subject to natural selection because it contributes to phenotypic variation.

The assumption that observed patterns of reproduction result from selection of individuals whose life history patterns most successfully convert energy into offspring has influenced research on the influence of environmental cues on rodent reproduction. The availability of food resources is temporally heterogeneous in many habitats. Ecological and reproductive physiologists have worked extensively to determine which environmental factors are exploited by mammals to coordinate reproduction with the availability of resources (Bronson, 1985, 1989; Negus and Berger, 1972, 1987, 1988; Sadleir, 1969).

The degree to which environmental heterogeneity is a problem for scheduling reproduction depends on the size and lifespan of the organism (Levins, 1968; Southwood, 1977). Cues may be categorized as either predicting future conditions or informing the animal of immediate conditions (Table 8.1). Predictive cues vary in the accuracy with which they identify future conditions. Information provided by immediate

TABLE 8.1
Factors that influence reproduction

Type of factor	Time scale	Example	Reference
External cues	Long-term	Photoperiod	Brady, 1982
	Intermediate	Plant chemistry	Bergber et al., 1977, 1981; Korn & Taitt, 1987; Rowsemitt & O'Connor, 1989
		Water	Breed, 1976; Christian, 1979; Janskey, 1986; Nelson & Desjardins, 1987; Soholt, 1977; Yahr & Kessler, 1975
	Immediate	Nutrition: calories, proteins	Dobson & Kjelgaard, 1985; Hamilton & Bronson, 1985; Perrigo, 1987; Spears & Clarke, 1987; Vandenbergh et al., 1972
		Social factors: pheromones	Brown & MacDonald, 1985; Vandenbergh, 1983
Physiological	Immediate	Age	Finch, 1987; Finch et al., 1984; Hagen & Forslund, 1979; Matt et al., 1987; Negus & Pinter, 1965; Petersen, 1986a, 1986b; Westlin & Gustafsson, 1984
		Body weight/ composition	Bushberg & Holmes, 1985; Dobson & Kjelgaard, 1985; Glass et al., 1987; Tuomi et al., 1983
		Previous reproduction	Matt et al., 1987; Negus & Pinter, 1965; Westlin & Gustafsson, 1983

cues will be accurate at the present, but may not allow sufficient time for animals to make major physiological adjustments necessary for reproduction. Animals may initiate development of reproductive structures in response to predictive cues; upon attaining a physiological state where reproduction is possible, the specific timing or energetic commitment can be made in response to the quality of the immediate environment and their physical condition (Kenagy and Barnes, 1984).

Evolutionary Concepts in Studies of Reproductive Physiology

The focus on variation in the preceding sections points out that, although an individual's ability to respond to an environmental cue has a genetic basis, the nature of the response may be modified by other factors. The study of variation is facilitated by using evolutionary concepts as a framework. The following examples illustrate how the recognition of ecological and evolutionary principles can facilitate an understanding of variation in physiological responses to environmental cues.

Responses to Pheromones

When forming hypotheses about the adaptive significance of pheromonal signaling systems one must consider both the emitter's and the recipient's perspectives (Drickamer, 1987; McClintock, 1981). For example, many laboratory studies demonstrate that male pheromones induce sexual maturation or estrus in females of several rodent species (Brown, 1985). Natural selection theory suggests that male signaling systems should evolve to attempt to induce females to breed whenever the costs to the male are minimal. The optimal timing for males frequently occurs before the onset of the most favorable conditions for females. Selection should favor females that delay reproduction until conditions would ensure their reproductive success. Females should ignore male stimuli when other cues indicate that conditions are suboptimal for females. Studies of females housed under less than optimal conditions (i.e., food restriction) may provide results demonstrating that the females respond to male stimuli (i.e., mate) only under good conditions. Females should track environmental and physiological cues relevant to enhancing their reproductive success. Selection may favor females that respond to male pheromones to assess the availability and quality of potential mates but not to obtain information regarding the optimal time to breed. Because of the different schedules of energy commitments in reproduction, females should be more selective in timing of breeding than males. This prediction extends Darwin's (1874) concept of female coyness to include not only greater selectivity in the quality of potential mates but also greater selectivity in the timing of breeding.

A great deal of evidence demonstrates that pheromones released by reproductively active females delay maturation of young conspecific females (Brown, 1985). Many researchers have interpreted this response as a mechanism of population control, with the adult females preventing young females from breeding. The implication is that it would be to the advantage of the young females to breed, but they cannot avoid the suppressive force of these pheromones. We argue that the role of these pheromones has been misinterpreted.

If it is truly to the young females' advantage to breed at a given time, and most of the young females are being suppressed by adult female pheromones to the advantage of the adult females, only one mutation would be necessary to alter the system at the population level. If one young female possessed a mutation rendering her insensitive to the inhibitory pheromone, she would breed while others of her generation were suppressed. In each succeeding generation, the percentage of young females insensitive to the pheromone would increase until the pheromone-insensitive phenotype dominated the population. Thus, a population consisting of young females that are being suppressed from breeding does not represent an ESS because this population can be invaded by a phenotype that is immune to pheromonal suppression. We suggest that this pheromonal system should not be viewed as a manipulative, suppressive tool of adult females, but rather as one cue that young females use to determine the appropriate time to breed. Only with this type of functional explanation of the pheromone does one obtain an ESS. The very close presence of an adult female provides a signal to the young female's neuroendocrine system documenting that she is either still in her mother's nest or that population densities are extremely high. The former situation would occur in some species where the young overwinter in the natal nest. This pheromone could be acting in concert with decreasing photoperiod and food supplies to provide cues that cause the young female's neuroendocrine system to delay maturation until the following spring. Evidence for such a possibility is provided by Drickamer (1987). Natural selection will favor phenotypes that respond to these pieces of information by delaying maturation. Those females that mature and breed when winter is approaching are more likely to be unsuccessful in producing offspring. Under such conditions, selection will favor the response to the adult female pheromone. Pheromones in a high-density population could also provide a cue that the young female should attempt to disperse to an area of lower population density before breeding.

This treatment of pheromonal responses demonstrates the utility of the ESS approach. The principles of natural selection can provide a coherent intellectual framework to allow us to understand why individuals often exhibit reproductive responses to pheromonal cues as well as other signals from conspecifics.

Photoperiod

Responses to photoperiod are known to vary depending on age, season of birth, or the reproductive state of the animal. All of these parameters may influence the probability that an individual will live to the next breeding season. Consider a polyestrous species with a lifespan of less than one year. If good conditions persist past the end of the normal breeding season, it may be advantageous for adults to continue to breed but for young to remain immature and wait for the onset of the next breeding season before maturing. This age difference would be favored if (1) the immature animals have a higher probability of living to the next breeding season than do animals that have expended resources on reproduction, (2) a delay of maturation allows for the delay of aggressive and dispersal behaviors that can decrease survival, and (3) an older animal's probability of living to the next breeding season is lower than the probability of successfully raising one last litter. These differences may be the evolutionary reason for the reduced sensitivity of the adults of some species to changes in photoperiod (Bell, 1980; Cole, 1954; Donham et al., 1989; Partridge and Harvey, 1988).

Changes in sensitivity to environmental cues are not always associated with senescence; sensitivity differences may be elicited in response to previous environmental conditions. A classic example of such an effect is photorefractoriness: animals become insensitive to a cue following prolonged exposure (Hoffmann, 1981). Neuroendocrine responses can also be influenced during the development of young animals. This effect may occur while the young are in utero. An interesting situation arises in rodent species that produce several litters each year. Young born at the beginning of the breeding season may have greater fitness if they mature rapidly and reproduce in the season of their birth, whereas young born at the end of the breeding season may benefit by waiting until the onset of the next breeding season. If young use photoperiod to determine whether they were born at the beginning or end of the breeding season, they may encounter ambiguous information. This ambiguity results from the fact that all but the longest and shortest photoperiods occur twice each year. If young are to use photoperiod as a cue, then responses to identical photoperiods must be different for young born at different times of the year. This change can be interpreted as an example of phenotypic plasticity.

Phenotypic plasticity can result in different responses to changing environmental conditions (Negus and Berger, 1988). This phenomenon was examined in the montane vole (*Microtus montanus*; Horton, 1984a, 1984b, 1985). Juvenile montane voles used changes in day length to regulate growth and sexual maturation. When laboratory-reared young were placed in short days (8L:16D; 8 h light:16 h dark per day), equivalent to midwinter day lengths, growth and reproductive development were inhibited. When reared in a long photoperiod (16L:8D), equivalent to midsummer day lengths, growth and reproductive development were stimulated (Petterborg, 1978; Pinter and Negus, 1965). Montane voles born in the spring and fall are exposed to ambiguous information because day lengths during these seasons are the same intermediate length. How, then, do voles respond to 14 hours of light per day? They can discriminate between 14L:10D occurring in the spring and fall on the basis of information received from their mothers during gestation. Young born to females exposed to 16L:8D during pregnancy were inhibited when raised in 14L:10D after birth, but young gestated in 8L:16D were stimulated by 14L:10D. These differences occurred despite the fact that the young had experienced no photoperiod other than 14L:10D from birth (Horton, 1984b, 1985). Similar maternal effects on photoperiodic responses of juveniles have been observed in meadow voles (*M. pennsylvanicus*) and the Siberian hamster (*Phodopus sungorus*) (Lee et al., 1987; Stetson et al., 1986). Extensive work with *P. sungorus* indicates that the maternal system produces a hormonal cue that programs the neuroendocrine development of her fetuses, altering their postnatal photoperiodic responses (Elliott and Goldman, 1989; Horton et al., 1989; Stetson et al., 1989; Weaver and Reppert, 1986). These maternal effects on photoperiodic responses are one of many examples in which the maternal environment has long-lasting effects on the behavioral and physiological development of young (see Huck et al., 1987).

Phenotypic plasticity in maturation rates is just one example of a phenomenon that can in part be understood by considering the probability of living to the next breeding opportunity. The evolution of the neuroendocrine system can involve pathways that integrate such factors as age and photoperiodic time measurement. With such an integrative mechanism, a rodent will either breed or not breed because of a combination of these factors.

Dietary Parameters

Cues from food resources may take the form of calories or other nutrients as well as nonnutritive compounds that provide information regarding the quality of the food resource. The amount of energy available in the environment may influence reproductive responses by acting either as a specific cue or by influencing the bodily composition of the animal. Thus, the role of nutrition in altering responses to other environmental cues requires careful and thorough consideration.

Differences in the patterns of energy requirements between male and female rodents may influence their responses to environmental cues. Field and laboratory studies have provided ample evidence that male and female rodents use different factors to time reproduction and have different amounts of control over the number of young produced. In a natural population of kangaroo rats (*Dipodomys merriami*), spermatozoa were present in the cauda epididymis of 17–100% of the males throughout most of the 2½-year study (Kenagy and Bartholomew, 1985). Females in the population were reproductively active only during limited periods from January to June. Other workers have found similar patterns in *D. merriami*, with a high degree of variability in timing of female reproductive activity (Reichman and Van De Graaff, 1975). The females, which can produce mature gametes in a few days, may be relying on cues provided by the sprouting of vegetation after sporadic rains (Reichman and Van De Graaff, 1975). Because testicular development and gametogenesis are more protracted, males cannot rely on cues provided by these sporadic rains; they remain reproductively competent throughout the year. Even if male reproductive development were not a prolonged process, males must be prepared to provide sperm at any time since they cannot predict when mating opportunities will occur in this system. A delay of only a few days when females are available would mean lost mating opportunities if other males were already competent.

Gender differences in reproductive responses to food supplies have also been shown to occur in both domestic and wild *Mus musculus* (Hamilton and Bronson, 1985, 1986). When food supplies are decreased, males will mature. Females will not mature, suggesting that the female neuroendocrine system is capable of estimating the probability of successfully rearing young. Natural selection favors male reproductive development under these conditions for two reasons: (1) if food availability improves rapidly, reproductively competent males will be the ones to obtain matings, and (2) in a heterogeneous habitat, there may be some females with sufficient caloric intake to breed. This sex difference is further supported by the fact that maintenance of reproductive capacity for a male is less costly than commitment to reproduction for a female. Under some conditions, a male may be able to maintain sperm production but limit territorial and aggressive behaviors. For a female, a commitment to reproduction is an on/off function with all of the subsequent energy expenses including gestation and lactation.

Laboratory studies with montane voles (*Microtus montanus*) have revealed gender differences in the ability of dietary factors other than calories to influence litter size as well as the frequency of litter production. Berger et al. (1987) administered a compound (6-methoxybenzoxazolinone; 6-MBOA) extracted from young grass shoots that stimulates reproduction when administered to mated pairs by means of constant-release implants; a 6-MBOA-containing implant was given to either the female only, male only, or both male and female. Only when the female was treated with 6-MBOA were increases found in the number of young per litter and frequency of litters. In contrast, implantation of either parent with 6-MBOA influenced the sex ratios of the litters: mated pairs without 6-MBOA produced male-biased litters; 6-MBOA treatment shifted the sex ratio to unity when either parent was treated with this compound. When both parents received 6-MBOA, the effect was additive, shifting the sex ratio to 1.00:1.25 (males: females). Although the adaptive significance of sex ratio is uncertain, such shifts would be expected to affect both parents equally (Myers, 1978; Trivers and Willard, 1973; Williams, 1975b).

The results of the 6-MBOA experiments discussed above (Berger et al., 1987) are consistent with an interpretation that suggests that the two sexes have evolved separate strategies in response to the cue. Exposure of males to 6-MBOA does not result in the production of larger or more-frequent litters, changes that would increase the costs of reproduction to females. Only when females are themselves exposed to 6-MBOA do litter size and frequency increase. In contrast, exposure of males to 6-MBOA can produce changes in the sex ratio of litters, changes that may result in little or no difference in the energy demand on females.

These examples demonstrate that the principles of natural selection can provide a coherent intellectual framework that enhances our understanding of why dietary cues elicit different responses from males and females.

Integration of Multiple Cues

Extensive studies of single ecological cues have led workers to propose ecological scenarios for the control of reproduction that are at odds with the concepts of natural selection. While each cue is indubitably important, consideration of a single variable in the laboratory can lead to an overestimation of the role of that variable in determining reproductive patterns in the field.

The factor that may have the most confounding influence is food. The availability of calories may greatly influence the likelihood of individuals to respond to other cues. With few exceptions, most studies of reproductive cues (i.e., photoperiod, pheromones) have been performed with ad libitum food (Pryor and Bronson, 1981). Recent work on golden hamsters (*Mesocricetus auratus*) suggests that the role of photoperiod in regulating reproductive function is overestimated in controlled laboratory studies. A shift to a higher-quality diet can decrease the rate of photoperiod-induced gonadal regression in both males and females of this species (Hoffman et al., 1987; Johnson and Hoffman, 1985). These results suggest that *M. auratus* may be more opportunistic in its breeding patterns than was previously believed (Hoffman et al., 1987).

Researchers using mammalian and nonmammalian species are beginning to recognize the importance of multiple cues (Bronson and Rissman, 1986; Drickamer, 1987; Kenagy and Barnes, 1984; Negus and Berger, 1987, 1988; Spears and Clarke, 1986, 1987; Wingfield and Moore, 1987). We believe that coupling principles of evolutionary biology, especially natural selection, with the recognition that many cues may be integrated will yield a clearer understanding of the role of the environment in regulating reproduction.

Studies of Population Biology

A major problem in ecology is the study of the dynamics of populations. Populations are made up of individuals. Yet as a consequence of the effort to simplify the mathematics, models developed to explain the growth of populations often fail to include the idea that individuals are different. The importance of individual variation to the ecology of populations is receiving increased awareness among population biologists (Lomnicki, 1988). The role of the environment in determining population dynamics is often controversial because of the lack of clearcut correlations between environmental perturbations and population growth and decline. However, if the physiological response of individuals to environmental cues varies as a function of age, gender, development, or other environmental factors, the effects of a single environmental change on the dynamics of populations may vary according to the initial composition of the population (Garsd and Howard, 1981; Lomnicki, 1988; Negus and Berger, 1988; Negus and Pinter, 1965).

Microtine rodents, whose population dynamics have attracted much attention for over 40 years, may provide an important example. Lidicker (1988) described the regulation of microtine populations as a multifactorial problem resulting from the interaction of reproduction, death, and dispersal. Limiting our focus to just one component of the problem, the control of reproduction, still yields a multifactorial equation. Voles of many species respond to many different environmental cues; a single species may respond to dietary, photoperiodic, and social cues (Table 8.2). If variation in the response to a single cue can result from the presence of other external environmental cues or the physiological condition of individuals, then the potential importance of individual variation is great. Additional studies of phenotypic variation in responsiveness to environmental cues might clarify the role of these cues in reproduction and, subsequently, in regulating the dynamics of populations.

We suggest that consideration of both the concepts of natural selection and the recognition of multiple cues should guide the design of future laboratory and field studies as well as their interpretations.

Conclusions

We have examined evolutionary theory, individual variation, and the role of environmental information in reproduction to clarify and in some cases alter the interpretation of physiological responses observed in the laboratory. We suggest that knowledge of individual variation in physiological responses to environmental information may facilitate

TABLE 8.2

Microtus species tested for responsiveness to environmental factors

Species	Photoperiod	Quality of diet	Temperature	Water	Pheromones
agrestis	Baker & Ranson, 1932; Clarke & Kennedy, 1967; Spears & Clarke, 1986, 1987	Spears & Clarke, 1987	Clarke & Kennedy, 1967; Spears & Clarke, 1987		Jemiolo et al., 1980; Spears & Clarke, 1986
arvalis	Lecyk, 1962		Daketse & Martinet, 1977		
californicus	Nelson et al., 1983	Gill, 1977; Nelson et al., 1983	Gill, 1977	Nelson et al., 1983	Heske, 1987; Heske & Nelson, 1984
montanus	Petterborg, 1978; Pinter & Negus, 1965	Berger et al., 1977, 1981, 1987			Berger & Negus, 1982; Gray et al., 1974
ochrogaster	Nelson, 1985; Nelson et al., 1989	Cole & Batzli, 1978	Nelson et al., 1989		Carter et al., 1980; Hasler & Conway, 1973; Hasler & Nalbandov, 1974
pennsylvanicus	Dark et al., 1983; Pistole & Cranford, 1982	Desy & Thompson, 1983			Baddaloo & Clulow, 1981; Pasley & McKinney, 1973; Watson et al., 1983
pinetorum	Lepri & Noden, 1984				Lepri & Vandenbergh, 1986; Schadler, 1981

our understanding of population biology. This chapter is not an exhaustive review of evolutionary theory and ecological considerations of reproduction or of the potential applications of these lines of thought. Reproductive physiologists, ecological physiologists, and population biologists are all interested in aspects of reproduction of animals. It would be helpful to our understanding of physiological and population processes if workers from each area possessed a greater understanding of the knowledge and thought processes that have been gained in the other areas. We believe that each field will reap substantial benefits if such cross-fertilization occurs. It is our hope that this treatment of the subject will encourage further exploration of the principles of natural selection by those who have not had formal training in the area. We also hope that this chapter will lead to more discussion between workers in these distinct but highly overlapping fields.

ACKNOWLEDGMENTS

This work was supported in part by USPHS grants 5R23HD18367 to C. N. Rowsemitt and F32HD06778 to T. H. Horton. We thank F. H. Bronson, J. J. Bull, E. L. Charnov, J. A. Endler, P. D. Heideman, L. D. Houck, N. C. Negus, and E. A. Rickart for their comments on the manuscript.

Literature Cited

Baddaloo, E. G. Y., and F. V. Clulow. 1981. Effects of the male on growth, sexual maturation, and ovulation of young female meadow voles, *Microtus pennsylvanicus*. Can. J. Zool., 59:415–421.

Baker, J. R. and R. M. Ranson. 1932. Factors affecting the breeding of the field mouse (*Microtus agrestis*). Part 1. Light. Proc. R. Soc. Lond., 110:313–322.

Bell, G. 1980. The costs of reproduction and their consequences. Am. Nat., 116:45–76.

Bennett, A. F. 1987. Interindividual variability: an underutilized resource. Pp. 147–165 *in* New Directions in Ecological Physiology. (M. E. Feder, A. F. Bennett, W. W. Burggren, and R. B. Huey, eds.). Cambridge University Press, New York.

Berger, J. 1989. Female reproductive potential and its apparent evaluation by male mammals. J. Mammal., 70:347–358.

Berger, P. J., and N. C. Negus. 1982. Stud male maintenance of pregnancy in *Microtus montanus*. J. Mammal., 63:148–151.

Berger, P. J., N. C. Negus, and C. N. Rowsemitt. 1987. Effect of 6-methoxybenzoxazolinone on sex ratio and breeding performance in *Microtus montanus*. Biol. Reprod., 36:255–260.

Berger, P. J., N. C. Negus, E. H. Sanders, and P. D. Gardner. 1981. Chemical triggering of reproduction in *Microtus montanus*. Science, 214:69–70.

Berger, P. J., E. H. Sanders, P. D. Gardner, and N. C. Negus. 1977. Phenolic plant compounds functioning as reproductive inhibitors in *Microtus montanus*. Science 195:575–577.

Boyce, M. S. 1988. Evolution of life histories: theory and patterns from mammals. Pp. 3–30 *in* Evolution of Life Histories of Mammals: Theory and Pattern. (M. S. Boyce, ed.). Yale University Press, New Haven.

Brady, J. 1982. Biological timekeeping. Society for Experimental Biology. Seminar Series 14. Cambridge University Press, Cambridge. 197 pp.

Brandon, R. 1978. Adaptation and evolutionary theory. Studies in the History and Philosophy of Science, 9:181–206. Reprinted in Sober, 1984.

Breed, W. G. 1976. Effect of environment on ovarian activity of wild hopping mice (*Notomys alexis*). J. Reprod. Fertil., 47:395–397.

Bronson, F. H. 1985. Mammalian reproduction: an ecological perspective. Biol. Reprod., 32:1–26.

——. 1989. Mammalian Reproductive Biology. University of Chicago Press, Chicago. 325 pp.

Bronson, F. H., and E. F. Rissman. 1986. The biology of puberty. Biol. Rev., 61:157–195.

Brown, R. E. 1985. The rodents I: effects of odours on reproductive physiology (primer effects). Pp. 245–344 *in* Social Odours in Mammals. Vol. 1 (R. E. Brown and D. W. MacDonald, eds.). Clarendon Press, Oxford. 506 pp.

Brown, R. E., and D. W. MacDonald. Social Odours in Mammals. Vol. 1. Clarendon Press, Oxford. 506 pp.

Bull, J. J. 1987. Evolution of phenotypic variance. Evolution, 41:303–315.

Bushberg, D. M., and W. G. Holmes. 1985. Sexual maturation in male Belding's ground squirrels: influence of body weight. Biol. Reprod., 33:302–308.

Carter, C. S., L. L. Getz, L. Gavish, J. L. McDermott, and P. Arnold. 1980. Male-related pheromones and the activation of female reproduction in the prairie vole (*Microtus ochrogaster*). Biol. Reprod., 23:1038–1045.

Caswell, H. 1983. Phenotypic plasticity in life-history traits: demographic effects and evolutionary consequences. Am. Zool., 23:35–46.

Charlesworth, B., and J. A. Leon. 1976. The relation of reproductive effort to age. Am. Nat., 110:449–459.

Christian, D. P. 1979. Comparative demography of three Namib desert rodents: responses to the provision of supplementary water. J. Mammal., 60:679–690.

Clarke, J. R., and J. P. Kennedy. 1967. Effect of light and temperature upon gonad activity in the vole (*Microtus agrestis*). Gen. Comp. Endocrinol., 8:474–488.

Clutton-Brock, T. H. ed. 1988. Reproductive Success: Studies of Individual Variation in Constrasting Breeding Systems. The University of Chicago Press, Chicago. 538 pp.

Clutton-Brock, T. H., F. E. Guinness, and S. D. Albon. 1982. Red Deer: Behavior and Ecology of Two Sexes. University of Chicago Press, Chicago. 378 pp.

Cole, F. R., and G. O. Batzli. 1978. Influence of supplemental feeding on a vole population. J. Mammal., 59:809–819.

Cole, L. C. 1954. The population consequences of life history phenomena. Q. Rev. Biol., 29:103–137.

Crews, D. 1987. Diversity and evolution of behavioral controlling mechanisms. Pp. 88–119 *in* Psychology of Reproductive Behavior: An Evolutionary Perspective (D. Crews, ed.). Prentice-Hall, Englewood Cliffs, N.J. 350 pp.

Daketse, M.-J., and L. Martinet. 1977. Effect of temperature on the growth and fertility of the field-vole, *Microtus arvalis*, raised in different daylength and feeding conditions. Ann. Biol. Anim. Biochim. Biophys., 17:713–721.

Dark, J., I. Zucker, and G. N. Wade. 1983. Photoperiodic regulation of body mass, food intake, and reproduction in meadow voles. Am. J. Physiol., 245:R334–R338.

Darwin, C. 1860. On the Origin of Species. A Facsimile of the First Edition. Harvard University Press, Cambridge, 1966. 502 pp.

——. 1874. The Descent of Man and Selection in Relation to Sex, 2d ed. Hurst and Co., New York. 705 p.

Desy, E. A., and C. F. Thompson. 1983. Effects of supplemental food on a *Microtus pennsylvanicus* population in central Illinois. J. Anim. Ecol., 52:127–140.

Dobson, F. S., and J. D. Kjelgaard. 1985. The influence of food resources on life history in Columbian ground squirrels. Can. J. Zool., 63:2105–2109.

Donham, R. S., T. H. Horton, and M. H. Stetson. 1989. Age, photoperiodic responses and pineal function in meadow voles, *Microtus pennsylvanicus*. J. Pineal Res., 7:243–252.

Drickamer, L. C. 1987. Seasonal variations in the effectiveness of urinary chemosignals influencing puberty in female house mice. J. Reprod. Fertil., 80:295–300.

Elliott, J. A., and B. D. Goldman. 1989. Reception of photoperiodic information by fetal Siberian hamsters: role of the mother's pineal gland. J. Exp. Zool., 252:237–244.

Falconer, D. S. 1981. Introduction to Quantitative Genetics, 2d ed. Longman, London. 340 pp.

Finch, C. E. 1987. Neural and endocrine determinants of senescence: investigation of causality and reversibility by laboratory and clinical interventions. Pp. 261–308 *in* Modern Biological Theories of Aging (H. R. Warner, R. N. Bulter, R. L. Sprott, and E. L. Schneider, eds.). Raven Press, New York. 324 pp.

Finch, C. E., L. S. Felicio, C. V. Mobbs, and J. F. Nelson. 1984. Ovarian and steroidal influences on neuroendocrine aging processes in female rodents. Endocr. Rev., 5: 467–497.

Fisher, R. A. 1958. The Genetical Theory of Natural Selection. 2d rev. ed. Dover, New York. 287 pp.

Gadgil M., and W. H. Bossert. 1970. Life historical consequences of natural selection. Am. Nat., 104:1–24.

Garsd, A., and W. E. Howard. 1981. A 19-year study of microtine population fluctuations using time-series analysis. Ecology, 62:930–937.

Gill, A. E. 1977. Polymorphism in an island population of the California vole, *Microtus californicus*. Heredity, 38:1–11.

Gittleman, J. L., and S. D. Thompson. 1988. Energy allocation in mammalian reproduction. Am. Zool., 28:863–875.

Glass, A. R., J. Anderson, and D. Herbert. 1987. Sexual maturation in underfed weight-matched rats. A test of the "critical body weight" theory of pubertal timing in males. J. Androl., 8:116–122.

Grafen, A. 1988. On the uses of data on lifetime reproductive success. Pp. 454–471 *in* Reproductive Success. Studies of Individual Variation in Constrasting Breeding Systems. (T. H. Clutton-Brock, ed.). University of Chicago Press, Chicago.

Gray, G. D., H. N. Davis, M. Zerylnick, and D. A. Dewsbury. 1974. Oestrus and induced ovulation in montane voles. J. Reprod. Fertil., 38:193–196.

Hagen, J. B., and L. G. Forslund. 1979. Comparative fertility of four age classes of female gray-tailed voles, *Microtus canicaudus*, in the laboratory. J. Mammal. 60:834–837.

Hamilton, G. D., and F. H. Bronson. 1985. Food restriction and reproductive development in wild house mice. Biol. Reprod., 32:773–778.

——. 1986. Food restriction and reproductive development: male and female mice and male rats. Am. J. Physiol., 250:R370–R376.

Hasler, M. J., and C. H. Conaway. 1973. The effect of males on the reproductive state of female *Microtus ochrogaster*. Biol. Reprod., 9:426–436.

Hasler, M. J., and A. V. Nalbandov. 1974. The effect of weanling and adult males on sexual maturation in female voles (*Microtus ochrogaster*). Gen. Comp. Endocrinol. 23:237–238.

Heske, E. J. 1987. Pregnancy interruption by strange males in the California vole. J. Mammal., 68:406–410.

Heske, E. J., and R. J. Nelson. 1984. Pregnancy interruption in *Microtus ochrogaster*: laboratory artifact or field phenomenon? Biol. Reprod., 31:97–103.

Hines, W. G. 1987. Evolutionary stable strategies: a review of basic theory. Theor. Pop. Biol. 31:195–272.

Hoffman, R. A., L. B. Johnson, M. K. Vaughan, and R. J. Reiter. 1987. Influence of diet on photoperiod-induced gonadal regression in female hamsters. Growth, 51:385–396.

Hoffmann, K. 1981. Photoperiodism in Vertebrates. Pp. 449–473 *in* Handbook of Behavioral Neurobiology. 4. Biological Rhythms (Jurgen Aschoff, ed.). Plenum, New York. 563 pp.

Horton, T. H. 1984a. Growth and maturation in *Microtus montanus*: effects of photoperiods before and after weaning. Can. J. Zool., 62:1741–1746.

——. 1984b. Growth and reproductive development of male *Microtus montanus* is affected by the prenatal photoperiod. Biol. Reprod., 31:499–504.

——. 1985. Cross-fostering of voles demonstrates in utero effect of photoperiod. Biol. Reprod., 33:934–939.

Horton, T. H., S. L. Ray, and M. H. Stetson. 1989. Maternal transfer of photoperiodic infomation in Siberian hamsters. III. Melatonin injections program postnatal reproductive development expressed in constant light. Biol. Reprod., 41:34–39.

Huck, U. W., J. B. Labov, and R. D. Lisk. 1987. Food-restricting first generation juvenile female hamsters (*Mesocricetus auratus*) affects sex ratio and growth of third generation offspring. Biol. Reprod., 37:612–617.

Jansky, L., G. Haddad, D. Pospisilova, and B. Dvorak. 1986. Effect of external factors on gonadal activity and body mass of male golden hamsters. (*Mesocricetus auratus*). J. Comp. Physiol. B, 156:717–725.

Jemiolo, B., A. Marchlewska-Koj, and A. Buchalczyk. 1980. Acceleration of ovarian follicle maturation of female caused by male in *Microtus agrestis* and *Clethrionomys glareolus*. Folia Biol., 28:269–272.

Johnson, L. B., and R. A. Hoffman. 1985. Interaction of diet and photoperiod on growth and reproduction in male golden hamsters. Growth, 49:380–399.

Kenagy, G. J. 1987. Energy allocation for reproduction in the golden-mantled ground squirrel. Symp. Zool. Soc. Lond., 57:259–273.

Kenagy, G. J., and B. M. Barnes. 1984. Environmental and endogenous control of reproductive function in the Great Basin pocket mouse *Perognathus parvus*. Biol. Reprod., 31:637–645.

Kenagy, G. J., and G. A. Bartholomew. 1985. Seasonal reproductive patterns in five coexisting California desert rodent species. Ecol. Monogr., 55:371–397.

Korn, H., and M. J. Taitt. 1987. Initiation of early breeding in a population of *Microtus townsendii* (Rodentia) with the secondary plant compound 6-MBOA. Oecologia, 71:593–596.

Lecyk, M. 1962. The effect of the length of daylight on reproduction in the field vole *Microtus arvalis* (Pall). Zool. Pol., 12:189–221.

Lee, T. M., L. Smale, I. Zucker, and J. Dark. 1987. Influence of daylength experienced by dams on post-natal development of young meadow voles (*Microtus pennsylvanicus*). J. Reprod. Fertil., 81:337–342.

Lepri, J. J., and P. F. Noden. 1984. Reproductive function is independent of photoperiod in adult male *Microtus pinetorum*. J. Mammal., 64:706–708.

Lepri, J. J., and Vandenbergh, J. G. 1986. Puberty in pine voles, *Microtus pinetorum*, and the influence of chemosignals on female reproduction. Biol. Reprod., 34:370–377.

Levins, R. 1968. Evolution in Changing Evironments: Some Theoretical Explorations. Princeton University, Princeton. 120 pp.

Lidicker, W. Z., Jr. 1988. Solving the enigma of microtine "cycles." J. Mammal., 69:225–235.

Lomnicki, A. 1988. Population Biology of Individuals. Princeton University Press, Princeton. 223 pp.

Matt, D. W., P. L. Sarver, and J. K. H. Lu. 1987. Relation of parity and estrus cyclicity to the biology of pregnancy in aging female rats. Biol. Reprod., 37:421–430.

Maynard Smith, J. 1982. Evolution and the Theory of Games. Cambridge University Press, Cambridge. 224 pp.

Mayr, E. 1976. Typologicial versus Population Thinking. Pp. 26–29 *in* Evolution and the Diversity of Life. Selected Essays. Belknap Press, Harvard University Press, Cambridge.

McClintock, M. K. 1981. Social control of the ovarian cycle and the function of estrous synchrony. Am. Zool., 21:243–256.

McClintock, M. K. 1987. A functional approach to the behavioral endocrinology of rodents. Pp. 176–203 *in* Psychobiology of Reproductive Behavior: An Evolutionary Perspective (D. Crews, ed.). Prentice-Hall, Englewood Cliffs, N.J. 350 pp.

Millar, J. S. 1977. Adaptive features of mammalian reproduction. Evolution, 31:370–386.

Mills, S., and J. Beatty. 1979. The propensity interpretation of fitness. Philos. Sci., 46:263–286.

Myers, J. H. 1978. Sex ratio adjustment under food stress: maximization of quality or numbers of offspring? Am. Nat., 112:381–388.

Negus, N. C., and P. J. Berger. 1972. Environmental factors and reproductive processes in mammalian populations. Pp. 89–98 *in* Biology of Reproduction: Basic and Clinical Studies (J. T. Velardo and B. A. Kasprow, eds.). Third Pan-American Congress of Anatomy Symposium, New Orleans.

——. 1987. Mammalian reproductive physiology: adaptive responses to changing environments. Pp. 149–173 *in* Current Mammalogy. Vol. 1. (H. H. Genoways, ed.). Plenum, New York.

——. 1988. Cohort analysis: environmental cues and diapause in microtine rodents. Pp. 65–74 *in* Evolution of Life Histories of Mammals: Theory and Pattern. (M. S. Boyce, ed.). Yale University Press, New Haven.

Negus, N. C., and A. J. Pinter. 1965. Litter sizes of *Microtus montanus* in the laboratory. J. Mammal., 46:434–437.

Nelson, R. J. 1985. Photoperiod influences reproduction in the prairie vole. *Microtus ochrogaster*. Biol. Reprod., 33:596–602.

Nelson, R. J., J. Dark, and I. Zucker. 1983. Influence of photoperiod, nutrition, and water availability on reproduction of male California voles. J. Reprod. Fertil., 69: 473–477.

Nelson, R. J., and C. Desjardins. 1987. Water availability affects reproduction in deer mice. Biol. Reprod. 37:257–260.

Nelson, R. J., D. Frank, L. Smale, and S. B. Willoughby. Photoperiod and temperature affect reproductive and nonreproductive functions in male prairie voles (*Microtus ochrogaster*). Biol. Reprod., 40:481–485.

Partridge, L., and P. H. Harvey. 1988. The ecological context of life history evolution. Science, 241:1449–1455.

Pasley, J. N., and T. D. McKinney. 1973. Grouping and ovulation in *Microtus pennsylvanicus*. J. Reprod. Fertil., 34:527–530.

Perrigo, G. 1987. Breeding and feeding strategies in deer mice and house mice when females are challenged to work for their food. Anim. Behav., 35:1298–1316.

Petersen, S. L. 1986a. Age- and hormone-related changes in vaginal smear patterns in the gray-tailed vole, *Microtus canicaudus*. J. Reprod. Fertil., 78:49–56.
——. 1986b. Age-related changes in plasma oestrogen concentration, behavioural responsiveness to oestrogen, and reproductive success in female gray-tailed voles, *Microtus canicaudus*. J. Reprod. Fertil., 78:57–64.
Petterborg, L. J. 1978. Effect of photoperiod on body weight in the vole *Microtus montanus*. Can. J. Zool., 56:431–435.
Pinter, A. J., and N. C. Negus. 1965. Effects of nutrition and photoperiod on reproductive physiology of *Microtus montanus*. Am. J. Physiol., 908:633–638.
Pistole, D. H., and J. A. Cranford. 1982. Photoperiodic effects on growth in *Microtus pennsylvanicus*. J. Mammal., 63:547–553.
Pryor, S., and F. H. Bronson. 1981. Relative and combined effects of low temperature, poor diet, and short daylength on the productivity of wild house mice. Biol. Reprod., 25:734–743.
Randolph, P. A., J. C. Randolph, K. Mattingly, and M. M. Foster. 1977. Energy costs of reproduction in the cotton rat *Sigmodon hispidus*. Ecology, 58:31–45.
Reichman, O. J., and K. M. Van De Graaff. 1975. Association between ingestion of green vegetation and desert rodent reproduction. J. Mammal., 56:503–506.
Reznick, D. 1985. Costs of reproduction: an evaluation of the empirical evidence. Oikos, 44:257–267.
Rowsemitt, C. N., and A. J. O'Connor. 1989. Reproductive function in *Dipodomys ordii* is stimulated by 6-methoxybenzoxazolinone. J. Mammal., 70:805–809.
Sadleir, R. M. F. S. 1969. The Ecology of Reproduction in Wild and Domestic Mammals. Methuen, London. 321 pp.
Schadler, M. H. 1981. Postimplantation abortion in pine voles (*Microtus pinetorum*) induced by strange males and pheromones of strange males. Biol. Reprod., 25:295–297.
Smith-Gill, S. J. 1983. Developmental plasticity: Developmental conversion versus phenotypic modulation. Am. Zool., 23:47–55.
Sober, E. ed. 1984. Conceptual Issues in Evolutionary Biology: An Anthology. MIT Press, Cambridge. 725 pp.
Soholt, L. F. 1977. Consumption of herbaceous vegetation and water during reproduction and development of Merriam's kangaroo rat, *Dipodomys merriami*. Am. Midl. Nat. 98:445–457.
Southwood, T. R. E. 1977. Habitat, the templet for ecological strategies? J. Anim. Ecol., 46:337–365.
Spears, N., and J. R. Clarke. 1986. Effect of male presence and of photoperiod on the sexual maturation of the field vole (*Microtus agrestis*). J. Reprod. Fertil., 78:231–238.
——. 1987. Effect of nutrition, temperature and photoperiod on the rate of sexual maturation of the field vole (*Microtus agrestis*). J. Reprod. Fertil., 80:175–181.
Stearns, S. C. 1976. Life-history tactics: a review of the ideas. Q. Rev. Biol., 51:3–47.
Stetson, M. H., J. A. Elliott, and B. D. Goldman. 1986. Maternal transfer of photoperiodic information influences the photoperiodic response of prepubertal Djungarian hamsters (*Phodopus sungorus sungorus*). Biol. Reprod., 34:663–670.
Stetson, M. H., S. L. Ray, N. Creyaufmiller, and T. H. Horton. 1989. Maternal transfer of photoperiodic information in Siberian hamsters II: the nature of the maternal signal, time of signal transfer and the effect of the maternal signal on peripubertal reproductive development in the absence of photoperiodic input. Biol. Reprod., 40:458–465.
Trivers, R. L., and D. E. Willard. 1973. Natural selection of parental ability to vary the sex ratio of offspring. Science, 179:90–92.

Tuomi, J., T. Hakala, and E. Haukioja. 1983. Alternative concepts of reproductive effort, costs of reproduction, and selection in life-history evolution. Am. Zool., 23:25–34.

Vandenbergh, J. G. 1983. Pheromonal regulation of puberty. Pp. 95–112 *in* Pheromones and Reproduction in Mammals (J. Vandenbergh, ed.). Academic Press, New York.

Vandenbergh, J. G., L. C. Drickamer, and D. R. Colby. 1972. Social and dietary factors in the sexual maturation of female mice. J. Reprod. Fertil., 28:397–405.

Via, S., and R. Lande, 1985. Genotype-environment interaction and the evolution of phenotypic plasticity. Evolution, 39:505–522.

Wallace, B. 1981. Basic Population Genetics. Columbia University Press, New York. 688 pp.

Watson, M., F. V. Clulow, and F. Mariotti. 1983. Influence of olfactory stimuli on pregnancy of the meadow vole, *Microtus pennsylvanicus*, in the laboratory. J. Mammal., 64:706–708.

Weaver, D. R., and S. M. Reppert. 1986. Maternal melatonin communicates daylength to the fetus in Djungarian hamsters. Endocrinology, 119:2861–2863.

Westlin, L. M., and T. O. Gustafsson. 1983. Influence of sexual experience and social environment on fertility and incidence of mating in young female bank voles (*Clethrionomys glareolus*). J. Reprod. Fertil., 69:173–177.

——. 1984. Influence of age and artificial vaginal stimulation on fertility in female bank voles (*Clethrionomys glareolus*). J. Reprod. Fertil., 71:103–106.

Williams, G. C. 1975a. Adaptation and Natural Selection: A Critique of Some Current Evolutionary Thought. Princeton University Press, Princeton. 307 pp.

——. 1975b. Sex and Evolution. Princeton University Press, Princeton. 200 pp.

Wingfield, J. C., and M. C. Moore. 1987. Hormonal, social, and environmental factors in the reproductive biology of freeliving male birds. Pp. 149–175 *in* Psychobiology of Reproductive Behavior: An Evolutionary Perspective. (D. Crews, ed.). Prentice-Hall, Englewood Cliffs. N.J.

Yahr, P., and S. Kessler. 1975. Suppression of reproduction in water-deprived Mongolian gerbils (*Meriones unguiculatus*). Biol. Reprod., 12:249–254.

James L. Blank

9. Phenotypic Variation in Physiological Response to Seasonal Environments

The study of mammalian life histories has as one of its goals the explanation of when and to what extent an individual should reproduce during its lifetime. One investigative approach to these questions, that of evolutionary biology, places a premium on determining what ultimate factors combine to regulate individual reproductive effort. Another approach, that of reproductive physiology, emphasizes a search for internal physiological mechanisms that regulate reproductive function. In large part these two approaches have developed separately. Although they both have merit, they have used different animal models, adopted different methodologies, and developed hypotheses that are tested at quite different levels of biological analysis. For example, many of the most significant advances in reproductive physiology have developed in the field of molecular endocrinology. Emphasis has been on characterizing how cellular and molecular mechanisms create a hormonal milieu in the organism appropriate for reproducing. Evolutionary approaches to life history questions have also been concerned with molecular phenomena, particularly at the level of the genome. But, the conceptual emphasis has often dealt less with measuring how specific mechanisms determine an individual's reproductive response and more with quantifying a theoretically predicted end result of reproduction, such as number and quality of offspring.

It can be argued that these different emphases have been caused by concentration on either theory or mechanism. This argument is misleading, however, because it implies that these two investigative ap-

Department of Biological Sciences, Kent State University, Kent, Ohio 44242.

proaches are unconnected, a view reinforced by the historical independence of both fields. In fact, each research strategy offers many conceptual and practical advantages to the other. As an example, both emphasize the central role of energy in regulating reproductive effort. Life history theory frequently incorporates energy as a selective force molding an optimal reproductive response to a fluctuating environment. Reproductive physiology, on the other hand, emphasizes how energy availability affects the function of specific physiological paths and how these paths, in turn, influence reproductive function. Both approaches ask the same question: How does energy limit reproduction? Both approaches stress the same investigative approach: identification and measurement of a physiological response, in this example that of the reproductive system.

One goal of this book is to evaluate how physiology constrains life history traits. In this chapter, I attempt to demonstrate how each of the disparate approaches of evolutionary biology and reproductive physiology offers special advantages toward answering questions concerning mammalian life histories. For this purpose I concentrate on recent laboratory studies of the reproductive and metabolic adjustments rodents make to physical factors characteristic of a winter environment. Experiments of this type have been useful for probing the regulatory mechanisms that respond to changes in the environment and that govern a wide variety of seasonal physiological adaptations. These mechanisms ultimately act through the nervous system, since environmental modalities are perceived and transduced into neuroendocrine signals that evoke seasonal adjustments by central integrative paths (Desjardins and Lopez, 1980; Hoffmann, 1973; Reiter, 1980).

I also emphasize the advantages of exploiting wild or outbred species as laboratory animal models. Historically, studies concerned with effects of environment on reproductive processes have predominantly exploited highly inbred animal stock such as the Syrian hamster (*Mesocricetus auratus*) (Elliot and Goldman, 1981; Hoffman and Reiter, 1965; Reiter, 1973; Turek and Campbell, 1979), Djungarian hamster (*Phodopus sungorus*) (Hoffmann, 1973; Yellon and Goldman, 1984), or white rat (*Rattus norvegicus*) (Nelson and Zucker, 1981; Reiter et al., 1968). These species have been useful because they typically exhibit robust physiological responses to easily manipulated environmental factors such as photoperiod or ambient temperature. Numerous generations of inbreeding have produced physiological responses that are uniform in scope and, relative to outbred stocks, uncomplicated by among-animal variation (Lynch et al., 1989). This uniformity has proven to be an advantage especially in studies designed to discriminate the neuroendocrine paths that respond to the environment and engage physiological adjustments. These paths are circuitous, and animal models that express a single overt physiological response exhibit lower unexplained variance in structural or functional measurements of underlying regulatory paths than do outbred animal models.

Emphasis on inbred animal models, however, creates investigative difficulties of equal importance. For example, one of the most characteristic features of animals living in natural populations is that individuals differ in type and extent of many physiological traits. As is argued below, these differences can serve as an important research tool to investigate underlying regulatory mechanisms. Studies that rely solely on inbred animal models for this purpose offer the least probable opportunity for identifying and studying mechanisms responsible for variation or its affect on life history phenomena. A second disadvantage inherent to inbreeding pertains to previous intentional or unintentional selection by the investigator for specific physiological traits. The extent to which inbreeding selectively modifies structural or functional aspects of regulatory paths that govern physiological responses to external environmental cues such as photoperiod is unknown. At an extreme, inbreeding may modify the appearance of a particular seasonal adjustment by means of a regulatory pathway not directly involved in regulating the same physiological response in outbred stocks. From an investigator's viewpoint the physiological response in each case is identical, but the mechanism causing the reponse is different.

Conclusions based on results from inbred stocks may, therefore, be accurate for a particular inbred species, or breeding stock, but not reflective of other breeding stocks or of mammals exposed to selection pressures in natural environments. Several excellent examples of this point are known. The time period over which metabolic (pelage) and reproductive characters (testis size) of the Syrian hamster undergo seasonal acclimitization to photoperiod has increased from about 4 to 8 or more weeks (Nelson, 1983; Reiter, 1980). This deceleration in testicular involution is difficult to reconcile in the context of a fixed time period (5 weeks) for replacement of the germinal epithelium in this species (Clermont, 1972). German breeding stocks of the Djungarian hamster exhibit a photoperiod-induced seasonal decline in body weight (Steinlechner et al., 1983) with little individual variation. American stocks, in contrast, exhibit comparatively little or no change in body weight (Lynch et al., 1989). Further, the proportion of males that exhibit testicular atrophy after exposute to short days varies from 20 to 90%, depending on the breeding stock (Heldmaier et al., 1985; G. Heldmaier, pers. comm., 1989). These differences exist despite the fact that the American stock was originally derived from the German stock, a strong indication that selection has occurred for different patterns of seasonal weight change (Lynch et al., 1989). What do these changes in response characteristics mean in terms of the underlying regulatory mechanisms that engage them? Which, if either, pattern of response represents the wild type effect? From which stock can conclusions be derived that can be generalized to other species?

These questions are particularly troublesome when asking how a particular physiological adjustment constrains or shapes an individual's ability to respond to its environment. The type and speed with which seasonal acclimation occurs are likely to determine how efficiently an individual copes with changes in proximate factors such as ambient temperature or food availability. Significantly, a growing body of laboratory evidence (Blank and Desjardins, 1986; Dark and Zucker, 1983; Desjardins and Lopez, 1983; Heath and Lynch, 1983), including that presented below, demonstrates that animals exhibit variability in many types of response characteristics and manifest mutiple physiological solutions to similar environmental demands. If such variability is more the rule than the exception, then study of only inbred animal models significantly reduces the opportunity to clarify what these solutions are, how they are regulated, what effect they have on individual ability to cope with environment, and how this variability acts as a constraint on life history characteristics. Phenotypic variation in animal models is, in this context, a useful research tool.

This chapter has three aims. The first is to describe individual variation in reproductive and metabolic adjustments when deer mice are exposed to ambient conditions that mimic the winter environment. These data provide an analysis of the types of individual variation seen in the laboratory when animals produced from free-living parents are exposed to seasonally changing cues such as photoperiod or ambient temperature. The second aim is to summarize current knowledge about the mechanistic basis for this variability. This analysis is reductionist and presented as a method for identifying the specific mechanisms that underlie phenotypic variation. The third aim is to discuss how different seasonal adjustments constrain the way in which an individual is able to respond to its environment. The overall purpose of all three aims is to demonstrate how physiological studies in the lab can lead to a better understanding of the determinants of mammalian life histories.

Materials and Methods

Results presented below were obtained from a laboratory population of deer mice (*Peromyscus maniculatus nebrascensis*) collected in the vicinity of Wind Cave National Park, South Dakota. Individuals used in all experiments were selected from the F_2 generation of an outbred F_1 breeding stock. F_1 breeders were selected from 280 individuals produced by 35 pairs of wild-caught parents (4 male and 4 female offspring per pair). These breeding techniques maintained a coefficient of relatedness between experimental subjects of 0.125 or less. Further details concerning breeding and housing techniques can be found in Blank and Desjardins, 1983, 1985, 1986. Technical details concerning specific methods concerned with reproductive and metabolic parameters can be found in citations in figure legends. Treatment effects were evaluated with ANOVA; differences between group means were evaluated using the Student-Neuman-Keuls test.

Results and Discussion

Photoperiod and Reproduction

North temperate populations of *Peromyscus* are like many rodent species in that breeding is restricted to the late spring, summer, and early fall. Onset and duration of the breeding season vary with year, latitude, species, population, age, and other factors (Desjardins and Lopez, 1980; Fairbairn, 1977; Lopez, 1981; Millar 1984, 1985; Taitt, 1981). In all cases, however, seasonality in breeding activity is a direct result of seasonality in functional aspects of neural paths that govern reproductive function. Specific environmental cues, principally photoperiod, have been coopted by animals as predictors of the suitability of the immediate and future environment for breeding (Elliot and Goldman, 1981). During the past 20 years considerable effort has been placed on tracing the paths via which the brain perceives, transduces, and translates predictive cues such as photoperiod into neuroendocrine effectors of reproductive adjustments (Glass, 1988). Nearly all research

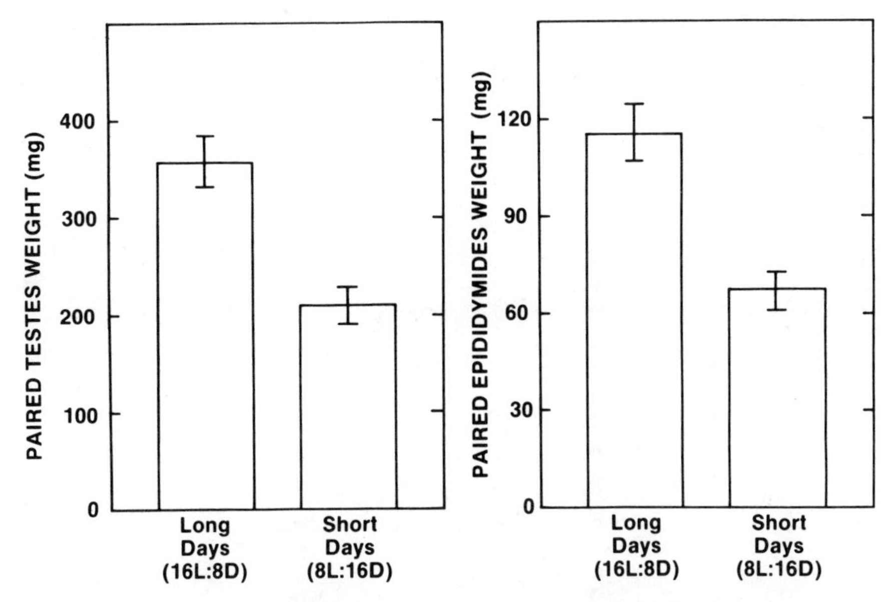

Fig. 9.1. (Left panel) Paired testes weight of deer mice exposed to long (16L:8D, N = 38) or short (8L:16D, N = 151) photoperiod. (Right panel) Paired epididymides weight for the same treatment groups. Bars, mean ±2 SE (vertical line) for the number of males designated above. Technical details can be found in Blank and Desjardins, 1985. (Data redrawn from Blank and Desjardins, 1986, by permission of *American Journal of Physiology*.)

effort has gone toward explaining how photoperiod serves in this capacity because it is the most error-free environmental signal available to synchronize breeding seasons.

Photoperiod has an unequivocal effect on gonadal function of deer mice. Males exposed to short day length (8L:16D) exhibited a 42% reduction ($P < 0.05$) in paired testes weight relative to breeding males maintained on long day length (16L:8D) (Fig. 9.1, left panel). Likewise paired epididymides underwent a 45% decline ($P < 0.05$) in weight (Fig. 9.1, right panel). The time required for attainment of minimum testis size is about 5 weeks, a period equal to the duration of the cycle of the germinal epithelium in mice (Clermont, 1972).

The function of the testis is twofold: to synthesize androgens, principally testosterone, and to support maturation of spermatozoa. By using either parameter to examine individual reproductive responses a more fine-grained characterization of the effect of photoperiod on testicular function can be obtained. Exposure to short day length caused a significant ($P < 0.05$) twofold reduction in the number of sperm present in paired epididymides (Fig. 9.2) and testes (Fig. 9.3). Inspection of indi-

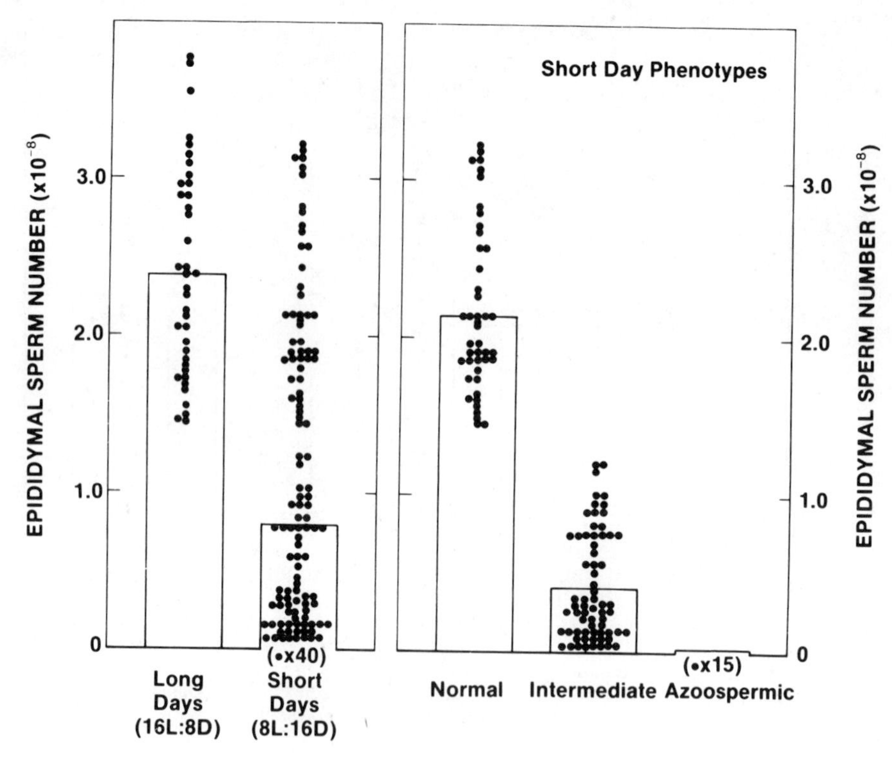

Fig. 9.2. Distribution of individual epididymidal sperm numbers (Blank and Desjardins, 1985; Desjardins and Lopez, 1983). Data were obtained from the same individuals presented in Figure 9.1. Bars, mean values; circles, individual values. (Left panel) Epididymal sperm number for deer mice exposed to long (16L:8D, N = 38) or short (8L:16D, N = 151) photoperiod. (Right panel) Epididymidal sperm number for mice categorized into one of three short day phenotypes based on individual spermatogenic responses to short photoperiod. Phenotypes are defined as follows: Normal (N = 44), males whose epididymal sperm numbers were equal to or greater than those of long day males; azoospermic (N = 40), deer mice with no spermatozoa present in their epididymides; and intermediate (N = 67), males whose epididymidal sperm numbers fell between the values used to identify normal and azoospermic phenotypes. Epididymal sperm numbers were used to assign males to each phenotype because this value represents the sperm available for ejaculation and fertilization of female mice. (Data redrawn from Blank and Desjardins, 1986, by permission of *American Journal of Physiology*.)

vidual spermatogenic responses for either organ, however, revealed considerable variation among males. On the one hand, fully 30% of all short day exposed males remained gonadally competent (normal phenotype) as assessed by finding normal sperm numbers in their testes and epididymides. Conversely, testes of an equal number of short day mice (azoospermic phenotype) contained only a few sperm and were judged infertile on the basis of coincident observation that their epididymides, which in breeding males contain sperm available for ejac-

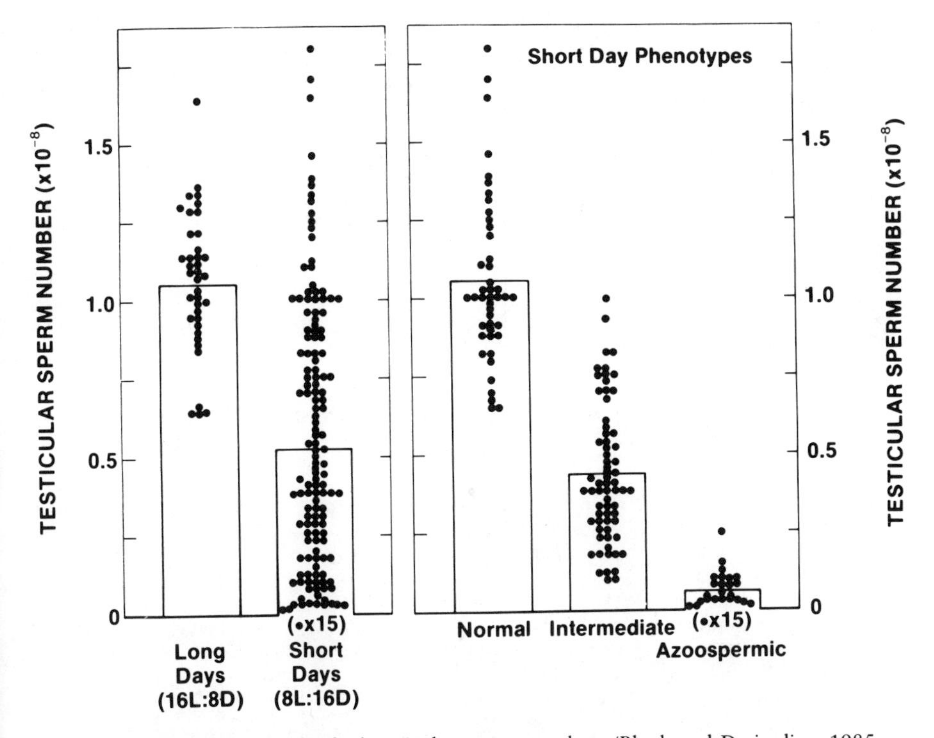

Fig. 9.3. Distribution of individual testicular sperm numbers (Blank and Desjardins, 1985; Desjardins and Lopez, 1983). Values were obtained from the same individuals for which data are described in Figure 9.2. (Left panel) Number of spermatozoa or spermatids shaped like them in deer mice exposed to either long (16L:8D, N = 38) or short photoperiod (8L:16D, N = 151). (Right panel) Individual testicular sperm number for the same phenotypic categories as described in Figure 9.2. Testes of azoospermic deer mice contain a few elongated spermatids, since the method used to quantify spermatogenesis does not differentiate between spermatozoa and spermatids shaped like them. (Data redrawn from Blank and Desjardins, 1986, by permission of *American Journal of Physiology*.)

ulation and fertilization, were devoid of sperm in this case. Testes of all remaining deer mice contained sperm at concentrations intermediate to these two extremes.

Significant differences among phenotypic categories also emerged when steroidogenic function was assessed by measuring plasma concentrations of testosterone (Fig. 9.4). Mice with normal sperm production had concentrations of testosterone that fell within the range of males exposed to long days. In contrast, plasma testosterone was significantly ($P < 0.05$) lower both among mice with intermediate numbers of sperm and those found to be azoospermic.

Therefore, there was considerable variability in spermatogenic and steroidogenic responses to short photoperiod among all tested deer mice. Although the array of responses is divided into three distinct sub-

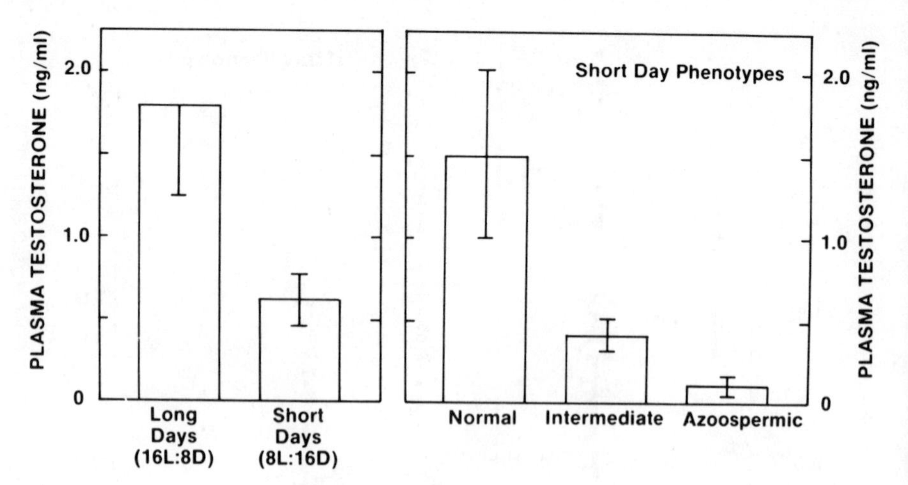

Fig. 9.4. Concentration of testosterone in plasma of adult male deer mice exposed to either long (16L:8D, N = 38) or short (8L:16D, N = 151) photoperiods (left panel). Bars, mean ±2 SE (vertical lines). Hormone values were sorted into short day phenotypes (right panel) as described for Figure 9.2. Testosterone was measured with a double-antibody radioimmunoassay (RIA) procedure verified for use with small volumes of plasma from male deer mice (Desjardins and Lopez, 1983). The within-assay coefficient of variation for five replicate determinations made on plasma from normal male deer mice was 13.6%. The minimum amount of testosterone detectable in unknown samples of deer mice blood was 0.025 ng/ml plasma. (Data redrawn from Blank and Desjardins, 1986, by permission of *American Journal of Physiology*.)

sets, this classification is arbitrary and used only for illustrative purposes. Only two short day phenotypes probably exist within a test population; those that are capable of inseminating females and those that are not. Mating tests to discriminate between those two classes were not performed.

Two important points concerning these data are noteworthy. First, individual variation in reproductive response to short photoperiod was not due to a latitudinal difference among subjects in site of origin; all deer mice used were derived from the same breeding population. This distinction is important because previous studies have demonstrated that populations of deer mice from northern latitudes contain proportionately more azoospermic males after short day exposure than do populations from southern latitudes (Desjardins and Lopez, 1983; Lopez, 1981). Similar latitudinal differences in reproductive characteristics have been demonstrated for the congeneric *Peromyscus leucopus* (Lynch et al., 1981).

Second, individual variation was not an artifact of conducting experiments on wild animal stocks within the laboratory environment. Two lines of experimental evidence support this statement. On the one hand, individual male deer mice of each short day phenotype respond in an identical manner to a second period of short day length exposure, following an intervening 16-week exposure (photorefractory period) to a stimulatory long day length (C. Desjardins, pers. comm., 1984; Lopez, 1981). On the other hand, the trait of either photoperiod responsiveness (short day-induced gonadal regression) or photoperiod nonresponsiveness (lack of short day-induced gonadal regression) responds rapidly to directional selection (Desjardins et al., 1986). The number of responsive or nonresponsive mice within a test population can be shifted by as much as 40% in only two generations. These measured differences in testicular function are, therefore, not likely to be caused by measurement errors or artifactual effects of laboratory housing. Taken together, these findings support the conclusion that individual variation in testicular response to photoperiod observed in the laboratory is representative of genetic differences among individuals living in natural habitats.

Multiple Environmental Cues

These results could have been predicted from studies on free-living rodents conducted as early as the 1930s and 1940s. Baker and Ransom (1932), for *Microtus agrestis*, and Whitaker (1940), for *Peromyscus leucopus*, found individuals of both sexes in breeding condition during the winter months. Indeed, winter breeding is one of the best described and most thoroughly documented physiological responses among winter populations of microtine rodents (Krebs and Meyers, 1974). These observations, surprisingly, have been overlooked by many laboratory physiologists interested in environmental regulation of reproductive function. Indeed, from the majority of lab studies, especially those exploiting the Syrian hamster as an animal model, field data have rarely been compared with laboratory data. In the case of the Syrian hamster this deficit results not from lack of appreciation for the importance of field data but rather from considerable lack of information about life history characteristics of this species in its natural habitat.

Baker and Ransom (1932) and Whitaker's (1940) data presaged results described here for deer mice. The presence of reproductively active males in winter populations demonstrates a failure of short day length or any other cue associated with the winter season to suppress

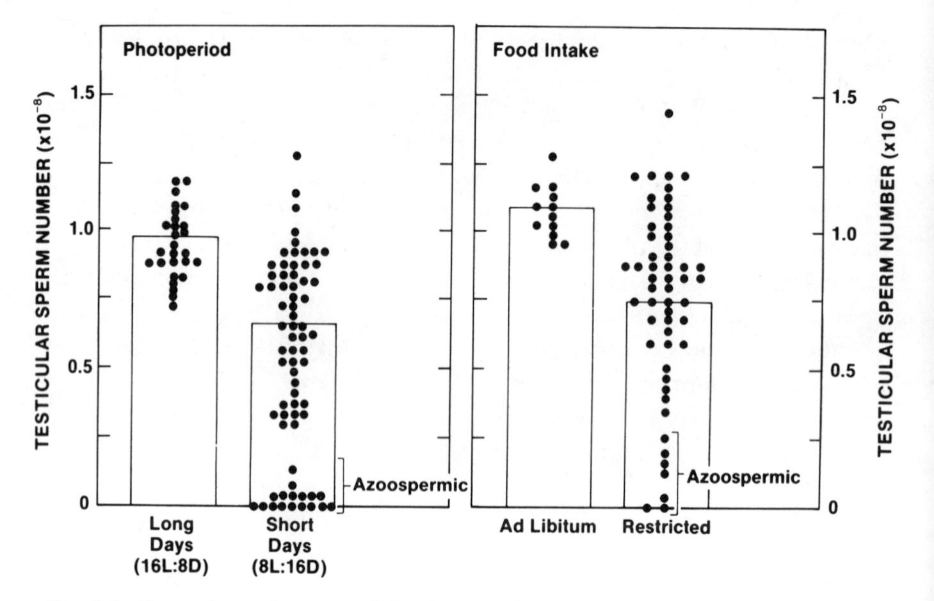

Fig. 9.5. Comparison of mean and distribution of testicular sperm number for deer mice exposed to either short photoperiod (8L:16D) or restricted food intake (70% of ad libitum). Data collected from long day–exposed males provided with ad libitum food served as controls in both instances. In each case, mean sperm number for all males exposed to the inhibitory cue was significantly ($P < 0.05$) lower than that of control males. (Data for food-restricted mice redrawn from Blank and Desjardins, 1983, by permission of *American Journal of Physiology*; data for photoperiod mice from unpublished results.)

reproductive function. In the present study, a minimum of 30% of all short day exposed deer mice were reproductively competent (Figs. 9.2 and 9.3); but estimates from field studies indicate that far fewer than 30% of all deer mice in natural populations are sexually active during the winter (Brown, 1966; Fairbairn, 1977; Sadleir, 1974). In hindsight, this inconsistency suggests that photoperiod is only one of several environmental factors that affect reproductive function in natural populations of deer mice. Additional laboratory experiments support this hypothesis. A mild restriction of food availability, for example, produces an array of testicular responses, nearly identical to those elicited by short photoperiod (Fig. 9.5). Other proximate cues have also been found to elicit testicular atrophy; these include water availability (Nelson and Desjardins, 1987) and cold ambient temperature (Desjardins and Lopez, 1983), the latter having an effect on testicular function only in combination with short photoperiod. In each case, individual deer mice differ in their gonadal response, with some males exhibiting azoospermia and others normal spermatogenesis. When mice are exposed simultaneously to three environmental factors, reproductive quiescence occurs in about 90%, but not 100%, of deer mice in laboratory populations (Desjardins and Lopez, 1983).

These data lead to two important generalities. On the one hand, individual deer mice respond differentially to a particular environmental cue. Some males are reproductively responsive and some are non-responsive after exposure to short photoperiod, food restriction, or some other factor. On the other hand, different individual deer mice exhibit the identical reproductive response to different environmental cues. In other words, testicular involution is elicited in some males by photoperiodic cues, in other males by food restriction, and in still others by some combination of these or other proximate factors. Thus, one can hypothesize that populations of deer mice are composed of subsets of mice that vary in their reproductive responses to a number of proximate cues, with a major difference among these subsets being their genetically programmed reproductive response to different environmental factors.

Neuroendocrine Regulation

Programmed gonadal responses to different environmental cues ultimately lie in genetic differences among individuals. Virtually nothing is known about how the genome regulates differential gonadal responses or about how the brain is organized to detect and respond to multiple environmental cues. With respect to the reproductive system the genome exerts its effects by modification of neuroendocrine signals from the hypothalmic-pituitary axis. The neural mechanisms that modify these signals are best described for one physiological system, reproduction, and one environmental cue, photoperiod. Detailed descriptions of the components of this pathway can be found elsewhere (Glass, 1988). Briefly, central paths that detect and respond to light and modify secretion of hormones of the hypothalamic-pituitary axis have been traced from the eye to their ultimate convergence in the hypothalamus. Transduction of day length information (i.e., lengthening or shortening) into a neurochemical signal is accomplished by the pineal gland, which, through its nocturnal release of melatonin within the 24-h day, conveys the photoperiodic message to neural loci residing in the hypothalamus. The anterior hypothalamus is thought to be particularly important in interpreting the pineal-mediated signal and in translating this signal into neuroendocrine effectors of pituitary gonadotropin secretion (Glass, 1988; Hastings et al., 1985). The measurements of testicular

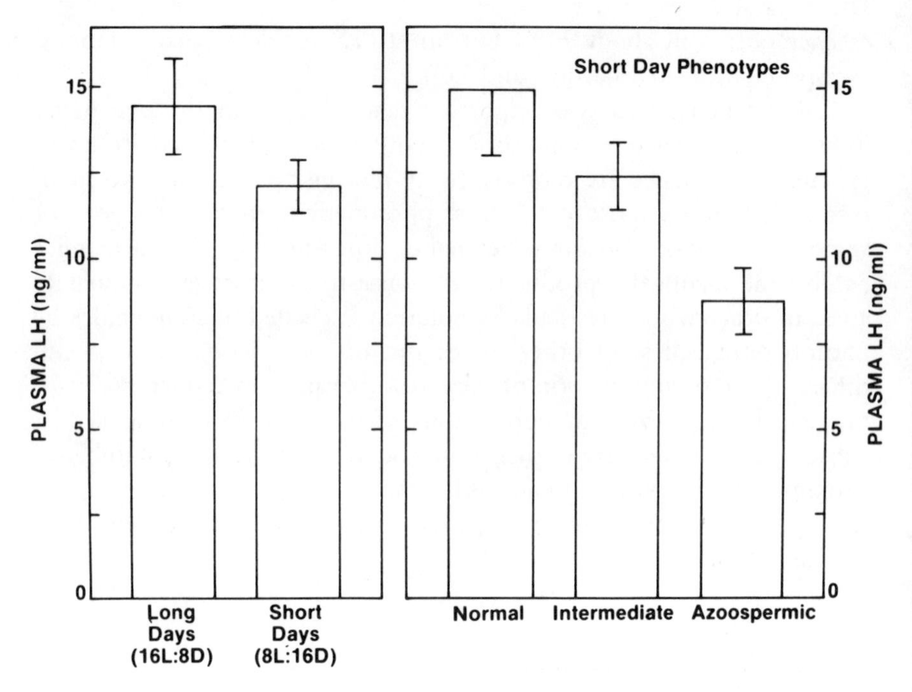

Fig. 9.6. Concentration of luteinizing hormone (LH) in blood plasma of adult male deer mice exposed to either long (16L:8D) or short (8L:16D) photoperiods (left panel). Bars, mean ±2 SE (vertical lines). Hormone values were sorted into short day phenotypes (right panel) as described for Figure 9.2. LH was measured with a double-antibody RIA validated for use in deer mice (Desjardins and Lopez, 1983). The within-assay coefficients of variation for four replicate determinations made on pools of plasma from normal intact and castrated male deer mice were 3.7 and 9.0%, respectively. The minimum amount of LH detectable in blood from unknown deer mice was 3.5 ng/ml plasma. (Data redrawn from Blank and Desjardins, 1986, by permission of *American Journal of Physiology*.)

function described above are a reliable index of these neuroendocrine adjustments.

The important physiological question is whether any part of the neuroendocrine system of deer mice responds to seasonal cues with lability equal to that of the gonad. With respect to two environmental cues, short photoperiod and reduced food availability, the answer is unequivocally yes; individual variation at the level of the testis is accompanied by individual variation at the level of the pituitary gland (see below).

Plasma luteinizing hormone (LH) concentrations, the putative pituitary hormone that stimulates synthesis of testosterone and sperm production, were normal in short day deer mice with normal sperm production and normal testosterone levels (Fig. 9.6). In contrast, plasma LH concentrations were significantly ($P < 0.05$) reduced in those deer mice with impaired sperm production and reduced testosterone levels

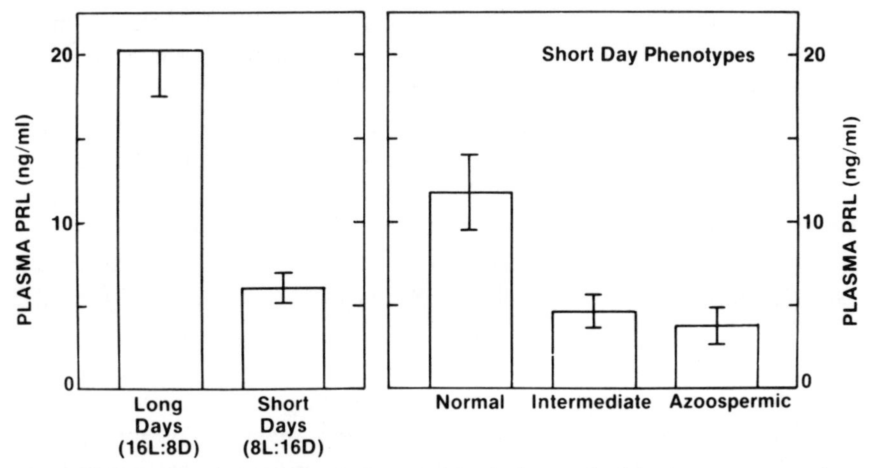

Fig. 9.7. Concentration of prolactin (PRL) in blood plasma of adult deer mice exposed to long (16L:8D) or short (8L:16D) photoperiods (left panel). Bars, mean ±2 SE (vertical lines). Hormone values were sorted into short day phenotypes (right panel) as described for Figure 9.2. PRL was measured by a double-antibody RIA validated for use in deer mice (Blank and Desjardins, 1986). The within-assay coefficients of variation for four replicate determinations made on pools of plasma from normal intact and castrated male deer mice were 2.8 and 3.3%, respectively. The minimum amount of PRL detectable in samples from unknown deer mice was 0.25 ng/10 μl plasma. (Data redrawn from Blank and Desjardins, 1986, by permission of *American Journal of Physiology*.)

(Fig. 9.4). Reduced food availability evokes similar relationships between sperm production, plasma testosterone, and LH concentrations among males, as does a combination of short photoperiod and cold ambient temperatures (Desjardins and Lopez, 1983).

Short photoperiod also caused a significant ($P < 0.05$) fourfold decline in plasma prolactin (PRL) levels (Fig. 9.7). Interestingly, plasma PRL was significantly ($P < 0.05$, in all cases) reduced among all phenotypic subsets of deer mice. This reduction included males that had normal spermatogenesis and normal levels of LH and testosterone after short day treatment.

Finding a positive correlation between blood concentrations of LH and testicular function is not surprising. These data are important, though, because they signify a normal functioning pituitary gland in all mice tested, thereby implicating higher neural centers as the primary cause of individual differences in testicular responses to photoperiod. Secretion of peptide hormones from the anterior pituitary is regulated by hormones synthesized and released from cell bodies residing in the hypothalamus. Individual differences among deer mice in LH or PRL release are probably mirrored by individual differences in synthesis or release of hypothalamic releasing hormones. This would appear likely given data available from other species (Glass, 1988) and the ultimate role of the hypothalamus as the final common path for modalities known to regulate pituitary hormone release.

Reduction in PRL secretion from the pituitary gland may be indicative of physiological adjustments in one or more nongonadal functions that depend on lower circulating concentrations of PRL. For example, short day–induced reduction in PRL is believed to direct changes in pelage (Duncan and Goldman, 1984a, 1984b) and metabolism (Dark and Zucker, 1983; Feist et al., 1988). Modification in PRL secretion demonstrates another important point. Prolactin release from the anterior pituitary is governed by neuroendocrine input from the hypothalamus (McNeilly, 1987). Changes in secretory activity of neurons residing in the anterior and basal medial hypothalamus direct changes in pituitary secretion of PRL. Therefore, the uniform reduction in PRL among all reproductive phenotypes of deer mice demonstrates that changes in photoperiod are detected in all males at the level of the hypothalamic neurons regulating pituitary function. This demonstration provides strong evidence that endocrine correlates of short day–induced gonadal phenotypes are attributable to differential operation of neurons of hypothalamic origin.

A second important conclusion that can be drawn from these data is that impaired testis function results in lowered sperm production and in reduced testosterone synthesis. Reduced testosterone not only eliminates sperm maturation but also drives a reduction in all testosterone-dependent or influenced functions (Desjardins, 1981; McEwen, 1980). For example, reduced testosterone synthesis causes cessation of all reproductive-related behaviors. These behaviors are not limited to solicitation of sex, but include male-male interactions (Kato, 1980), territory formation and maintenance (Kenagy, 1985), and general activity (Ellis and Turek, 1983). A reduction in testosterone is a recognized prerequisite in some rodent species for development of seasonal metabolic adjustments such as pelage molt, hibernation, or daily torpor (Barnes et al., 1987; Feist et al., 1988; Goldman et al., 1985; Vitale et al., 1985). Thus, individual differences in neural function are expressed as phenotypic differences in numerous physiological functions, not only at the gonadal level but also at the organismal, and especially, behavioral levels. From a strictly reductionist viewpoint, reproductive phenotypes can accurately be termed neuroendocrine phenotypes. This viewpoint is conceptually important because it suggests that differences among individuals at the organismal level can be attributed to differ ences in operation of specific neural loci in the hypothalamus.

Seasonal Metabolic Adjustments

Reproductive adjustments are not the only or even the most obvious physiological accommodation made on a seasonal basis. Free-living animals engage numerous adaptations that allow them to cope with environmental exigencies during all periods of the year (Desjardins and Lopez, 1981; Hoffmann, 1973; Lynch and Epstein, 1976). During the winter months animals engage numerous metabolic and thermoregulatory adjustments (Heldmaier et al., 1985; Heldmaier and Lynch, 1986). Accommodations take place, for example, in characteristics of the pelage, tissue components, and body size. Many species also exhibit adaptive hypothermia (either on a daily or longer basis). Metabolic adjustments of deer mice, including those mentioned above, are evoked by the same environmental cues that regulate reproductive function, and these adjustments also exhibit considerable variation in appearance. One question is addressed here: Do different reproductive phenotypes display unique sets of metabolic adjustments?

Two categories of seasonal metabolic adjustments are particularly distinct and have been investigated in parallel with the reproductive adjustments detailed above; these are, use of daily torpor and proliferation and cellular adaptation of brown adipose tissue (BAT). These adjustments produce opposite metabolic states. Daily torpor reduces energy required to maintain high body temperature and thereby serves to reduce energy expenditure (Heldmaier and Steinlechner, 1981; Vogt and Lynch, 1982). In a process termed nonshivering thermogenesis, brown adipose tissue consumes energy in a biochemically futile cycle that liberates energy from breakdown of substrate as heat instead of fixing this energy into molecules of ATP (see Wunder, this volume). Torpor was assessed by continuously monitoring body temperature during exposure to various environmental cues (Fig. 9.8). The capacity for endogenous heat production by BAT was measured by assessing biochemical changes in BAT tissue that are known to be related to nonshivering thermogenesis (Fig. 9.10).

Individual Variation in Metabolic and Reproductive Responses

The appearance of daily torpor and proliferation of BAT tissue and function vary considerably among deer mice exposed to winter condi-

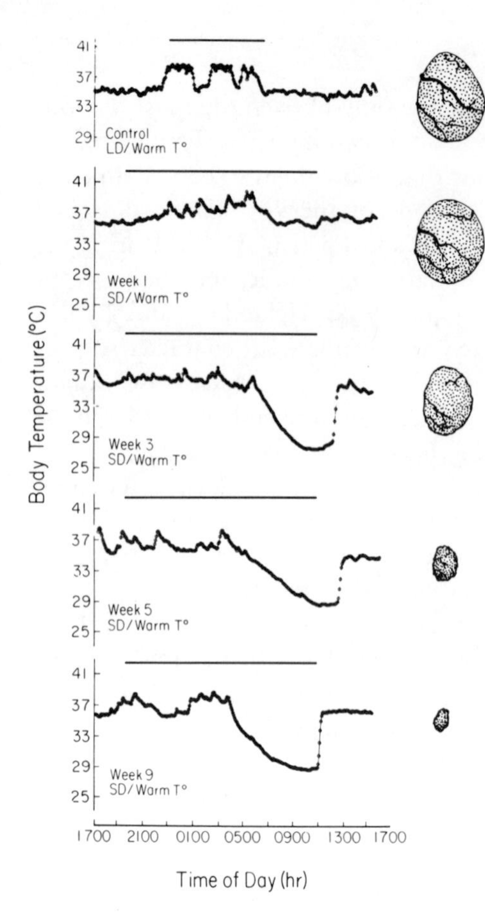

Fig. 9.8. Typical 24-h pattern of body temperature (Tb) changes in one representative adult male deer mouse exposed to a long day length (16L:8D) at 23°C. (top record) and to a subsequent 9-week period of short day length (8L:16D) at 23°C (bottom 4 records). Tb was measured using temperature-sensitive transmitters (Model X-M, Minimitter Inc., Sunriver, OR) implanted intraperitoneally (Blank and Desjardins, 1985). Tb was subsequently measured at 4-min intervals with a computer-assisted data-aquisition system. As illustrated by the gradual reduction in testis size shown in this figure, all males that displayed daily torpor eventually exhibited short day–induced testicular atrophy. Furthermore, when observed, torpor appeared in mice 1–14 days after initial exposure to short, warm days and on at least 90% of all days in mice that displayed this behavior.

tions in the lab (short photoperiod and cold ambient temperature). The type of response seen is contingent upon the reproductive response. Only individuals that exhibit gonadal regression and cessation of spermatogenesis develop daily torpor, and these same individuals develop the greatest capacity to produce heat via nonshivering thermogenesis.

Daily torpor is evoked only by short day length; cold ambient temperature elicits a lower minimum body temperature reached during a torpor bout but does not elicit daily torpor in the absence of simultaneous exposure to short day lengths. Figures 9.8 and 9.9 illustrate this point for two representative deer mice exposed for 9 weeks to short day length (8L:16D) and warm (23°C) ambient temperature (Fig. 9.8) or short day length and cold (2°C) ambient temperature (Fig. 9.9). Torpor appeared in male deer mice during the first 2 weeks, and as early as the first day, following exposure to either environmental treat-

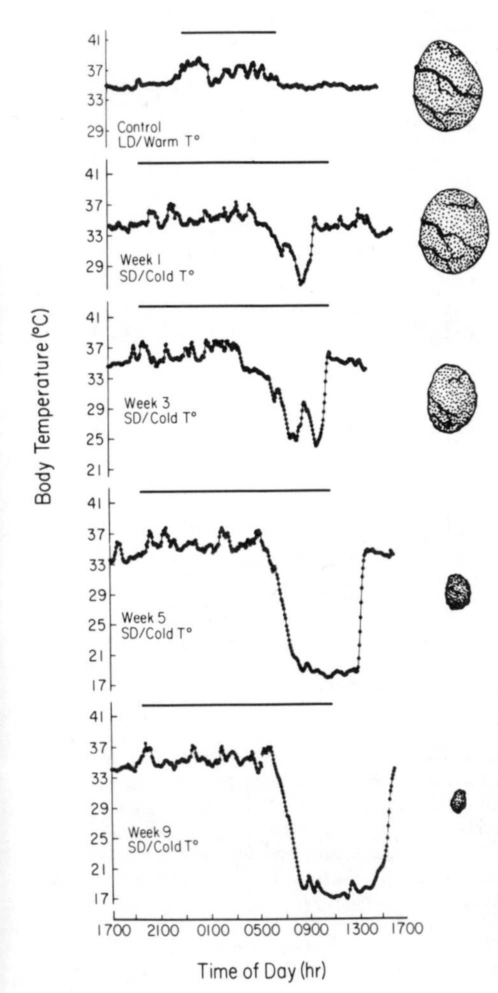

Fig. 9.9. Typical 24-h pattern of body temperature (Tb) changes in one representative adult male deer mouse exposed to a long day length at 23°C (top record) and during a 9-week exposure to a short day length (8L:16D) at 2°C (bottom 4 records). Tb was measured and plotted as described for Figure 9.8. As with short, warm day exposure, daily torpor appeared 1–14 days after short, cold day exposure, on at least 90% of all days once observed, and only in males that eventually exhibited testicular regression.

ment, and it increased in depth and duration until a steady pattern was reached by weeks 6–8. Daily torpor appeared in only about 12% of all individuals tested but in 50% of those categorized as azoospermic (epididymides devoid of sperm).

Proliferation and thermogenic properties of the interscapular BAT pad were assessed in the interscapular BAT pad of deer mice exposed to short (8L:16D), cold (2°C) days. Thermogenic activity was estimated by measuring the activity of cytochrome oxidase, an enzyme of the mitochondrial electron transport chain. An increase in activity of this enzyme indicates an increase in mitochondrial protein and is indicative of an increase in capacity for nonshivering thermogenesis (Rafael

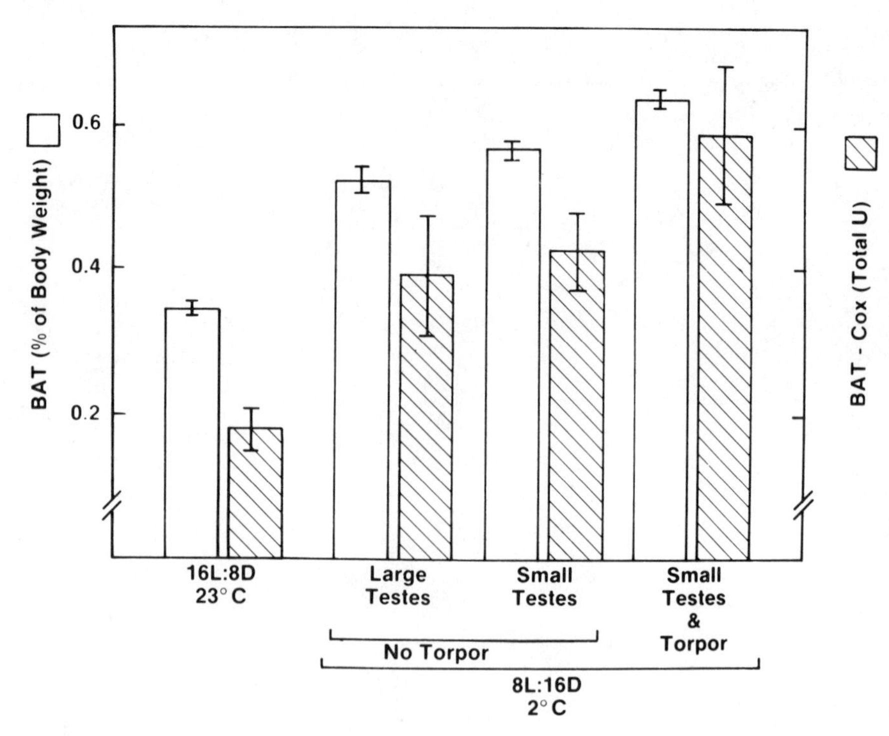

Fig. 9.10. Interscapular brown adipose tissue (BAT) pad weight as a percent of body weight (left y axis) and thermogenic activity of brown fat (BAT) obtained from the interscapular BAT pad (right y axis) for male deer mice exposed to long days (16L:8D) at 23°C or short days (8L:16D) at 2°C. Thermogenic activity was assessed by measuring the activity of cytochrome oxidase (Cox) in BAT mitochondria according to methods described by Rafael et al. (1985) and Blank et al. (1988). Bars, mean ±2 SE. Short, cold day–exposed mice were categorized according to gonad size (large, i.e., normal phenotype; small, i.e., azoospermic phenotype) and use of daily torpor. All short, cold groups showed a significant ($P < 0.05$) increase in total units of activity per BAT pad. Activity in nontorpid males with large and small gonads did not differ from each other ($P > 0.05$) and exhibited a 30 and 40% increase in activity, respectively. Atrophic males that also used torpor had significantly ($P < 0.05$) higher units of activity per pad than all other short, cold day mice. Total units equal the number of micromoles of O_2 consumed per minute per fat pad at 25°C. (Data from Blank et al., 1988.)

et al., 1985). Short, cold days caused increased proliferation of BAT and greater thermogenic activity in all deer mice; but, the magnitude of these changes varied with reproductive phenotype and use of daily torpor (Fig. 9.10). BAT weight in short, cold day–exposed males with normal-sized testes more than doubled, while total cytochrome oxidase activity increased by 30%, relative to long day breeding males. In contrast, short, cold day–exposed deer mice with atrophic testes that also used torpor exhibited a 64% increase in BAT weight and a 100% increase in total cytochrome oxidase activity, compared with long day males. Cytochrome oxidase activity in nontorpid deer mice with atrophic testes was intermediate to these two groups.

Taken together, these data emphasize that neuroendocrine adjustments in reproduction and metabolism are, like most traits, phenotypically, and probably genotypically variable within populations. Contemporary evolutionary theory would assume that the purpose of these adjustments is to increase propagation of an individual's genes. Thus, one can hypothesize that variation in composition of each suite of adjustments would represent different strategies toward increasing survivorship or reproductive success or both. With available evidence, making generalizations about the survival or reproductive value of these adjustments for free-living individuals and measuring their effect on fitness are difficult. The hypothesis that a common regulatory mechanism underlies a suite of physiological adjustments is attractive since it suggests that phenotypic-level differences between individuals result from a small reduced number of regulatory paths; the different suites of seasonal adjustments displayed by male deer mice may be controlled by a definable set of neural structures. Identification of these paths and their functions would greatly add to the ability to determine the source and lability of phenotypic variation among free-living animals.

Conclusions

Implicit in the question of how physiology constrains an individual's life history characteristics is the assumption that expression of life history traits reflects expression of underlying regulatory mechanisms. Further, the manner by which an individual adapts to its environment is limited by the type and extent of the regulatory mechanisms that can exist. Unfortunately, more is known about the types of life history traits that are expressed by individuals than of the regulatory mechanisms that underlie them. To fully appreciate how an individual's physiological makeup constrains its life history characteristics one must make connections between regulatory mechanisms and functional responses.

Life history adaptations by which deer mice cope with demands of their winter environment are numerous and have been described in both field and lab. Data presented in this chapter provide strong evidence that the seasonal reproductive and metabolic adjustments deer mice make to their winter environment occur in specific sets. These data do not address the issue of cause and effect between the constituent responses of these sets. Nevertheless, each set may provide differing abilities to cope with the cold temperatures and reduced food supplies of the winter environment. In this manner, each set limits or constrains the possible range of responses an individual can make to its environment, thereby affecting its fitness. In this context, two areas of investigation deserve additional attention.

First, the presence of phenotypic differences implies that individual deer mice living within a single population use several strategies to meet the demands of the environment. What these strategies are and how they influence individual fitness remain uncertain. A combination of the respective investigative strengths of life history studies and laboratory physiology could provide meaningful answers to these questions. Field research, for example, demonstrates deer mice breed during mild winters (Brown, 1966; Fairbairn, 1977; Sadleir, 1974) or when supplemental food is experimentally provided (Taitt, 1981). These data suggest that, with respect to reproduction, energy is limiting; but, some individuals are capable of exploiting windows of increased energy availability. Speculation that these deer mice are reproductively nonresponsive to environmental cues such as photoperiod or food restriction is tempting. Investigations that explore this possibility can exploit techniques of laboratory physiology that assess reproductive (Glass et al., 1988) and metabolic function (Blank et al., 1988) in an attempt to uncover advantages or disadvantages in retaining reproductive function during the winter months.

As a modest example of this approach, data collected on laboratory-reared deer mice provide preliminary evidence of a thermoregulatory advantage for reproductive quiescence. Figure 9.11 illustrates a small but consistent difference among reproductive phenotypes in the ability to withstand cold ambient temperatures. Males responding to short day lengths with testicular involution maintain body temperature at an ambient temperature 5°C colder than that at which reproductively active short day males can maintain body temperature. This small difference may, during some years, be of considerable importance. At intervals of roughly 8 years between 1910 and 1960, the minimum temperature reached during the coldest month of the year (January) fell below the maximum cold resistance of reproductively active short day males but above that of reproductively quiescent males (Fig. 9.12). Al-

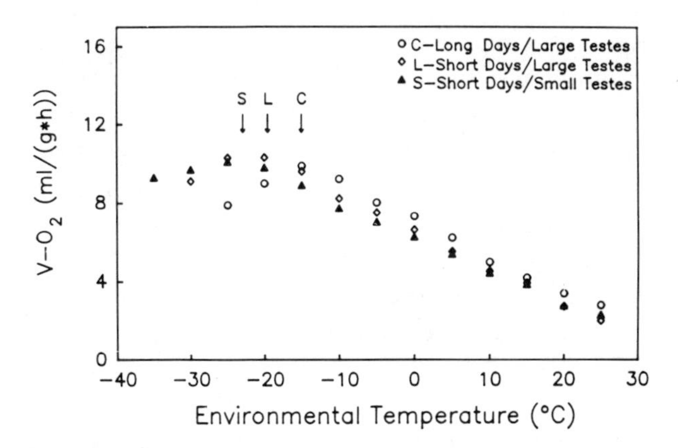

Fig. 9.11. Cold-tolerance test for males of three different deer mice phenotypes. Cold resistance is defined here as the ambient temperature (Ta) at which metabolic heat production can no longer be maintained. For the present study, this temperature was determined by placing individual males at progressively lower Ta, in steps of 10°C, beginning at thermoneutrality. At each temperature level, mice were maintained for approximately 1 h until a minimum oxygen consumption rate was obtained. Cold limit was − 22.8°C for short day–exposed deer mice (8L:16D for 8 wks, N = 4) with fully regressed (small) testes; −17.9°C for short day–exposed deer mice with nonregressed (large) testes (N = 5); and −13.5°C for long day–exposed controls (16L:8D, N = 8).

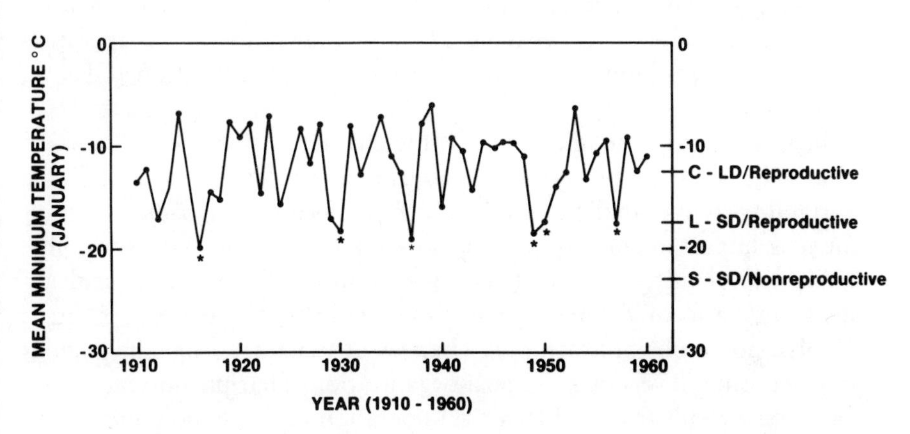

Fig. 9.12. Minimum ambient temperature (Ta; °C) measured on the coldest day of the coldest month (January) from 1910 to 1960 at Wind Cave National Park, S.D., in the vicinity of the collection site for deer mice used in this experiment. Data were provided by Richard Klukas, Wind Cave National Park, Hot Springs, S.D. Note that the cold limit (the lowest Ta at which mice can maintain heat production) for reproductive short day deer mice phenotypes is located above the lowest Ta in some years (asterisks). The cold limit for nonreproductive males is lower than the lowest Ta in any year.

though they cannot reproduce, reproductively quiescent males may have greater survivorship even during the most severe winters. Individuals of the reproductively active phenotype, therefore, may possess greater reproductive output during mild winters, but greater mortality during severe winters. In a simplistic case, one can envision maintenance of a balanced polymorphism between the two phenotypes, with their relative fitness varying from year to year depending on local climatic conditions. This type of scenario begs for experimental approaches that combine laboratory and field efforts.

A second investigatory area deserving attention is that concerned with the primary neural paths that regulate the suite of seasonal reproductive and metabolic responses. The observations that reproductive and metabolic adjustments occur in particular sets and that adjustments in the two systems occur in response to the identical environmental cues, such as photoperiod, provide strong evidence that a common neural regulatory system exists. As mentioned above, the hypothalamus serves as the site of this common regulatory path. Indeed, a few neural loci within this portion of the brain regulate reproductive function (Glass, 1988) as well as body temperature regulation (Benveniste, 1989; Wunnenberg et al., 1978) and BAT function (Imai-Matsumura and Nakayama, 1987). From a proximate point of view, seasonal reproductive and metabolic adjustments may occur together because the hypothalamic centers governing these adjustments involve common neural effector systems. Little is known about the operation of these loci in animals undergoing seasonal acclimitization (Glass et al., 1988).

In this regard, a major tenet of this chapter is that individual variability in seasonal physiological adjustments is driven by individual variability in hypothalamic function. While seemingly trivial, this argument is important in providing a proximate basis for explaining phenotypic-level differences in seasonal responses. The physiological adjustments affecting the change from summer to winter condition involve numerous and complex changes in function, from the cellular to organismal levels of biological organization. That phenotypic differences may result from a differential operation of a common and identifiable neural path is an exciting concept. The major challenge in investigating this concept is to make functional connections between neural effector mechanisms and whole-organism responses. Little is known of these control mechanisms, primarily because of the lack of methodology to provide a dynamic assessment of neural function. With the advent of techniques such as microdialysis sampling, the possibility exists to continuously assess and manipulate neurotransmitter activity within specific hypothalamic regions (Rea et al., 1989).

The key concept in this view of how physiology constrains life history is flexibility of an individual's response to its environment. As applied to this concept, life history theory can define what physiological adjustments should occur under specific environmental conditions on the basis of the effects of these adjustments on individual fitness. Laboratory physiology can identify what regulatory mechanisms cause these same adjustments. Together, these two approaches offer the opportunity for a comprehensive assessment of the mechanisms by which genomic and environmental effects are transduced into physiological adaptation.

ACKNOWLEDGMENTS

I thank Lester McClanahan and Richard Klukas, U.S. Department of the Interior, Wind Cave National Park, Hot Springs, S.D., for assistance in collecting parental stocks of deer mice. I also thank Don Carroll and Raylene Hensley for assistance in all aspects of this investigation. Many of the studies were conducted with the help and cooperation of Claude Desjardins, for which I am very grateful. I thank D. Glass, R. Dorman, and L. Orr for helpful comments on the manuscript.

This research was supported, in part, by National Institutes of Health and Human Development Grant HS-13470, National Research Service Award HD-06431, and by an Alexander von Humboldt-Stiftung Research Fellowship provided by the Federal Republic of Germany.

Literature Cited

Baker, J. R., and R. M. Ransom. 1932. Factors affecting the breeding of the field mouse (*Microtus agrestis*). Part 1. Light. Proc. R. Soc. Lond., 110:313–322.

Barnes, B. M., P. Light, and I. Zucker. 1987. Temperature dependence of in vitro androgen production in testes from hibernating ground squirrels, *Spermophilus lateralis*. Can. J. Zool., 65:3020–3023.

Benveniste, H. 1989. Brain microdialysis. J. Neurochem., 52:1667–1979.

Blank, J. L., and C. Desjardins. 1983. Supermatogenesis is modified by food intake in mice. Biol. Reprod., 30:410–415.

——. 1985. Differential effects of food restriction on pituitary-testicular function in mice. Am. J. Physiol., 248:R181–R189.

——. 1986. Photic cues induce multiple neuroendocrine adjustments in testicular function. Am. J. Physiol., 250:R199–R206.

Blank, J. L., R. J. Nelson, and A. Buchberger. 1988. Cytochrome oxidase activity in brown fat varies with reproductive response and use of torpor in deer mice. Physiol. & Behav., 43:301–306.

Brown, L. N. 1966. Reproduction of *Peromyscus maniculatus* in the Laramie Basin of Wyoming. Am. Midl. Nat., 76:183–189.

Clermont. Y. 1972. Kinetics of spermatogenesis in mammals: seminiferous epithelium cycle and spermatogonial renewal. Physiol. Rev., 52:198–236.

Dark, J., P. G. Johnston, M. Healy, and I. Zucker. 1983. Latitude of origin influences photoperiodic control of reproduction of deer mice (*Peromyscus maniculatus*). Biol. Reprod., 28:213–220.

Dark, J., and I. Zucker. 1983. Short photoperiods reduce winter energy requirements of the meadow vole, *Microtus pennsylvanicus*. Physiol. & Behav., 31:699–702.

Desjardins, C. 1981. Endocrine signaling and male reproduction. Biol. Reprod., 24:1–21.

Desjardins, C., F. H. Bronson, and J. L. Blank. 1986. Genetic selection for reproductive photoresponsiveness in deer mice. Nature, 322:172–173.

Desjardins, C., and M. J. Lopez. 1980. Sensory and nonsensory modulation of testicular function. Pp. 381–393 *in* Testicular Development, Structure and Function (A. Steinberger and E. Steinberger, eds.). Raven Press, New York. 523 pp.

——. 1983. Environmental cues evoke differential responses in pituitary-testicular function in deer mice. Endocrinology., 112:1398–1406.

Duncan, M. J., and B. D. Goldman. 1984a. Hormonal regulation of the annual pelage color cycle in the Djungarian hamster, *Phodopus sungorus*. I. Role of gonads and the pituitary. J. Exp. Zool., 230:89–96.

——. 1984b. Hormonal regulation of the annual pelage color cycle in the Djungarian hamster, *Phodopus sungorus*. II. Role of prolactin. J. Exp. Zool., 230:97–102.

Elliot, J. A., and B. D. Goldman. 1981. Seasonal reproduction, photoperiodism and biological clocks. Pp. 377–419 *in* Neuroendocrinology of Reproduction (N. A. Adler, ed.). Plenum Press, New York. 556 pp.

Ellis, G. B., and F. W. Turek. 1983. Testosterone and photoperiod interact to regulate locomotor activity in male hamsters. Horm. Behav., 17:66–75.

Fairbairn, D. J. 1977. Why breed early? A study of reproductive tactics in *Peromyscus*. Can. J. Zool., 55:862–871.

Feist, C. F., D. D. Feist, and G. R. Lynch. 1988. The effects of castration and testosterone on thermogenesis and pelage condition in white-footed mice (*Peromyscus leucopus*) at different photoperiods and temperatures. Physiol. Zool., 6:26–33.

Glass, J. D. 1988. Neuroendocrine regultion of seasonal reproduction by the pineal gland and melatonin. Pp. 219–259 *in* Pineal Research Reviews, vol. 6 (R. J. Reiter, ed.). Liss, New York. 311 pp.

Glass, J. D., S. Ferreira, and D. R. Deaver. 1988. Photoperiodic adjustments in hypothalamic amines, gonadotropin-releasing hormone, and beta-endorphin in the white-footed mouse. Endocrinology, 123:1119–1127.

Goldman, B. D., J. M. Darrow, M. J. Duncan, and L. Yogev. 1985. Photoperiod, reproductive hormones, and winter torpor in three hamster species. Pp. 341–350 *in* Living in the Cold: Physiological and Biochemical Adaptations (H. C. Heller, X. J. Musacchia, L. C. H. Wang, eds.). Elsevier, New York. 587 pp.

Hastings, M. H., A. C. Roberts, and J. Herbert. 1985. Neurotoxic lesions of the anterior hypothalamus disrupt the photoperiodic but not the circadian system in the Syrian hamster. Neuroendocrinology, 40:316–34.

Heath, H. W., and G. R. Lynch. 1983. Intraspecific differences in use of photoperiod and temperature as environmental cues in white-footed mice, *Peromyscus leucopus*. Physiol. Zool., 56:506–512.

Heldmaier, G., H. Bockler, A. Buchberger, G. R. Lynch, W. Puchalski, S. Steinlechner, and H. Wiesinger. 1985. Seasonal acclimation and thermogenesis. Pp. 490–501 *in* Circulation, Respiration, and Metabolism (R. Giles, ed.). Springer-Verlag, Berlin. 672 pp.

Heldmaier, G., and G. R. Lynch. 1986. Pineal involvement in thermoregulation and acclimatization. Pineal Res. Rev., 4:97–139.

Heldmaier, G., and S. Steinlechner. 1981. Seasonal pattern and energetics of short daily torpor in the Djungarian hamster, *Phodopus sungorus*. Oecologia, 48:265–270.

Hoffman, R. A., and R. J. Reiter. 1965. Pineal gland: influence of gonads on male hamsters. Science, 142:1609–1611.

Hoffmann, K. 1973. The influence of photoperiod and melatonin on testis size, body weight, and pelage colour in the Djungarian hamster (*Phodopus sungorus*). J. Comp. Physiol., 85:267–282.

Imar-Matsumura and T. Nakayama. 1987. The central mechanism of brown adipose tissue thermogenesis induced by preoptic cooling. Can. J. Physiol. Pharmacol., 65: 1299–1303.

Kato, J. 1980. Steroid hormone receptors during brain development. Pp. 389–413 *in* Development of Responsiveness to Steroid Hormones (A. M. Kaye and M. Kaye, eds.). Pergamon, Elmford, New York. 589 pp.

Kenagy, G. J. 1985. Strategies and mechanisms for timing of reproduction and hibernation in ground squirrels. Pp. 383–392 *in* Living in the Cold: Physiological and Ecological Adaptations (H. C. Heller, X. J. Musacchia, L. C. H. Wang, eds.). Elsevier, New York. 587 pp.

Lopez, M. J. 1981. Reproductive and temporal adaptations to seasonal change in deermouse, *Peromyscus maniculatus*. Ph.D. dissertation, University of Texas, Austin.

Krebs, C. J., and J. H. Meyers. 1974. Population cycles in small mammals. Adv. Ecol. Res., 8:268–389.

Lynch, G. R., and A. L. Epstein. 1976. Melatonin induced changes in gonads, pelage, and thermogenic characters in the white-footed mouse, *Peromyscus leucopus*. J. Comp. Physiol., 125:156–163.

Lynch, G. R., H. W. Heath, and C. M. Johnston. 1981. Effect of geographic origin on the photoperiodic control of reproduction in the white-footed mouse, *Peromyscus leucopus*. Biol. Reprod., 25:475–480.

Lynch, G. R., C. B. Lynch, and R. M. Kliman. 1989. Genetic analyses of photoresponsiveness in the Djungarian hamster, *Phodopus sungorus*. J. Comp. Physiol., 164:475–481.

McEwen, B. S. 1980. Gonadal steroids: humoral modulators of nerve-cell function. Mol. Cell. Endocrinol., 18:151–164.

McNeilly, A. S. 1987. Prolactin and the control of gonadotrophin secretion. J. Endocrinol., 115:1–5.

Millar, J. S. 1984. Reproduction and survival of *Peromyscus* in seasonal environments. Pp. 253–266 *in* Winter Ecology of Small Mammals (J. F. Merritt, ed.). Carnegie Museum of Natural History Publication No. 10. Pittsburgh, Penn. 352 pp.

——. 1985. Life history characteristics of *Peromyscus maniculatus nebrascensis*. Can. J. Zool., 63:1280–1284.

Nelson, R. J. 1983. Photoperiodic regulation of reproductive development in male prairie voles: influence of laboratory breeding. Biol. Reprod., 33:418–422.

Nelson, R. J., and C. Desjardins. 1987. Water availability affects reproduction in deer mice. Biol. Reprod., 37:257–260.

Nelson, R. J., and I. Zucker. 1981. Photoperiodic control of reproduction in olfactory bulbectomized rats. Neuroendocrinology, 32:266–271.

Rafael, J., P. Vsiansky, and G. Heldmaier. 1985. Seasonal adaptation of brown adipose tissue in the Djungarian hamster. J. Comp. Physiol., B155:521–528.

Rea, M. A., J. L. Blank, S. Ferreira, D. M. Terrian, and J. D. Glass. 1989. In vivo microdialysis of amino acids in the suprachiasmatic nuclei. *In* Proceedings of the 19th Annual Meeting of the Society for Neuroscience, Phoenix, Arizona. Neurosci. Abstr. 293.10. Society for Neuroscience, Washington, D.C.

Reiter, R. J. 1973. Pineal control of a seasonal reproductive rhythm in male golden hamsters exposed to natural daylight and temperature. Endocrinology, 92:423–430.

——. 1980. The pineal and its hormones in the control of reproduction in mammals. Endocrinol. Rev., 1:109–131.

Reiter, R. J., J. C. Hoffmann, and P. H. Rubin. 1968. Pineal gland: influence on gonads of male rats treated with androgen three days after birth. Science, 160:420–421.

Sadleir, R. M. F. S. 1974. The ecology of the deer mouse *Peromyscus maniculatus* in a coastal coniferous forest. II. Reproduction. Can. J. Zool., 52:119–131.

Steinlechner, S., G. Heldmaier, and H. Becker. 1983. The seasonal cycle of body weight in the Djungarian hamster: photoperiodic control and the influence of starvation and melatonin. Oecologia, 60:401–405.

Taitt, M. J. 1981. The effect of extra food on small rodent populations: I. Deermice. (*Peromyscus maniculatus*). J. Anim. Ecol., 50:111–124.

Taylor, C. R., and E. R. Weibel. 1981. Design of the mammalian respiratory system. I Problem and strategy. Resp. Physiol., 44:1–10.

Turek, F. W., and C. S. Campbell. 1979. Photoperiodic regulation of neuroendocrine-gonadal activity. Biol. Reprod., 20:32–49.

Vitale, P. M., J. M. Darrow, M. J. Duncan, C. A. Shustak, and B. D. Goldman. 1985. Effects of photoperiod, pinealectomy and castration on body weight and daily torpor in Djungarian hamsters (*Phodopus sungorus*). J. Endocrinol., 106:367–375.

Vogt, F. D., and G. R. Lynch, 1982. Influence of ambient temperature, nest availability, huddling, and daily torpor on energy expenditure in the white-footed mouse, *Peromyscus leucopus*. Physiol. Zool., 55:56–63.

Weibel, E. R., C. R. Taylor, P. Gehr, H. Hoppeler, O. Mathieu, and G. M. O. Maloiy. 1981. Design of the mammalian respiratory system. IX. Functional and structural limits for oxygen flow. Resp. Physiol., 44:151–164.

Whitaker, W. L. 1940. Some effects of artificial illumination on reproduction in the white-footed mouse, *Peromyscus leucopus noveboracensis*. J. Exp. Zool., 83:33–60.

Wunnenberg, W., G. Kunen, and R. Laschefski-Sievers. 1986. CNS regulation of body temperature in hibernators and non-hibernators. Pp. 185–192 *in* Living in the Cold: Physiological and Biochemical Adaptations (H. S. Heller, X. J. Musacchia, L. C. H. Wang, eds.). Elsevier, New York. 587 pp.

Yellon, S. M., and B. D. Goldman. 1984. Photoperiod control of reproductive development in the male Djungarian hamster (*Phodopus sungorus*). Endocrinology, 114: 664–670.

Steven D. Thompson

10. Gestation and Lactation in Small Mammals: Basal Metabolic Rate and the Limits of Energy Use

Mammalian reproduction is energetically expensive (Gittleman and Thompson, 1988; Loudon and Racey, 1987), and limits on a female's abilities to allocate energy to reproduction are thought to have profound effects on the variation of mammalian behavior and reproductive patterns. Aspects of social structure and group size (Coelho, 1986; Contreras, 1984; Fa and Southwick, 1987; Gittleman and Harvey, 1982; Harvey and Clutton-Brock, 1981; Smuts et al., 1987), mating and breeding strategies (Altmann, 1983; Austad and Sunquist, 1986; Berman, 1988; Caley et al., 1988; Clutton-Brock et al., 1982; Goldizen, 1987a, 1987b; Lee and Moss, 1986; Oliveras and Novak, 1986; Ortiz et al., 1984; Trillmich, 1986), and life history strategies or tactics (Harvey, 1986; Harvey and Clutton-Brock, 1985; Hennemann, 1983, 1984; Lee, 1987, 1988; Martin, 1984a, 1984b; McNab, 1980; Sacher and Staffeldt, 1974) are frequently attributed to compromises necessitated by the high energetic demands of reproduction (Brody, 1964; Hanwell and Peaker, 1977; Loudon and Racey, 1987; Peaker et al., 1984). Moreover, energetics have played a key role in the development of evolutionary scenarios depicting both the origin of eutherian mammals and the subsequent evolutionary pattern of mammalian reproduction (Hayssen et al., 1985; Lillegraven et al., 1987; Low, 1978; McNab, 1980, 1986a; Morton et al., 1982; Parker, 1977). Of particular importance are arguments that the seemingly graded series of energetic characteristics (e.g., thermoregulation, basal metabolic rate) among monotremes, marsupials, and eutherian mammals may reflect differences in reproduction, rather than a phylogenetic progression of (or phylogenetic constraints on) thermoregulatory abilities per se (Lillegraven et al., 1987; McNab, 1978, 1986a).

Lincoln Park Zoological Gardens, Chicago, Illinois 60614.

Although the literature on energy requirements for reproduction in domesticated mammals is voluminous (Brockway et al., 1963; Brody, 1964; Bruce and Clark, 1979; Clark, 1981; Graham, 1964; Hadjipieris and Holmes, 1966; Kleiber, 1975; Mount, 1979; Myrcha et al., 1969; Oftedal, 1985), energetic strategies of wild mammals may differ greatly from those of livestock and laboratory mammals (Glazier, 1985a; Nicoll and Thompson, 1987; Oftedal et al., 1987; Oftedal and Gittleman, 1988; Randolph et al., 1977; Speakman and Racey, 1987). Moreover, while among wild mammals there is substantial intraspecific and interspecific variability in gross patterns of energy use during reproduction, by comparison the reproductive energetics of domestic mammals seem relatively invariate. Much of the variation in wild mammals seems to be the result of remarkable maternal abilities to compensate for limitations on energy use or availability during reproduction (Carl and Robbins, 1988; Gittleman and Thompson, 1988; McClure, 1987; Mulder, 1987). However, neither the extent of nor the bases for functional limitations on energy use during reproduction are well documented among wild mammals (Gittleman and Thompson, 1988; Glazier, 1985a; McClure, 1987; Oftedal and Gittleman, 1988; Randolph et al., 1977; Thompson and Nicoll, 1986).

This chapter explores the patterns of, and possible limits to, energy intake and expenditure during gestation and lactation as they relate to (1) interspecific variation of energy requirements for reproduction, (2) allocation of energy within reproduction, and (3) the relationships between nonreproductive measures of energy use, such as the basal metabolic rate, and reproductive and population parameters. This chapter also addresses contradictions between correlative analyses of reproductive and energetic parameters (Glazier, 1985a, 1985b; McNab, 1980, 1986a, 1987) and the mechanism of interaction between nonreproductive energetic strategies and energy expenditure during reproduction (Hayssen, 1984; McNab, 1986b, 1987; Nicoll and Thompson, 1987).

Definitions and Data

Definitions

Nonreproductive energy use is herein referred to as either maintenance—which denotes the sum of the basal metabolic rate, digestive costs (e.g., SDA, specific dynamic action), and all behavior and thermoregulatory expenditures over a 24-h day as determined by food consumption of captive animals—or basal costs, which represent the calculated energy use for 24-h at the basal metabolic rate (BMR). Basal metabolic rates include only those measurements that conform to the standard criteria proposed by Kleiber (1975) and reinforced by Lavigne et al. (1986): measurements of oxygen consumption for nonreproductive adults, at rest in the zone of thermoneutrality, postabsorptive, and measured during the resting phase of the day. Resting metabolic rate (RMR) refers to any measurement of oxygen consumption made under conditions that violate one or more of the established criteria for BMR (except "at rest"). Measurements of pregnant or lactating females under otherwise standard conditions, for example, are referred to collectively as maternal resting metabolic rates (RMR_m; more specifically, RMR_g during gestation and RMR_l during lactation). Other deviations from BMR criteria are denoted through similar qualifiers, for example, age, reproductive state, or thermal stress; comparisons of RMR are made only when measurement conditions are similar for all criteria. The energy requirement (ER) for reproduction is the total amount of energy necessary for reproduction in addition to energy that would be used by a nonreproductive adult female kept at a constant body mass, and housed under the same environmental conditions (i.e., air temperature and relative humidity), for the same period of time (i.e., conception to weaning). The mean daily energy requirement (MER) is ER divided by the duration of reproduction or of a reproductive phase (gestation or lactation). ER and MER may be expressed either as (1) multiples of maintenance or basal energy requirements or (2) absolute energy values (in kilojoules). ER may be expressed with respect to either daily or total (gestation and lactation combined) values. Reproductive effort (RE) is herein defined as the ratio of ER or MER to total energy use or mean total daily energy use during reproduction (i.e., physiological RE: Randolph et al., 1977). Because there are few data for gestation, and because lactation is by far the most expensive aspect of reproduction (Loudon and Racey, 1987; Oftedal, 1985; Peaker et al., 1984), partial REs are calculated for lactation (RE_{lact}) as (MER × days of lactation)/(conception to weaning time). Efficiency of net production refers to the ratio of energy deposited as offspring tissues to the total ER.

Data

Because energy use changes as reproduction proceeds, (Kurta et al., 1987; Oftedal, 1984; Ortiz et al., 1984; Randolph et al., 1977; Thompson and Nicoll, 1986; Weiner, 1987), the primary criteria for inclusion in quantitative analyses were whether energy use during gestation or lactation or both was (1) monitored longitudinally or (2) determined either by food consumption or respirometric measurements on captive individuals. Data that were clearly collected as a cross section of a population (Bowen et al., 1985; Robbins et al., 1981; Weiner, 1977) or data that were not verifiably longitudinal (Portman, 1970) were not included. This criterion eliminated virtually all milk-output studies, the cross-sectional design of which makes them difficult to compare among studies. Milk-output or suckling-frequency studies may also seriously underestimate maternal expenditures for maintenance and production of milk (Lucas et al., 1987; Oftedal and Gittleman, 1988). Data from domestic mammals, including laboratory strains of *Mus musculus* (Myrcha et al., 1969; Studier, 1979), *Rattus norvegicus* (Morrison, 1955, 1956), and *Mesocricetus auratus* (Fleming, 1978), were omitted from quantative comparisons and statistical analyses because (1) livestock and laboratory mammals have been subjected to considerable artificial selection, either intentionally or incidental to general husbandry practices; (2) experimental designs and methods of data collection and presentation often prohibit direct comparisons of data for such mammals with that for wild species without an excessive number of assumptions and recalculations; and (3) heat production from fermentation and increased ingestion complicates comparisons of data for large herbivores with data for many small mammals (Oddy et al., 1984; Verhagen et al., 1986). However, since the majority of experimental work has been done on domesticated mammals, reference to these species is essential in considerations of reproductive flexibility and the causal mechanisms behind variation in energy use during mammalian reproduction. Several studies of wild mammals (Havera, 1979; Krol, 1985; Mover et al., 1988; Perrin and Clarke, 1987) were also omitted from quantitative analyses (Table 10.1) because data did not permit calculation of relative levels of energy use during reproduction.

Data extracted from the literature were converted to kilojoules per day assuming 4.1868 kJ/Kcal and 20.097 J/ml O_2 ($RQ \approx 0.79$: Bartholomew, 1977). All values were adjusted for urinary loss of digested

TABLE 10.1
Species used in energetic analyses

Species[a]	Mass (g)	BMR		MER[c]		Reference
		kJ/d	f_m[b]	Gest.	Lact.	
1. *Clethrionomys gapperi*	23.4	22.4	128		557	Innes & Millar, 1981; McNab, 1988
2. *C. glareolus*	25.0	28.0	152	327	502	Kaczmarski, 1966; McNab, 1988
3. *Echinops telfairi*	140.6	18.8	28	366	1078	Thompson & Nicoll, 1986
4. *Elephantulus rufescens*	84.1	37.5	82	317	478	Thompson & Nicoll, 1986
5. *Microtus arvalis*	23.2	26.1	150	207	470	Trojan & Wojciechowska, 1967; Migula, 1969
6. *M. pennsylvanicus*	23.4	24.5	140	397	595	Innes & Millar, 1981; McNab, 1988
7. *M. pinetorum*	25.4	28.5	153		496	Lochmiller et al., 1982, 1983; B. K. McNab, pers. comm., 1988
8. *Monodelphis domestica*	82.0	28.7	64	239	600	Thompson & Nicoll, 1986
9. *Neotoma floridanus*	184.0	67.5	82	299	404	McClure, 1987
10. *N. floridanus* (f)	184.0	67.5	82	253	473	McClure, 1987
11. *Peromyscus eremicus*	21.5	16.6	101		490	Glazier, 1985a
12. *P. floridana*	42.0	21.5	79		523	Glazier, 1985a
13. *P. leucopus*	21.6	20.0	121	283	563	Glazier, 1985a; Millar, 1975, 1978
14. *P. leucopus*	21.0	19.6	121		577	Glazier, 1985a
15. *P. maniculatus*	29.0	33.2	161		428	Glazier, 1985a; Stebbins, 1977
16. *P. maniculatus*	14.5	19.7	161		559	Glazier, 1985a
17. *P. maniculatus*	17.0	22.2	161		519	Glazier, 1985a; Millar, 1979
18. *P. polionotus*	13.4	12.4	108		569	Glazier, 1985a
19. *Phodopus sungorus*	30.2	22.1	104	237	426	Weiner, 1987; Weiner and Heldmaier, 1987
20. *Sigmodon hispidus* (KA)	175.0	67.4	85	280	496	Mattingly & McClure, 1982; Randolph et al., 1977
21. *S. hispidus* (TX)	135.0	67.2	103	271	378	Mattingly & McClure, 1982; Randolph et al., 1977
22. *S. hispidus* (TXf)	135.0	62.0	95	279	438	Mattingly & McClure, 1982; Randolph et al., 1977
23. *Suncus murinus*	30.2	29.5	139		496	Dryden et al., 1974

[a]f = fat stored during gestation added to lactation costs; KA = Kansas population; TX = Texas population.
[b]f_m = percent of Kleiber's (1975) predicted BMR.
[c]Mean daily energy requirement as percent of observed BMR.

energy (Glazier, 1985a; Randolph et al., 1977). Respirometric values of MER were not adjusted for digestive or assimilative efficiencies, behavior, or thermoregulatory costs, and thus reflect only the minimal cost of reproduction (Nicoll and Thompson, 1987). Kleiber's (1975) predicted BMR was converted to kilojoules per day as 293 $M^{0.75}$, where M = mass in kilograms. Although statistical analyses of ratios have recently received justifiable criticisms (Packard and Boardman, 1988), comparison to an intuitively satisfying allometric standard (i.e., a ratio) remains one of the more lucid means for conveying information about relative values of resting and basal metabolic rate (Harvey and Clutton-Brock, 1985; McNab, 1988). For ease of discussion and presentation of results (McNab, 1978), measurements were often converted to multiples of the BMR predicted from Kleiber's (1975) allometric equation of BMR on body mass. All statistical analyses were performed with the SYSTAT statistical package (Wilkinson, 1986) on energy use as kilojoules per animal per day and with body mass as a covariate.

BMR and Mammalian Reproduction

Field and laboratory studies have focused on two sets of factors that may interact to determine the limits on energy use during reproduction. On the one hand, correlations between ecological factors, such as indices of resource availability and reproductive activity, suggest that the ability to locate and acquire energy limits allocation of energy to reproduction in free-living populations (Austad and Sunquist, 1986; Bronson, 1979; Clutton-Brock et al., 1982; Dunbar, 1985; Jolly, 1984; Kenagy and Bartholomew, 1985; Lee and Moss, 1986; Lochmiller et al., 1987; Pianka, 1976; Richard and Nicoll, 1987; Silk, 1986; Smuts et al., 1987; Stenseth et al., 1980; Stewart, 1986; Thompson and Gittleman, 1988; Townsend and Calow, 1981). On the other hand, considerations of form and function have suggested that morphophysiological limits on food intake, rate of energy extraction from food, or cellular capacities for energy processing are equally important constraints on the rate at which energy may be used during reproduction (Demment, 1983; Demment and Van Soest, 1985; Gross et al., 1985; Heasley, 1983; Kirkwood, 1983; Kleiber, 1975; McNab, 1980; Sibley, 1981; Wunder, this volume). The extent of food restriction in free-living populations is difficult to quantify (King and Murphy,

1985; Watson, 1970). For captive mammals the maximum sustained rate of energy intake during lactation is four to five times the basal metabolic rate (Brody, 1964; Kirkwood, 1983; Kleiber, 1975); this observation supports the hypothesis that a functional upper limit on energy use exists and can be reached under conditions of ad libitum food (Kenagy, 1987). However, because many field studies suggest that free-living populations seldom have unlimited access to food (Fa and Southwick, 1988; Watson, 1970), ecological limits to energy use during reproduction might often be expected to (1) prevent free-living individuals from approaching their functional limits on energy use and (2) vary greatly in time and space (Kenagy and Bartholomew, 1985; Lochmiller et al., 1987; Loy, 1988).

Although many invertebrates, fishes, amphibians, reptiles, and some birds make extensive use of maternal tissues to fuel the energy demands of reproduction (Calow, 1979, 1981; Congdon et al., 1982; Millar, 1975; Pianka, 1976; Tuomi et al., 1983; Tytler and Calow, 1985), most female mammals meet the energy demands of reproduction through an increase in food intake and not by a sacrifice of their own structural tissues or corporal reserves (Glazier, 1985a; Heasley, 1983; Innes and Millar, 1981; Kaczmarski, 1966; Kenagy, 1987; Loveridge, 1986; Mattingly and McClure, 1982, 1985; McClure, 1987; Migula, 1969; Millar, 1978; Randolph et al., 1977; Smith and McManus, 1975). Reliance on increased food consumption has focused attention on gut size as one factor that may limit the rate at which mammals can utilize energy during reproduction. Guts of many small herbivores become enlarged during reproduction. This increased gut size has been interpreted as a result of selection to increase the rate of energy absorption in response to the high energy demands of reproduction (Demment, 1983; Demment and Van Soest, 1985; Gebczynska and Gebczynski, 1971; Gross et al., 1985; Myrcha, 1962, 1965; Sibley, 1981). Whether temporary enlargement of the gastrointestinal tract during reproduction (1) actually facilitates increased food intake or is merely a consequence of that intake (i.e., work hypertrophy), (2) increases the rate or efficiency of nutrient absorption, or (3) is important among either larger herbivores or mammals such as frugivores, omnivores, and carnivores, which have relatively high-energy diets, is unclear (Campbell and Fell, 1964; Cripps and Williams, 1975; Fell et al., 1963; Gross et al., 1985; Kenagy, 1987; Smith and Baldwin, 1974). Although a logical and attractive hypothesis (Demment, 1983), few data support the notion that gut size per se limits the ability of most mammals to consume or process energy during reproduction (Kenagy, 1987; Leon et al., 1983; Peterson and Baumgardt, 1971; Wunder, this volume). Moreover, even if gut size is potentially limiting, temporary shifts to higher-quality diets or extension of daily feeding time during reproduction could increase energy intake without the necessity of an increase in gut size (Kenagy, 1987; Lee, 1987; Leon and Woodside, 1983; Sibley, 1981; Silk, 1986).

In comparison to studies of gut size, much more attention has been given to the relationships between ecological constraints on nonreproductive components of a mammal's energetic strategy (sensu McNab, 1978, 1979) and the effects (constraints) of those aspects on capabilities for allocating energy to reproduction (Hennemann, 1983; Lillegraven et al., 1987; McNab, 1980, 1986a; Schmitz and Lavigne, 1984). Physiological components of the energetic strategy such as BMR, RMR, body temperature (T_b), precision of thermoregulation, thermal conductance (C = heat transfer coefficient, now more commonly referred to as h), evaporative water loss (EWL), and use of torpor or hibernation, as well as the degree of seasonal shifts (acclimation) in these parameters, vary both interspecifically and intraspecifically with body size and climate (Hinds and MacMillen, 1985; Hulbert and Dawson, 1974; McNab, 1979, 1983, 1988; McNab and Morrison, 1963; Scheck, 1982; Thompson, 1985). These parameters, along with behavior and the thermal environment, proscribe the minimum and reflect average daily energy use by a nonreproductive individual (Yousef, 1985a, 1985b). Because most mammalian females appear to satisfy their nonreproductive energy demands (BMR, thermoregulatory costs, and supportive behaviors) before they allocate any energy to reproduction (Heasley, 1983; Leon and Woodside, 1983; Mattingly and McClure, 1985; McClure, 1981; Millar, 1975; Mount, 1979; Wunder, 1978), the combination of behavioral and physiological tactics by which energy balance is maintained outside the breeding season could have direct effects on the amount of energy that a female can possibly allocate to reproduction (Clark, 1981; Leon and Woodside, 1983; Leon et al., 1983; Mount, 1979). In this regard, one of the most striking interspecific patterns in mammalian energetics is the correlation between food habits and BMR.

Folivores, large insectivores, and frugivores have relatively low BMRs ($\leq$85% of that predicted by Kleiber's 1975 allometric equation), while meat eaters and grazers of all sizes have relatively high BMRs ($\geq$95% of that predicted by Kleiber, 1975; Hennemann et al.,

1983; Kurland and Pearson, 1986; McNab, 1980, 1983, 1986a, 1988). These correlations have been attributed to size-related and other intrinsic limitations on abilities to acquire or process a particular type of food or combination of foods (Gaulin, 1979; McNab, 1980). According to this "food habits hypothesis," interspecific differences in BMR reflect evolutionary adjustments to balance daily energy requirements for a given body mass against net gain in energy available from a given diet (Demment, 1983; Gaulin, 1979; Kurland and Pearson, 1986; McNab, 1978, 1980, 1983, 1986b; Sailer et al., 1985). Despite the large number of studies in support of the food habits hypothesis, it still entails considerable controversy (Derrickson, 1989; Elgar and Harvey, 1987; Hinds and MacMillen, 1984; McNab, 1987; Pagel and Harvey, 1988a).

Because BMR is responsible for at least 30% of the daily energy budget of most wild endotherms (Goldstein, 1988; Goldstein and Nagy, 1985; Nagy, 1987; Williams and Nagy, 1984) it has appreciable ecological significance. Low BMR could, for example, permit use of food types, or combinations of food types, from which the net rate of energy gain is relatively low (McNab, 1980, 1983, 1986b, 1988). The latter might be accomplished either by a net reduction in daily energy requirements (Gaulin, 1979; Goldstein and Nagy, 1985; Kurland and Pearson, 1986; McNab, 1974, 1983, 1986b; Williams and Nagy, 1984) or through compensation for allocation of large amounts of energy to a particular (specialized) activity, such as foraging. Moreover, although the underlying basis of the BMR is still elusive (Heusner, 1987; McNab, 1983, 1988, this volume; Racey and Speakman, 1987), it has great value as a standardized measure of energy use that can be used in direct comparisons both within and among species (McNab, 1988).

Because BMR should reflect the rate of biosynthesis (Hammel, 1976; Stevens, 1973; Taigen, 1983; Whittam, 1964), constraints imposed by food habits could indirectly limit the deposition of new tissue during gestation and the rate or quality of milk production during lactation (Glazier, 1985a, 1985b; Hennemann, 1983, 1984; Hofman, 1983; Martin, 1984a, 1984b; McNab, 1980, 1986b, 1987). This "BMR-reproduction hypothesis" assumes that interspecific differences in BMR are maintained during reproduction (Fig. 10.1a) as maternal resting metabolic rate (RMR_m, a generic term for metabolic rate during either lactation or gestation); thus, variation in BMR is considered to be directly responsible for variation in mammalian reproductive parameters.

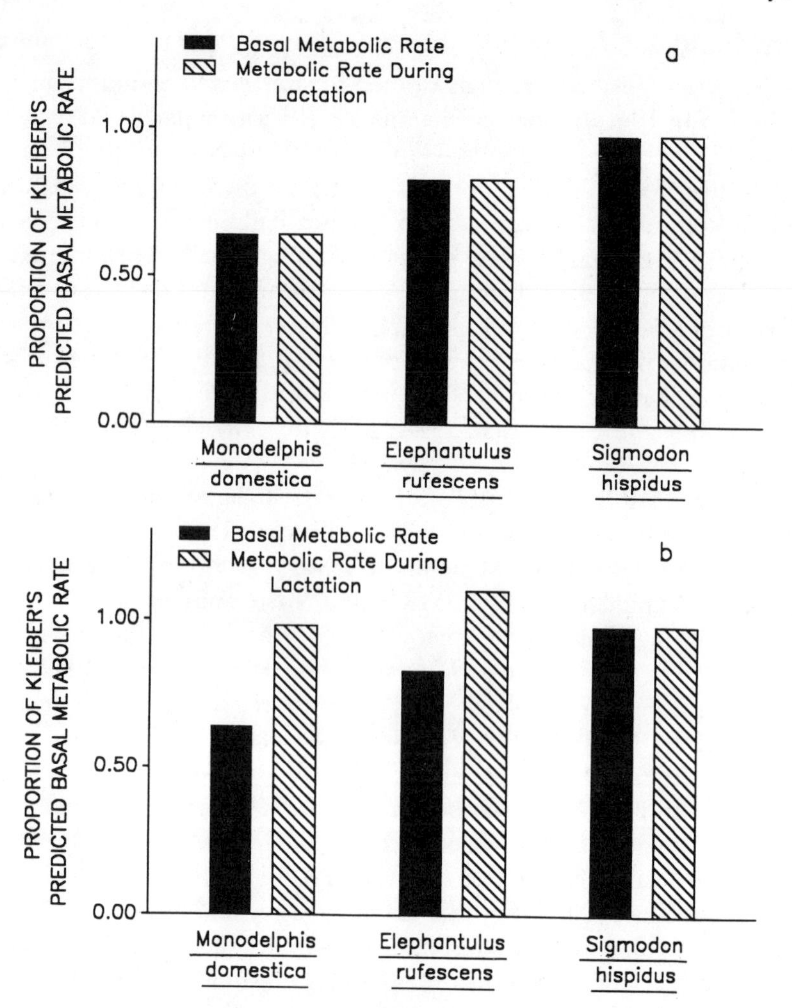

Fig. 10.1. (a) Hypothesized relationships between BMR and resting metabolic rate during lactation (RMR_l). (b) Actual relationships determined using respirometry. (Data for *Monodelphis domestica* [Marsupialia: Didelphidae] and *Elephantulus rufescens* [Macroscelidia: Macroscelididae] from Thompson and Nicoll, 1986. Data for *Sigmodon hispidus* [Rodentia: Cricetidae] from Randolph et al., 1977.)

Species with high BMRs should have high RMR_ms and higher rates of biosynthesis during reproduction, and thus should be capable of allocating energy to reproduction faster than species with relatively low BMRs (Glazier, 1985b; Hofman, 1983; Martin, 1984a, 1984b; McNab 1980, 1986a; Pagel and Harvey, 1988a; Rasmussen and Izard, 1988; Sacher and Staffeldt, 1974). Because a greater speed of reproduction has obvious advantages for interspecific competition (Low, 1978;

McNab, 1980; Stearns, 1976; Thompson, 1987), research has focused on the prediction that as relative BMR increases through evolutionary time, species will hold the total energy allocated to reproduction constant and reduce the duration of reproduction (I call this the BMR-speed prediction). An equally likely alternative (Glazier, 1985b) is that high-BMR species will hold the duration of reproduction more or less constant and "opt" for an evolutionary increase in the total energy expended for reproduction (I call this the BMR-reproductive effort prediction).

Existing support for hypotheses linking BMR and reproduction is almost entirely correlative. The BMR-speed prediction is supported by (1) inverse correlations of BMR with length of gestation (Hayssen et al., 1985; Martin, 1984a, 1984b; McNab, 1980, 1986b, 1987; Pagel and Harvey, 1988; Sacher and Staffeldt, 1974) and duration of parental care (Hennemann, 1984; McNab, 1986a) and (2) positive correlations of BMR with growth rates (McNab, 1980) and reproductive potential (Hennemann, 1983; Lillegraven et al., 1987; McNab, 1980; Rasmussen and Izard, 1988; Schmitz and Lavigne, 1984). Interpretation of these correlations, many of which are based on small sample sizes and involve data not adjusted for differences in adult body mass, is extremely difficult (Pagel and Harvey, 1988a).

The BMR-reproductive effort prediction is supported by positive correlations of BMR with total energy allocated during lactation (Glazier, 1985b) and annual fecundity (McNab 1980, 1986a, 1987). As yet, only a few studies have considered the speed and effort predictions as alternative strategies, and no study has assessed the possibility that within any diverse sample of mammalian species, some high-BMR species will have opted for greater speed, others will have opted for greater effort, and some may have opted for moderate increases in both speed and effort (Hofman, 1983; Pagel and Harvey, 1988a; Rasmussen and Izard, 1988).

One major shortcoming of the BMR-reproduction hypothesis is the presumption that (RMR_m), during either gestation (RMR_g) or lactation (RMR_l), remains the same as, or similar to, the BMR (Hayssen, 1984; Louden, 1987; Thompson and Nicoll, 1986). Based primarily on data from livestock and humans, RMR_g is generally thought to be much greater than the BMR (Brody, 1938, 1964; Brody et al., 1938; Kleiber, 1975; Hayssen, 1984; Mover et al., 1988; Racey and Speakman, 1987; Rahn, 1980, Sandiford and Wheeler, 1924; Sandiford et al., 1931). This increase in metabolic rate has been attributed to some combination of increased biosynthesis associated with fetal growth (primarily in the fetal tissues) and an increase in the relative amount of other metabolically active maternal tissue such as gut, liver, or pancreas. The RMR_m of many small mammals with average or relatively high BMRs, however, is not elevated with respect to BMR (e.g., *Sigmodon hispidus*: Randolph et al., 1977; *Neotoma floridana:* McClure and Randolph, 1980; *Reithrodontomys megalotis*: Nicoll and Thompson, 1987). Data for several high-BMR species even suggest that RMR_g is lower than BMR; this difference may persist at least into early or midlactation (e.g., *Microtus arvalis*: Trojan and Wojciechowska, 1967; *Mus musculus*: Studier, 1979; humans: Prentice and Whitehead, 1987; Prentice and Prentice, 1988). Other studies, however, suggest just the opposite (Mover et al., 1988; Partridge et al., 1986; Weiner, 1977, 1987). In contrast, wild mammals with low BMRs typically have RMR_ms substantially greater than their BMR (Figs. 10.1b, 10.2; Fleming et al., 1981; McClure, 1987; Nicoll and Thompson, 1987; Perrin and

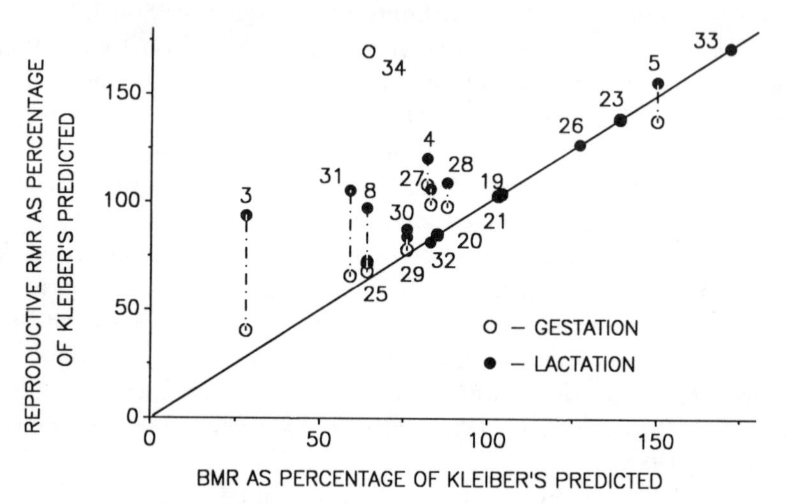

Fig. 10.2. Variation in maternal RMR during gestation (RMR_g) and lactation (RMR_l) as a function of BMR. Solid line is BMR = RMR_m. Species and numbers from Table 10.1: *Echinops telfairi* (3), *Elephantulus rufescens* (4), *Microtus arvalis* (5), *Monodelphis domestica* (8), *Phodopus sungorus* (19), *Sigmodon hispidus* (20, 21), *Suncus murinus* (23). Other species are *Pseudocheirus peregrinus* (25), *Reithrodontomys megalotis* (26), *Thrichomys apereoides* (27), *Hoplomys gymnurus* (28), *Potorous apicalis* (29), *Didelphis virginiana* (30), *Praomys natalensis* (31), *Bettongia gaimardi* (32), *Nesomys rufus* (33), *Hemecentetes semispinosus* (34). (Data for [25–29] from Nicoll and Thompson, 1987; [30] from Fleming et al., 1981; [31] from Haim and Fourie, 1980; Perrin and Clarke, 1987; [32] from Rose, 1987; [33, 34] from M. E. Nicoll, pers. comm., 1986.)

Clarke, 1987; Richard and Nicoll, 1987; Rose, 1987; Thompson and Nicoll, 1986).

The basis for the discrepancy between the BMR to RMR_m shift attributed to domesticated mammals and the absence of this shift in small wild mammals is unclear. Some of the discrepancies seem to be attributable to the manner in which data have been reported. Earlier studies of humans or domesticated mammals often reported only the peak increase in total, as opposed to mass-specific, energy use during gestation without correction for increased energy demands attributable to the increase in maternal body mass (mother and fetus combined), and few such studies report longitudinal data for RMR_m throughout either gestation or lactation. Moreover, while the burst of development in the final stage of gestation might be expected to demand increased metabolic activity by maternal organs, placental structures, and fetal tissues, much of this burst is confounded by metabolic and physical activities associated with parturition per se (Gittleman and Thompson, 1988). There are additional complications. (1) At least some of the earlier RMR_m data for livestock (e.g., horses: Brody, 1964) may have been elevated because mothers were young and still growing, thus they were probably allocating energy to their own structural growth and corporal reserves, and a portion of their high RMR_m may be attributable to the elevated RMR typically associated with growth (Kleiber, 1975; McClure and Randolph, 1980; Poczopko, 1979; Poo et al., 1939). (2) In other cases, because animals were allowed free access to food prior to metabolic measurements it is possible that specific dynamic action (SDA), resulting from increased gut content (Oddy et al., 1984; Verhagen et al., 1986), was partially responsible for the elevated RMR_m. Reanalysis of the human data indicate that, contrary to earlier reports, healthy women do not have markedly increased rates of metabolism during either gestation or lactation (Prentice and Whitehead, 1987). Moreover, when RMR_m data from livestock and humans are expressed as the mean for gestation (Thompson and Nicoll, 1986), they seldom show increases of more than a few percent over BMR (Brody, 1964; Hoversland et al., 1974; Pommerenke et al., 1930). Finally, livestock with BMRs lower than predicted by Kleiber (1975) increase RMR_m in accordance with the pattern seen in wild species (goats, cattle: Brody, 1964; sheep: Hoversland et al., 1974).

For most mammals, the lower the BMR the greater the difference between RMR_m and BMR. At least there is a tendency for the RMR_g

of low-BMR species to be somewhat lower than the RMR_l (Loudon, 1987; Nicoll and Thompson, 1987; White et al., 1988). As a consequence, over a wide interspecific range of relative BMRs, RMR_m is usually near or slightly above the BMR predicted by Kleiber for a nonreproductive female of similar body mass (Figs. 10.1, 10.2).

Although there are several possible explanations for a shift from BMR to a higher RMR_m, none can readily account for differential shifts by high- and low-BMR species. Early studies presumed that high levels of biosynthesis in the fetus or high fetal surface areas (or both) were responsible for elevated RMR_gs (Murlin, 1910; Pommerenke et al., 1930). Surface area arguments, however, have been thoroughly discounted (Huesner, 1987; McNab, 1983, 1988). Although direct measurements of fetal metabolic rate give conflicting results, at least some studies suggest that fetal metabolic rate is not high enough to produce significant changes in RMR_m (Bissonette et al., 1980; Cotter et al., 1969; Kleiber et al., 1943; Pike, 1981). More recently, hypertrophy and hyperplasia of the metabolically demanding tissues such as the gut, liver, pancreas, uterine, and mammary tissues, as well as the increasing size and complexity of the placenta itself have emerged as potential causes of elevated RMR_m (Canas et al., 1982; Kirkwood, 1983; Poo et al., 1939; Smith and Baldwin, 1974; Wunder, this volume). Unfortunately, because these changes have been studied only in domesticated mammals with relatively high BMRs and little or no difference between BMR and RMR_m, their significance for low-BMR species remains unclear. Absence of a drastically elevated RMR_m in high-BMR species, however, implies either an internal reallocation of metabolic activity to enlarged tissues (organs)—at the expense of other tissues—or a compensatory increase in hydration or an increase in the size of metabolically inert tissues to the extent that metabolic changes in more-active tissues are obscured. This might be termed an apparent, rather than a true, lowering of the relative energy demands of hypertrophied organs (Illingworth et al., 1986). One hypothesis with considerable behavioral support is that a temporary restructuring of the relative metabolic demands of various body components might be necessary to prevent maternal overheating (Croskerry et al., 1978; Leon et al., 1978, 1983). Despite evidence for organ hypertrophy, attempts to directly determine the relative increases in energy demands of various organs during reproduction have been relatively unsuccessful (Oddy et al., 1984). Indirect studies, however, suggest that increased protein synthesis occurs in some enlarged organs (Grigor et al., 1987; Jolicoeur et al., 1980). Unfortunately, there are no data on the relative body composition or organ energy demands of any low-BMR species; thus it is impossible to address the question of whether the BMR-RMR_m shift in low-BMR species is related to changes in body composition or metabolic activity of specific organs. Nevertheless, available data indicate that BMR per se is not causally responsible for correlations with reproductive parameters. Efforts should be made to identify and examine functional and autocorrelates of BMR that may have direct affects on reproductive parameters.

One important correlate of BMR is the maximal capacity to use energy. This capacity could be limited either by abilities to consume or to expend energy. Under acute energy demands, the maximum rate at which energy can be expended is about 3–7 times BMR (i.e., the factoral aerobic scope) during cold stress and 5–15 times BMR during locomotion (Chappell, 1984; Dawson et al., 1986; Hinds and MacMillen, 1985; Koteja, 1987; Lechner, 1978; MacMillen and Hinds, this volume, Taigen, 1983; Taylor et al., 1981). If a constant factoral scope exists for the prolonged high energy demands of mid and late lactation, then the capacity for the maximal rate of energy intake and expenditure during reproduction might be the underlying basis for correlations between the BMR and reproductive parameters. Support for a BMR-factoral scope-reproduction link comes from reports that (1) mean energy intakes during lactation, for a variety of domesticated and a few wild animals, cluster around five times BMR (Brody, 1964; Kirkwood, 1983; Kleiber, 1975) and (2) energy expenditures are 4–6 times BMR among free-living female birds feeding nestlings (Drent and Daan, 1980; King, 1974; Roby and Ricklefs, 1986). Although a direct link may exist between BMR and maximum cellular capacities for energy turnover (Stevens, 1973), BMR may be only an autocorrelate of a limit imposed by the interaction of dietary quality and availability, gut size, and body mass on maximal abilities to process energy during reproduction (Demment, 1983; Kirkwood, 1983; McNab, 1986b).

Energy Use During Reproduction

For small wild mammals there were few data on energy use during gestation and only a moderate number on energy use in lactation. Throughout the following discussions, unless otherwise cited, generalizations refer to studies in Table 10.1. A few thorough, well-designed studies such as those on *Sigmodon hispidus* (Mattingly and McClure, 1982, 1985; McClure, 1987; McClure and Randolph, 1980), *Phodopus sungorus* (Weiner, 1987), and *Peromyscus* (Glazier, 1985a, 1985b; Millar, 1975, 1979) remain the basis for much of our knowledge on the reproductive energetics of small wild mammals. The ecological and taxonomic bias in the sample of species in Table 10.1, most of which are microtine or cricetine rodents, may result in similar bias concerning patterns of energy use during reproduction (cf. Pagel and Harvey, 1988a).

Temporal Patterns

For most homeothermic mammals, energy use during reproduction is generally a monotonically increasing function of time (Fig. 10.3). However, intraspecific and interspecific variability in both the shape and magnitude of the energy use curve as a function of time and stage of development is quite high (Kurta et al., 1987; McClure, 1987; Nicoll and Thompson, 1987; Randolph et al., 1977; Speakman and Racey, 1987; Thompson and Nicoll, 1986; Weiner, 1987). This level of variation greatly complicates comparison of cross-sectional studies of reproductive energetics. Human females are a particularly good example in

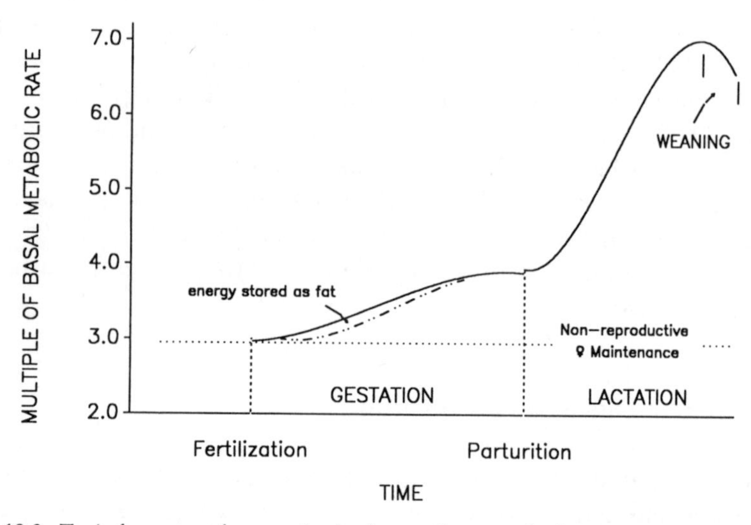

Fig. 10.3. Typical pattern of energy intake for small mammals during gestation and lactation. Area above dashed and dotted line during gestation reflects fat stored for use during lactation.

that they display a remarkable variety of patterns associated with factors such as social and nutritional history or overall dietary quality (Prentice and Prentice, 1988; Prentice and Whitehead, 1987). Energy use by some high-BMR species such as *Microtus arvalis* and *Crocidura russula* appears to decline during early gestation (from nonreproductive levels). Many other species (e.g., *Microtus pennsylvanicus, Clethrionomys glareolus, C. gapperi, Elephantulus rufescens*) show little increase in energy use until relatively late (after 70% or more) in gestation. For most mammals, a dramatic increase in maternal size and energy use is evident during the last 20% of gestation; peak energy use during gestation always occurs during the days immediately preceding parturition. Following peak ER in late gestation and parturition, some species display a pronounced drop in energy use during the first few days of lactation (Gittleman and Thompson, 1988); other species show a continued increase (from the gestational peak) in energy use to a peak ER late in lactation. Unlike the peak in gestation, which predictably anticipates parturition, the peak in lactation may vary greatly in its relative temporal relationship to weaning (Oftedal, 1984).

Mammals that enter torpor during gestation or lactation may display either highly variable patterns of energy use corresponding to torpor events (e.g., *Echinops telfairi*: Thompson and Nicoll, 1986; bats: Racey, 1973, 1981; Racey and Speakman, 1987) or a seeming constant use of energy (e.g., *Myotis lucifugous*: Kurta et al., 1987; Studier et al., 1973, Studier and O'Farrell, 1976). In bats, torpor prolongs gestation (Audet and Fenton, 1988; Racey, 1973, 1981; Racey and Swift, 1981) and may provide essential compensation for the high flight costs associated with increased foraging necessary to meet the energy demands of reproduction (Audet and Fenton, 1988; Racey and Speakman, 1987; Speakman and Racey, 1987). Although Thompson and Nicoll (1986) reported one instance of torpor in a lactating tenrec, neither the frequency nor the impact of torpor on temporal patterns of energy use during mammalian reproduction are well known outside the Microchiroptera (Racey, 1981).

Relative shapes of the question and lactation portions of the energy use curve may be also complicated by accumulation of energy stores (e.g., lipids) during gestation for later use during lactation (see below). Many species store energy as fat either prior to conception or during gestation for use later during reproduction (e.g., rodents: McClure, 1987; ungulates: Oftedal, 1985; Tyler, 1987; pinnipeds: Anderson and Fedak, 1987; Oftedal et al., 1987). Moreover, the relative importance of these fat stores may vary greatly: they are essential for ice breeding pinnipeds (Oftedal et al., 1987) and some ungulates (Oftedal, 1985), important prerequisites for man and other species (Falk and Millar, 1987; Frisch, 1984), and perhaps only safety margins or buffers for other ungulates and small rodents (Mattingly and McClure, 1982, 1985; McClure, 1987; Tyler, 1987). Despite an abundant literature, much is left to be learned about the role of energy storage in mammalian reproduction (Pond, 1983, 1984). In most studies, whether these transitory stores are laid down and used continuously or only during certain portions of gestation and lactation is unclear (Weiner, 1987). Although changes in maternal mass remain the most prevalent means of assessing storage, such changes are not sufficient indicators of storage because, in live animals, these changes are often indistinguishable from either hyperplasic or hypertrophic changes in the size of other tissues (e.g., liver, gut). Furthermore, although this pattern of storage is not uncommon, whether these stores are intended for use on a daily basis to buffer short-term shortages in energy availability (e.g., low foraging success due to weather) or to supplement inadequacies of energy processing (e.g., limitations of gut size) solely during peak lactation is unknown.

The body mass of some lactating females may decrease precipitously, albeit at a fairly even rate, during lactation, and maternal mass immediately postweaning may be 10–15% less than before conception. For primiparous females this loss may result in the establishment of a new, lower, nonreproductive body mass that is returned to after subsequent reproductions (Bercovitch, 1984; Roberts et al., 1984; Studier, 1979). Evaluations of maternal fat reserves in small rodents, bats, and even some ungulates have tended to downplay their importance because they typically are not large enough to provide more than a few days total daily energy requirements (Millar, 1987; Racey and Speakman, 1987; Speakman and Racey, 1987; Tyler, 1987) or because reproduction may proceed normally when food restriction during gestation prevents deposition of said reserves (Mattingly and McClure, 1982). However, because of the great energy demands incurred near the end of lactation (see below), even small augmentations of lipid stores may be essential either to supplement limitations on the rate of intake and digestion at peak energy demands or to buffer short-term foraging failures and low food intake throughout lactation (Tyler, 1987).

The shape of the lactation portion of the energy use curve may vary with maintenance costs and rate of development of the nursing offspring as well as with the relative durations of gestation and lactation. Offspring respiration is a substantial portion of MER (Nicoll and Thompson, 1987; Randolph et al., 1977), and variations in the ontogenetic pattern of the nursling's capacity to thermoregulate (Hill, 1976, 1983, this volume; McClure and Randolph, 1980; McNab, 1983), their thermal environment (Barnett and Dickman, 1984; Leon et al., 1983), and initial body size (McNab, 1983; Tracy, 1977) affect the pattern of increasing energy allocation to offspring respiration and MER during lactation (Gittleman and Thompson, 1988; McClure and Randolph, 1980; Nicoll and Thompson, 1987). Since lactation is more expensive, a shift of energy demand to gestation (e.g., a slight increase in duration or investment during gestation) might actually decrease

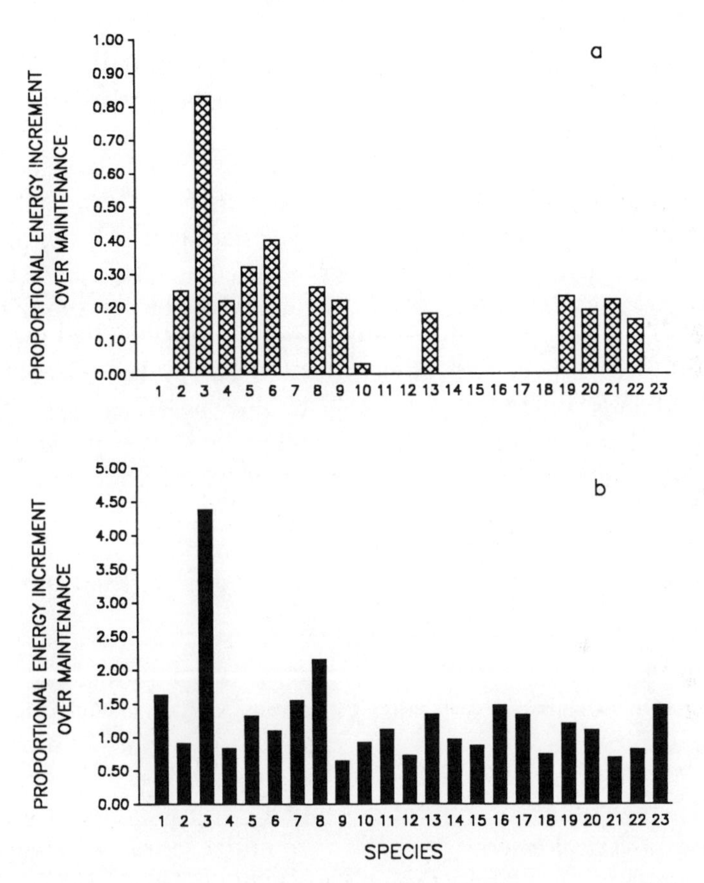

Fig. 10.4. Incremental mean energy requirement (MER) for (a) gestation and (b) lactation as a function of maintenance energy requirements for nonreproductive adult females housed under similar conditions. Note difference in scales. Species as in Table 10.1.

mean daily or peak levels of energy stress (requirements) during lactation.

Gestation

Comparative data on energy use during gestation are available for only a handful of small wild mammals (Table 10.1). Those data suggest that gestation is relatively inexpensive: for most species, the range of MER is 18–25% of maintenance costs (Fig. 10.4a) and 125–300% of measured BMR. MER for gestation seems largely independent of litter size, although this apparent independence may be due to either small sample size or low variation in the cost of gestation. Although there are few reported values for peak ER during gestation, existing data indicate that peak costs during gestation are only slightly greater than MER (Table 10.2; Krol, 1985) and generally well below MER and peak ER during lactation.

As noted above, for many species much of the cost of gestation is devoted to storage of lipids for later use during lactation. MER of *Sigmodon hispidus* from Kansas during gestation was 25% of maintenance requirements. When the energy stored as fat (to be used during lactation) was subtracted from the total MER of gestation, however, the net MER was only 16% of maintenance: thus, 36% of the cost of gestation was actually in anticipation of energy use during lactation (Randolph et al., 1977). Fat storage by the eastern woodrat (*Neotoma floridana*) is even more striking (McClure, 1987). *N. floridana*'s MER during gestation is 22% above maintenance; however, about 86% of that energy is stored as fat (Fig. 10.4a) leaving the actual MER at about 3% of maintenance. A similarly low MER of about 3% has also been reported for pregnancy in *Myotis lucifugus* (T. Kunz, pers. comm.,

TABLE 10.2
Peak energy use and mean daily energy requirement (MER) as multiples of BMR

Species	Gestation		Lactation		Reference
	Peak	MER	Peak	MER	
Peromyscus maniculatus	—	—	7.66	4.28	Stebbins, 1977
	—	—	6.19	5.19	Millar, 1979
P. leucopus	—	—	6.88	5.63	Millar, 1975, 1978
Clethrionomys glareolus	4.10	3.27	6.20	5.02	Kaczmarski, 1966
Microtus arvalis	2.56	2.07	8.06	4.70	Migula, 1969

1988; Kurta et al., 1987) although higher values are reported for other bats (Speakman and Racey, 1987). Because both bats and woodrats have relatively long gestations (Millar, 1977), their low MER during gestation suggests that the low cost of gestation may be advantageous by permitting either significant net production when energy availability is low or by permitting the augmentation of energy reserves for later use during a relatively short, energy-demanding lactation (Mattingly and McClure, 1982; McClure, 1987). While lipid storage during gestation is well documented for *N. floridana, Sigmodon hispidus*, and *Phodopus sungorus*, storage does not appear to be a common strategy either in other small herbivores (e.g., *Peromyscus, Microtus, Clethrionomys*: Glazer, 1985a; Millar, 1978, 1987; *Spermophilus saturatus*: Kenagy, 1987) or in small insectivorous mammals (*Elephantulus rufescens*: Nicoll and Thompson, 1987; *Suncus murinus*: Dryden et al., 1974). Moreover, fat storage during gestation is often not obligatory, as at least some *Sigmodon hispidus* produce litters of normal size and mass when fat deposition is prevented (or limited) by restriction of food intake (Mattingly and McClure, 1982; McClure, 1987). However, among many small mammals, severe food restriction during reproduction results either in smaller or fewer offspring or both per reproductive attempt or in an extension of gestation or lactation rather than substantial usage of prereproductive maternal tissues and corporal reserves (Mattingly and McClure, 1982, 1985; McClure, 1981; Merson and Kirkpatrick, 1981; Millar, 1975; Oswald and McClure, 1987; Woodside et al., 1981).

One exception to the general pattern of low gestational MER is *Echinops telfairi* (Fig. 10.4a). This tenrec has an extremely labile body temperature (T_b) and extremely low maintenance requirements (RMR = 28% of Kleiber's predicted at T_b = 30°C): during gestation both maternal RMR and T_b are substantially elevated from nonreproductive values (Table 10.1). The high gestational MER here is due largely to the substantial elevation of maternal metabolism (Nicoll and Thompson, 1987) and suggests that a high T_b and RMR_m are essential during gestation.

Lactation

For small mammals, lactation is two to three times as expensive as gestation (Fig. 10.4b). With the exception of *E. telfairi*, MER ranges from 65 to 210% of maintenance requirements, with most values fall-

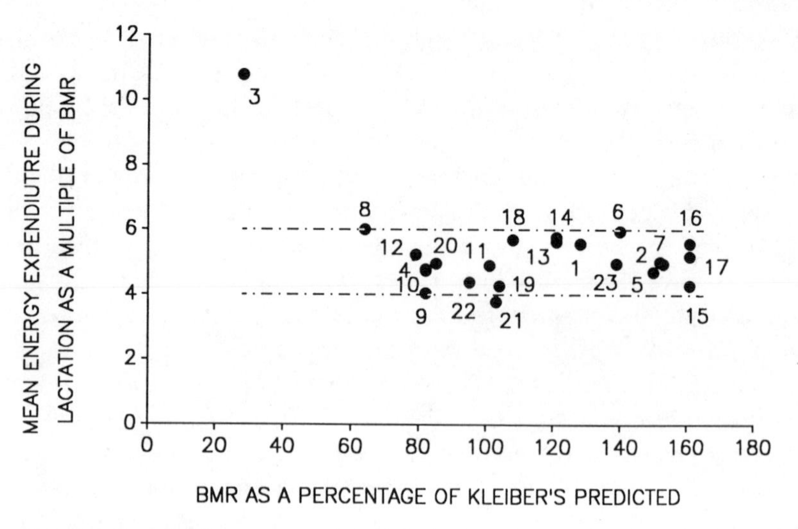

Fig. 10.5. Mean total daily energy use (MER) during lactation as a function of relative BMR. Species and data from Table 10.1.

ing between 85 and 160%. Moreover, although among most eutherian mammals gestation is much longer than lactation (Thompson, 1987), 65–85% of the total ER is expended during lactation (Table 10.1; Oftedal, 1985; Nicoll and Thompson, 1987; Thompson and Nicoll, 1986). As with previous reports for other wild mammals, domesticated mammals, and free-living birds the range of lactational MER for the species in Table 10.1 is four to six times BMR (Fig. 10.5). Birds, for which the costs of feeding nestlings remain fairly constant throughout parental care (Roby and Ricklefs, 1986), differ from mammals in that mammalian costs reach a definite peak of six to eight times BMR near the time of weaning (Table 10.2). The durations of peak energy demands are unknown, but most workers have speculated that these peaks last only a few days (Oftedal, 1984). If these peaks denote morphophysiological limits to intake and absorption of energy (e.g., gut size), use of fat stores may be particularly important in supplementing energy use at this point in time.

Much of the variation in MER during lactation has been attributed to differences in litter size. Many studies indicate that the cost of producing each neonate is relatively constant among litter sizes both within (Deag et al., 1987; Kenagy, 1987; Knight et al., 1986; Lochmiller et al., 1982; Loveridge, 1986; Mattingly and McClure, 1982; Millar, 1978; Randolph et al., 1977; Smith and McManus, 1975) and between species (Mattingly and McClure, 1982; Nicoll and Thompson, 1987;

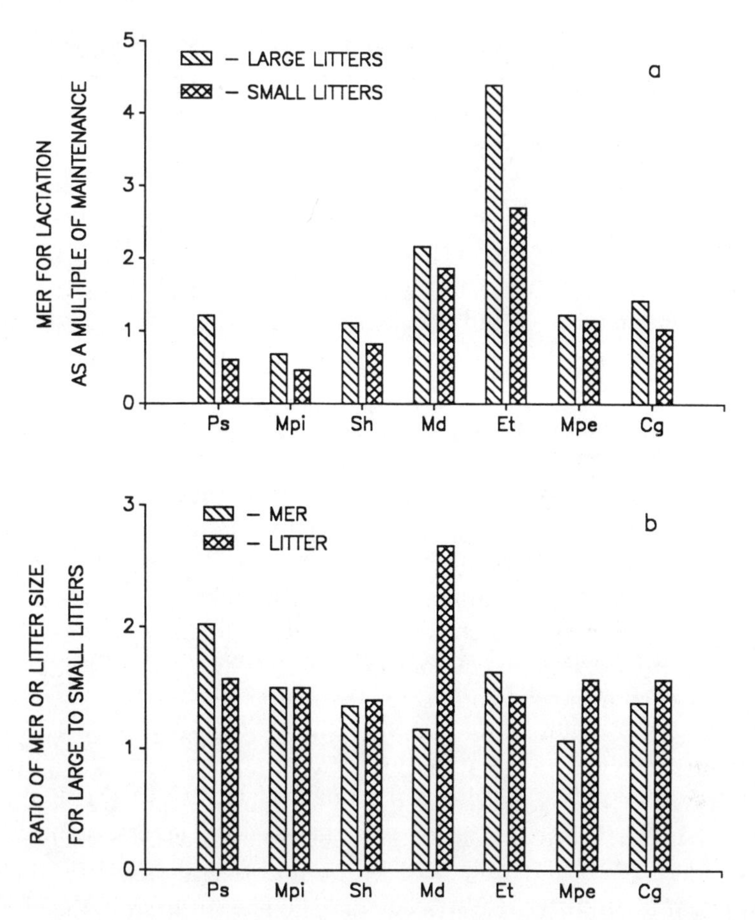

Fig. 10.6. Mean energy requirement (MER) as a function of (a) large or small litter size and (b) proportional change in MER compared with the proportional change in litter size. Proportional changes are the ratio of MER for large to small litters and the ratio of the corresponding large to small litter size. Ps, *Phodopus sungorus* (Weiner, 1987); Mpi, *Microtus pinetorum* (Lochmiller et al., 1982); Sh, *Sigmodon hispidus* (Mattingly and McClure, 1982, 1985); Md, *Monodelphis domestica* (Nicoll and Thompson, 1987); Et, *Echinops telfairi* (Nicoll and Thompson, 1987); Mpe, *Microtus pennsylvanicus* (Innes and Millar, 1981); Cg, *Clethrionomys gapperi* (Innes and MIllar, 1981).

Thompson and Nicoll, 1986). Others report considerable interspecific variation in the energy costs associated with differences in litter size (Fig. 10.6a; Leon and Woodside, 1983; Leon et al., 1983). Although larger litters generally demand more energy, differences in litter size are not always proportional to differences in MER (Fig. 10.6b; Heasley, 1983; Innes and Millar, 1981; Lodge, 1957; Lodge and Heaney, 1973; Mattingly and McClure, 1985; Ota and Yokohama, 1967; Sadlier,

1982). For example, there is relatively little difference between the cost of rearing *Microtus pennsylvanicus* litters of four or five and five or six, but similar differences in *Clethrionomys gapperi* litter size produce substantial differences in MER (Fig. 10.6; Innes and Millar, 1982).

Lactating females have several options for adjusting their energy use to compensate for changes in litter size. Perhaps the best-documented tactic is to vary the energy invested per young such that individual offspring mass is an inverse function of litter size (Millar, 1977; Ota and Yokohama, 1967), but this is not always the case (Mattingly and McClure, 1982, 1985). More subtly, parental behavior may alter the thermoregulatory costs (heat loss) for nurslings by adjustments of the parent's time in the nest, nest insulation, or nest location (Hill, 1983, this volume; Leon et al., 1978).

The intraspecific relationship between litter size and MER may also be complicated by dichotomous strategies among breeding females. Within some species, a tendency exists for larger mothers to have both larger litters and larger young than smaller mothers (*Peromyscus maniculatus*: Myers and Master, 1983; *Kerodon rupestris*: Roberts et al. 1984; *Rattus norvegicus*: Woodside et al., 1981); this pattern is typically reversed in interspecific comparisons (Millar, 1977). Pronounced effects of litter size may be found readily in comparisons of large and small litters (where litter sizes differ by at least 33%), but even this relationship is not consistent (Fig. 10.6). Large litters are also more costly than small litters in low-BMR species such as *Monodelphis domestica* and *Phodopus sungorus*. Moreover, several studies imply that the dichotomy in litter size strategy may be present even in the absence of differences in maternal body mass (Oswald and McClure, 1987; Pond and Mersmann, 1974; Weiner, 1987); this could make it exceedingly difficult to predict reproductive costs on the basis of maternal body mass alone (Mattingly and McClure, 1982).

Efficiency of Gestation and Lactation

Because of the relatively direct transfer of energy and nutrients across membranes, the lack of direct thermoregulatory stress on the fetus (i.e., cold), and the low level of activity of the fetus, one would intuitively expect conversion of energy into offspring tissue (net production) to be more efficient during gestation than during lactation. In contrast, during lactation energy must be obtained, digested, assimi-

TABLE 10.3
Conversion efficiencies from energy use to net production of fetal or nursling tissues

Species[a]	Gestation	Lactation
2. *Clethrionomys glareolus*	.11	.15
5. *Microtus arvalis*	.12	.15
7. *M. pinetorum*	—	.26
10. *Neotoma floridana*	.12	.14
11. *Peromyscus eremicus*	—	.20
12. *P. floridanus*	—	.20
14. *P. leucopus*	—	.21
16. *P. maniculatus*	—	.22
18. *P. polionotus*	—	.18
19. *Phodopus sungorus*	.15	.46
20. *Sigmodon hispidus*	.15	.33
21. *S. hispidus*	.26	.31

[a]Species numbers and references from Table 10.1.

lated, synthesized into milk, then transferred to the offspring who must then allocate energy among maintenance, thermoregulatory costs, activity, and growth. Despite these obvious differences only 12–26% of the ER for gestation is converted into fetal tissue, while 12–46% may be converted during lactation (Table 10.3). These efficiencies of gestation may be deceptively low and may rise considerably when corrected for energy allocated to fat deposition, expansion of the uterus, the placenta, and mammary tissues. Tissier et al. (1979), for example, found that the efficiency of net production was 39% during gestation when new maternal tissues were included as net production (Oftedal, 1985; Rattray et al., 1974; Robinson et al., 1980). Moreover, lactation efficiencies often decline when corrected for the use of fat deposited during gestation (McClure, 1987; Randolph et al., 1977; Weiner, 1987), although they generally remain remarkably high. Despite greater efficiency under optimal conditions, because minimal energy requirements are much higher for lactation (due to thermoregulatory costs and activity of nursing young, see below), lactation requires far more energy per day before its highest levels of lactational efficiency are attained.

Allocation of Energy during Reproduction

Within reproduction, numerous options for allocation of energy both between gestation and lactation and within gestation and lactation exist. Evolutionary shifts of development between gestation and lactation seem to reflect trade-offs of speed versus daily energy requirements (McClure, 1987). Gestation generally is less energetically demanding on a daily basis, but prolonged gestation implies both slower reproduction and greater total ER (for this reason, partial REs must be adjusted for the duration of any given reproductive phase; Payne and Wheeler, 1967).

The most energy-demanding component of reproduction is usually respiration of nursing young. Small body size and poor insulation of many neonates require either high rates of maintenance energy for thermoregulation or a warm thermal environment (Hill, 1983, this volume; Leon et al., 1978; Markussen et al., 1985). In addition to behavioral strategies, mammals seem to adjust the costs of nursling thermoregulation by varying patterns of thermoregulatory development. In precocial species, for which thermoregulatory capacities and behavioral competence are well developed at or soon after birth, thermoregulatory activity costs of nursing young may be substantial, although growth rates may be relatively high and nutritional independence reached at an early age (Hill, 1983, this volume; Markussen et al., 1985; McClure and Randolph, 1980). While those costs may be less for altricial species, Nicoll and Thompson (1987) estimated that offspring respiration may reach 65% of total ER. Although direct comparisions of the ontogeny of thermoregulation in altricial versus precocial species are rare (McClure and Randolph, 1980; Waldschmidt and Müller, 1988), several authors have suggested that relative to precocial species, some altricial species may reduce daily energy demands, even at the price of increased total ER, by delaying the onset of thermoregulation (Hill, this volume; McClure and Randolph, 1980) or lowering the trajectory of the ontogenetic "hump" in juvenile RMR (McClure and Randolph, 1980; McNab, 1983). Marsupials, for example, appear to both delay the onset of thermoregulation and reduce the trajectory of juvenile RMR (Gemmell and Johnston, 1985; Morrison and Petajan, 1962; Rose, 1987; Shield, 1966); this strategy has the effect of reducing daily energy use during lactation (Lillegraven et al., 1987; McNab, 1983, 1986a; Morton et al., 1982).

Except in somewhat unusual cases such as *Echinops* or *Elephantulus*, energy allocated to increase mass-specific RMR_m from BMR is generally only a small portion of the total ER (Nicoll and Thompson, 1987).

Regardless of whether the limits to a female's abilities to allocate energy to reproduction are functional aspects of her biological design or the result of local resource availability, each female's ability to direct energy to reproduction is ultimately limited. Within those limits, certain fixed requirements that most females appear to defend, even to the detriment of their offspring, exist. These requirements include energy for maintenance, thermoregulatory costs, and deposition of energy reserves necessary for future survival (e.g., during hibernation). Although there is some flexibility in the minimal values of these requirements, for the most part this flexibility seems limited to compensatory shifts in behavior that reduce thermal stress and reduce energy spent on activity (e.g., locomotion). Once these minimal survival requirements are met, whatever additional energy can be obtained and processed, up to the prevailing ecological or morphophysiological limit, can be allocated to reproduction. Thus, the absolute amount of energy a typical mammalian female can possibly allocate to reproduction is dependent on both the limits of her abilities to obtain or process energy and her abilities to minimize her own requirements. While changes in availability of, or access to, food have frequently been held responsible for decreased reproductive performance in wild mammals, increased thermoregulatory costs or levels of physical activity (locomotion) have more frequently, and often more convincingly, been shown to detract directly from growth rates and net production in domesticated mammals (Mount, 1979; Pond and Maner, 1974; Yousef, 1985b).

Thermoregulation

During reproduction there may be some expansion of the maternal thermal neutral zone toward lower T_as and a slight reduction in thermal conductance (Verhagen et al., 1986). While increased heat production from an elevated RMR_m does not typically have either a predictable or substantial compensatory effect for cold stress due to low T_a or high convective heat loss (Clark, 1981; Leon and Woodside, 1978; Maust et al., 1972; Pond and Maner, 1974; Yousef, 1985a, 1985b), the generality of this pattern is unclear (Porter and McClure 1984; Roberts and Coward, 1985). Larger mammals seem most affected by energy allocated to compensate for heat loss during reproduction (Clark, 1981; Verhagen et al., 1986; Yousef, 1985b), but overheating can also reduce the energy available for reproduction in small mammals as well. Female laboratory rats may suspend the thermogenic capacities of brown fat or restrict physical contact with their litters to prevent unacceptable increases in maternal T_b (Croskerry et al., 1978; Leon et al., 1978; Trayhurn et al., 1982). The widespread importance of thermoregulatory stress as a drain on energy allocated to reproduction implies that species with (1) broad zones of thermoneutrality, (2) relatively low or high upper and lower critical temperatures (T_{lc}, T_{uc}), respectively, and (3) low thermal conductances (heat transfer coefficients) should be less susceptible to thermal stress during reproduction (cf. Bruce and Clark, 1979; Porter and McClure, 1984; Verhagen et al., 1986; Yousef, 1985a, 1985b). Unfortunately, although most small captive mammals are bred at temperatures well below thermoneutrality, the effects of thermal stress on the energetics of reproduction have not been studied in small wild mammals.

The effects of thermoregulatory stress on nurslings seem to have an even greater energetic significance than those on the mother. The response to thermal stress is a complex function of thermoregulatory competence, thermoregulatory costs, and maintenance costs of nursing young. For larger animals, there may be little recourse, other than an increase in heat production and thermoregulatory cost, for combating thermal stress, and such stress typically leads to decreased growth rates and lower adult size (Bruce and Clark, 1979; Mount, 1960, 1979; Verhagen et al., 1987a, 1987b). Among small mammals, maternal or paternal attendance at the nest, direct heat transfer between parent and offspring and among offspring, indirect heating of the nest environment by parental heat loss, and location and integrity of the nest itself are malleable factors that can play a major compensatory role in reducing thermoregulatory expenses of nursing offspring (Leon and Woodside, 1983). Moreover, trade-offs involving thermoregulatory costs of the young may have profound effects on life history features such as litter size and rate of development (McClure and Randolph, 1980; Hill, this volume).

Activity

Maternal activity (e.g., locomotion, standing, exposure to thermal stress) is generally higher for free-living than for captive mammals. Reduced activity in domesticated mammals is well documented as a means of increasing net productivity and growth rates (Clark, 1981; Mount, 1979; Pond and Maner, 1974); thus the types and levels of activity should be viewed as important determinants of the energy available for reproduction in wild mammals. By reducing activity to only that necessary to obtain food and perform minimal social behavior, females may actually reduce normal maintenance costs during reproduction below those of their nonreproductive state (Anderson and Harwood, 1985). Through these behavioral compensations, more energy can be channeled to reproduction. In a classic example, captive lab rats that made extensive use of running wheels prior to reproduction did not increase their food intake during reproduction. The ER of reproduction was met almost entirely by a decrease in wheel running (Morrison, 1955, 1956; Slonaker 1924; Wang, 1925). This behavioral compensation may have similar consequences in, for example, monogamous or cooperative breeders in which individuals other than the mother expend energy to gather food or carry young during lactation, thus permitting the mother to devote less energy to activity and more to the production of milk (Goldizen, 1987a, 1987b).

Reproductive Consequences of a Low BMR

If the BMR is linked to capacities for maximal energy processing during reproduction, then species with low BMRs would seem to have two distinct disadvantages. First, because most species with low BMR have an elevated RMR_m (Fig. 10.2; Thompson and Nicoll, 1986), the amount of ER allocated to RMR_m is inversely proportional to BMR. If this energy is unavailable for trade-offs involving other aspects of ER (e.g., net production), net production of species with low BMRs should be less efficient than those species with high BMRs. However, the elevation of RMR_m may often be a relatively small component of the total ER (Nicoll and Thompson, 1987). The second, and more important, drawback of a low BMR is that a high BMR should confer the capacity to allocate a greater absolute amount of energy to reproduction. Consider, for example, two species of the same body size, one with a BMR that is 50% and the other with a BMR that is 100% of Kleiber's predicted (Fig. 10.7). Then, assuming that their capacities for sustained energy intake during reproduction are about five times BMR (Table 10.1; Fig. 10.5), the low-BMR species would be able to expend energy at a rate of up to only 250% of Kleiber's predicted BMR, while the high-BMR species could be capable of up to 500% of Kleiber's predicted. If minimal maternal maintenance costs remained a constant fac-

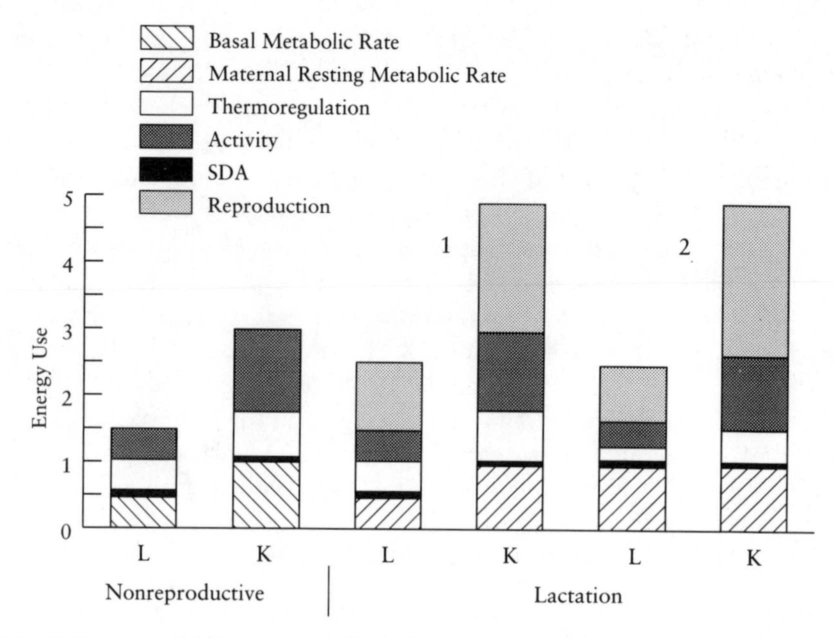

Fig. 10.7. Potential differences in abilities of mammals with low (L) and high (K) BMRs to allocate energy to reproduction. Energy use: 1.0 = Kleiber's predicted BMR. (1) Condition when BMR = RMR_l for low-BMR species. (2) Both reduced thermoregulatory costs for the high-BMR species (Kenagy, 1987) and $RMR_l >$ BMR for the low-BMR species. BMR of low-BMR species is one-half that of high-BMR species.

tor of BMR (maintenance is typically two times BMR in captive individuals), then in the above example the abilities of the two species to allocate energy to reproduction would differ in direct proportion to the difference in their BMRs (150 vs. 300% of Kleiber's predicted available for production). However, if, as seems more likely, maternal maintenance is a function of RMR_m, then for the low-BMR species RMR_l would be ≈ 100% of Kleiber's predicted, maternal maintenance would be 200% of Kleiber's predicted, and the energy available for reproduction would be only 250 − 200 = 50% of Kleiber's predicted, or about 17% of the capacity of the high-BMR species. While this pattern could vary greatly as a function of the key components (i.e., maternal maintenance, RMR_l, and the actual factoral relationships), the pattern does provide an underlying mechanism for correlations between BMR and reproductive parameters. Perhaps because the range of relative BMRs is so small, available data (Table 10.1) do not really support these hypotheses. Species with moderate BMRs (65–90% of predicted) are variable both in the allocation of energy to RMR_m (Fig. 10.2) and their maximum energy use. However, when contrasted with a single high-BMR species (*Microtus arvalis*), it would appear that high BMR does confer a great energy advantage during reproduction (Table 10.1). Clearly, more data on the energetics of high- and low-BMR species are needed to address this question.

The Energetic Limits to Reproduction

The focus of this chapter has been on either captive wild mammals with food and water available ad libitum or on domesticated mammals, primarily the laboratory rat, under some experimental regime of restricted food availability. How data for captive individuals generalize to free-living populations is unknown. As noted above, a plethora of field studies have attributed variation in reproductive patterns to an underlying variation in access to or abundance of food sources. Unfortunately, the few direct studies of energy use by free-living wild mammals provide a confusing picture of reproductive energetics. Nagy's (1987) comprehensive review of daily energy use (field metabolic rates, FMR) by nonreproductive mammals (both male and female) suggests that most free-living mammals operate at a daily energy use of about three times the BMR. If functional limits to energy turnover are really four to six times the BMR, under the best of conditions most wild mammals have the capacity to allocate only about two to three times the BMR to reproduction. Field studies on energy use by small marsupials (*Antechinus stuartii*: Nagy et al., 1978; *Gymnobelideus leadbeateri*: Smith et al., 1982; *Petaurus breviceps*: Nagy and Suckling, 1985) found that FMRs of lactating females were only slightly higher than those of either males or nonreproductive females. Similarly, Montgomery and Nagy (1980) reported no difference among FMRs of nonreproductive, pregnant, and lactating female sloths (*Bradypus variagatus*). However, Lee and Nagy (cited in Lee and Cockburn, 1985) reported that in the FMRs of another small marsupial, *Antechinus swainsoni*, during late lactation was 175% of that for nonreproductive females. Racey and Speakman (1987) and Speakman and Racey (1987) also reported high ERs for free-living bats. Moreover, in the latter cases, torpor was apparently employed as a means of minimizing the increase in total energy intake for reproduction by reducing maternal maintenance requirements. In the most detailed study to date, Kenagy (1987) found that during late lactation, an average female golden-mantled ground squirrel (*Spermophilus saturatus*) has an FMR 82% greater than at mating. While some of this increase might be due to increased body mass, thermoregulatory expenses are greatly reduced from mating to late lactation. Kenagy estimated BMR at 95% of Kleiber's predicted and reported that maximal energy turnover by mothers near peak lactation was 3.1 times BMR (G. J. Kenagy, pers. comm., 1985). Since most hibernators, and ground-dwelling squirrels in particular, tend to have BMRs nearer 75% of Keliber's predicted (McNab, 1988), peak lactation in free-living golden-mantled squirrels could easily be nearer four to five times BMR. Similar consideration of Lee and Nagy's data for *A. swainsoni* (75% increase over a "typical" nonreproductive FMR of three times BMR) also suggests that FMR may approach four to six times BMR during lactation. Finally, the similarities between avian FMRs during reproduction and those attained by captive mammals fed ad libitum suggest that during the breeding season, many mammals are limited by functional rather than ecological aspects of energy use (Kenagy, 1987). More FMRs of free-living small mammals during lactation must be accumulated before it will be possible to adequately address the question of whether functional limits to energy processing are regularly attained during reproduction by free-living mammals.

There is now enough circumstantial evidence for a link between BMR and reproduction to warrant at least a simple reconsideration of the "BMR-speed" and "BMR-effort" predictions. Using direct assessments of energy use during reproduction, and assuming that (1) the time from conception to weaning is an appropriate measure of the speed of reproduction (Lee and Cockburn, 1985; McNab, 1986a; Thompson, 1987) and (2) MER during lactation is a reasonable index of total ER for gestation and lactation, then the prediction that variations in BMR lead directly, or indirectly, to variations in reproductive effort or speed of reproduction or both, can be addressed using the data from Table 10.1 (one value for each species). Conception to weaning time (CW) is in days, BMR is in kilojoules per day, and body mass is in grams. As in previous studies, strong correlations exist between relative BMR and both RE and CW (Fig. 10.8) because the latter are strongly correlated with body mass. When effects of body mass were removed using either partial correlation or comparisons of the respective residuals (Harvey and Clutton-Brock, 1985; Thompson, 1987), BMR was not significantly correlated with CW (Table 10.4). With CW time and body mass held constant, however, BMR was significantly corretated with RE_{lact}. These analyses are tantalizing in that

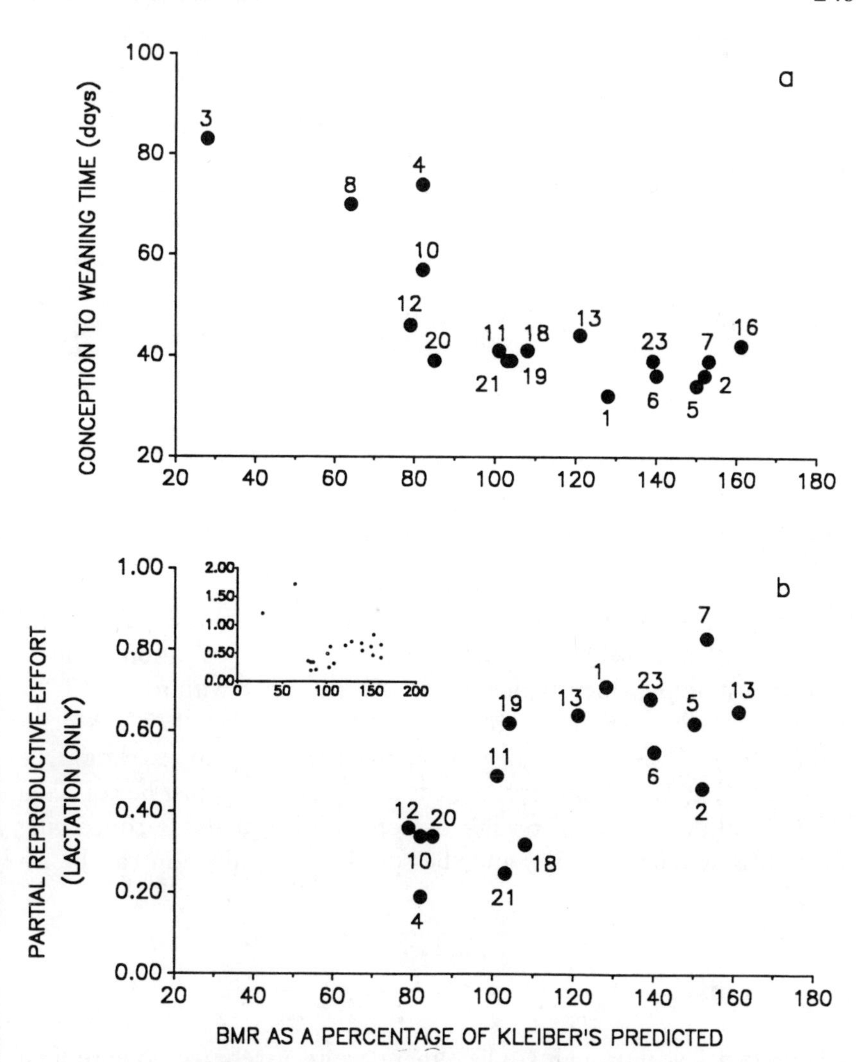

Fig. 10.8. (a) conception to weaning time. (b) reproductive effort as functions of relative BMR (without *Monodelphis domestica* and *Echinops telfairi*). Inset in (b) shows all 17 species from Table 10.1.

they suggest that correlations of CW time with BMR (McNab, 1986a) may reflect an autocorrelation between CW time and RE (or RE_{lact}).

While these results suggest that BMR may be more closely related to RE than speed of reproduction per se, the present data set contains a large number of closely related small rodents (*Peromyscus, Microtus*), and thus poses myriad complications with respect to both phylogenetic

TABLE 10.4
Partial correlation matrixes for basal metabolic rate (BMR), conception to weaning time (CW), and reproductive effort during lactation (RE_{lact})

	CW	RE_{lact}
BMR	−.292	.542*
CW		−.386
	CW (RE_{lact} constant)	
BMR	−.106	
	RE_{lact} (CW constant)	
BMR	.487**	

Note: Body mass held constant. N = 17: species 1–8, 10–13, 16, 18, 19, 21, 23 from Table 10.1.
**Significant at P ≤ .025.
*Significant at P ≤ .050.

bias and level of analysis (Harvey and Clutton-Brock, 1985; Pagel and Harvey 1988a, 1988b). One such bias might be low variation in duration of gestation, which typically varies little either within a genus or between closely related genera (Gould, 1977; Millar, 1977; Moelhman, 1986). Thus, while suggestive, the present analyses should be considered highly preliminary; a great deal more data must be collected before confidence in the results of comparative analyses concerning how BMR is related to RE and the speed of reproduction can be increased.

Conclusions

Despite a few thorough studies, there are relatively few longitudinal data on the diversity of strategies for allocation of energy to reproduction. Data exist for domestic mammals, a moderate number of rodents and insectivores, and a few bats. Many studies have focused on either gestation or lactation, while the cross-sectional data of most studies present only a glimpse of energy use during reproduction.

Nevertheless, available data make clear that, even among closely related taxa, a variety of options for managing the allocation of energy to reproduction exist. This level of variation and the potential bias of the sample composition indicate that a great deal more descriptive and experimental data must be assembled before generalized trends will be clearly recognizable. Moreover, observed variation in energetic strategies, even among conspecifics, points to great dangers of using indirect estimates of energy use or energy investment (e.g., birth mass, litter size, duration of reproduction) in comparative analyses of reproductive behaviors and life history strategies. The relatively low daily energy cost and small proportion of total ER allocated to gestation serve to emphasize the pitfalls of such indirect estimates of energy use.

The existing evidence encourages further investigations into a link between BMR and maximal capacities for energy use, and limits on energy use during reproduction. More data are sorely needed, on a wide variety of species, to examine the interrelationships of BMR, gut size, and energy use during reproduction. Moreover, attempts must be made to determine the relative importance of functional and ecological limits on reproductive energy use in free-living individuals and populations. In particular, it is critical that studies of energy use in free-living mammals focus on reproductive energy use and the quantitative importance of behavioral compensation in increasing a female's abilities to allocate energy to reproduction.

ACKNOWLEDGMENTS

The manuscript was greatly improved by comments and criticisms from T. Tomasi, O. Oftedal, K. Fischer, B. K. McNab, and D. Boness. This work was supported by the Friends of the National Zoo and a career development fellowship from the National Zoological Park.

Literature Cited

Altmann, J. 1983. Costs of reproduction in baboons. Pp. 67–77 *in* Behavioral Energetics: The Costs of Survival (W. P. Aspey and S. I. Lustick, eds.). Ohio State University Press, Columbus.

Anderson, S. S., and M. A. Fedak. 1987. The energetics of sexual success of gray seals and comparison with the costs of reproduction in other pinnipeds. Symp. Zool. Soc. Lond., 57:319–327.

Anderson, S. S., and J. Harwood. 1985. Time budgets and topography: how energy reserves and terrain determine the breeding behavior of grey seals. Anim. Behav., 33:1343–1348.

Audet, D., and M. B. Fenton. 1988. Heterothermy and the use of torpor by the bat *Eptesicus fuscus* (Chiroptera: Vespertilionidae): a field study. Physiol. Zool., 61:197–204.

Austad, S. N., and M. E. Sunquist. 1986. Sex-ratio manipulation in the common opossum. Nature, 324:58–60.

Barnett, S. A., and R. G. Dickman. 1984. Milk production and consumption and growth of young of wild mice after ten generations in a cold environment. J. Physiol., 346:409–417.

Bartholomew, G. A., Jr. 1977. Energy metabolism. Pp. 57–110 *in* Animal Physiology: Principles and Adaptations. (M. S. Gordon, ed.). Macmillan, New York. 699 pp.

Bercovitch, F. B. 1984. The energetic costs of lactation in olive baboons. Am. J. Phys. Anthropol., 84:138.

Berman, C. M. 1988. Maternal condition and offspring sex ratio in a group of free-ranging rhesus monkeys: an eleven-year study. Am. Nat., 131:307–328.

Bissonette, J. M., J. Metcalfe, A. R. Hohimer, and M. L. Pernoll. 1980. Uterine oxygen uptake in the pregnant pigmy goat. Resp. Physiol., 42:373–381.

Bowen, W. D., O. T. Oftedal, and D. J. Boness. 1985. Birth to weaning in four days: remarkable growth in the hooded seal, *Systophora cristata*. Can. J. Zool., 63:2841–2846.

Brockway, J. M., J. D. McDonald, and J. D. Pullar. 1963. The energy cost of reproduction in sheep. J. Physiol., 167:318–327.

Bronson, F. H. 1979. The reproductive ecology of the house mouse. Q. Rev. Biol., 54:265–299.

——. 1985. Mammalian reproduction: an ecological perspective. Biol. Reprod., 32:1–26.

Brody, S. 1938. Relation between heat increment of gestation and birth weight. Bull. Univ. Mo. Agric. Exp. Stn., 283:1–28.

——. 1964. Bioenergetics and growth, with special reference to the efficiency complex in domestic animals. Hafner, New York.

Brody, S., J. Riggs, K. Kaufman, and V. Herring. 1938. Energy-metabolism levels during gestation, lactation, and post-lactation rest. Bull. Univ. of Mo. Agric. Exp. Stn., 281:1–43.

Bruce, J. M., and J. J. Clark. 1979. Models of heat production and critical temperature for growing pigs. Anim. Prod., 28:353–369.

Caley, M. J., S. Boutin, and R. A. Moses. 1988. Male biased reproduction and sex-ratio adjustment in muskrats. Oecologia, 74:501–506.

Calow, P. 1979. The cost of reproduction—a physiological approach. Biol. Rev., 54:23–40.

——. 1981. Resource utilization and reproduction. Pp. 245–270 *in* Physiological Ecology: An Evolutionary Approach to Resource Use (C. R. Townsend and P. Calow, eds.). Sinauer, Sunderland, Mass. 393 pp.

Campbell, R. M., and B. F. Fell. 1964. Gastrointestinal hypertrophy in the lactating rat and its relation to food intake. J. Physiol., 171:90–98.

Canas, R., J. Romero, and R. L. Baldwin. 1982. Maintenance energy requirements during lactation in rats. J. Nutr., 112:1876–1180.

Carl, G., and C. T. Robbins. 1988. The energetic cost of predation avoidance in neonatal ungulates: hiding versus following. Can. J. Zool., 66:239–246.

Chappell, M. A. 1984. Maximum oxygen consumption during exercise and cold exposure in deer mice, *Peromyscus maniculatus*. Resp. Physiol., 55:367–377.

Clark, J. A. 1981. Environmental aspects of housing for animal production. Butterworths, London.

Clutton-Brock, T. H., F. E. Guinness, and S. D. Albon. 1982. Red Deer: Behavior and Ecology of Two Sexes. University of Chicago Press, Chicago.

Coelho, A. M., Jr. 1986. Time and energy budgets. Comp. Primate Biol., 2A:141–146.

Congdon, J. D., A. E. Dunham, and D. W. Tinkle. 1982. Energy budgets and life histories of reptiles. Pp. 233–271 *in* Biology of the Reptilia, vol. 13 (C. Gans and F. H. Pough, eds.). Academic Press, London.

Contreras, L. C. 1984. Bioenergetics of huddling: test of a psycho-physiological hypothesis. J. Mammal., 65:256–262.

Cotter, J. R., J. N. Blecher, and H. Prystowsky. 1969. Blood flow and oxygen consumption of pregnant goats. Am. J. Obstet. Gynecol., 103:1098–1101.

Cripps, A. W., and V. J. Williams. 1975. The effect of pregnancy and lactation on food intake, gastrointestinal anatomy and the absorptive capacity of the small intestine in the albino rat. Br. J. Nutr., 53:17–32.

Croskerry, P. G., G. K. Smith, and M. Leon. 1978. Thermoregulation and the maternal behavior of the rat. Nature, 273:299–300.

Dawson, T. J., B. K. Smith, and W. R. Dawson. 1986. Metabolic capabilities of Australian marsupials in the cold. Pp. 478–483 *in* Living in the Cold: Physiological and Biochemical Adaptations (H. C. Heller, X. J. Musacchia, L. C. H. Wang, eds.). Elsevier, New York.

Deag, J. M., C. E. Lawrence, and A. Manning. 1987. The consequences of differences in litter size for the nursing cat and her kittens. J. Zool., 213:153–179.

Demment, M. W. 1983. Feeding ecology and the evolution of body size in baboons. Afr. J. Wildl. Ecol., 21:219–233.

Demment, M. W., and P. J. Van Soest. 1985. A nutritional explanation for body-size patterns of ruminant and nonruminant herbivores. Am. Nat., 125:641–672.

Derrickson, E. M. 1989. The comparative method of Elgar and Harvey: silent ammunition for McNab. Funct. Ecol., 3:123–125.

Drent, R. H., and S. Daan. 1980. The prudent parent: energetic adjustments in avian breeding. Ardea, 68:225–252.

Dryden, G. L., M. Gebczynski, and E. L. Douglas. 1974. Oxygen consumption by nursling and adult musk shrews. Acta Theriol., 19:453–461.

Dunbar, R. I. M. 1985. Reproductive Decisions: An Economic Analysis of Gelada Baboon Social Strategies. Princeton University Press, Princeton.

Elgar M. A., and P. H. Harvey. 1987. Basal metabolic rates in mammals: allometry, phylogeny and ecology. Funct. Ecol., 1:25–36.

Fa, J. E., and C. H. Southwick. 1987. Ecology and behavior of food-enhanced primate groups. A. R. Liss, New York.

Falk, J. W., and J. S. Millar. 1987. Reproduction by female *Zapus princeps* in relation to age, size, and body fat. Can. J. Zool., 65:568–571.

Fell, B. F., K. A. Smith, and R. M. Campbell. 1963. Hypertrophic and hyperplastic changes in the alimentary canal of the lactating rat. J. Pathol. Bacteriol., 85:179–188.

Fleming, A. F. 1978. Food intake and body weight regulation during the reproductive cycle of the golden hamster (*Mesocricetus auratus*). Behav. Biol., 24:291–306.

Fleming, M. W., J. D. Harder, and J. J. Wukie. 1981. Reproductive energetics of the Virginia opossum compared with some eutherians. Comp. Biochem. Physiol., 70B: 645–648.

Frisch, R. E. 1984. Body fat, puberty, and fertility. Biol. Rev., 59:161–188.

Gaulin, S. J. 1979. A Jarman/Bell model of primate feeding niches. Human Ecol., 7:1–20.

Gebczynska, Z., and M. Gebczynski. 1971. Length and weight of the alimentary tract of the root vole. Acta Theriol., 16:359–369.

Gemmell, R. T., and G. Johnston. 1985. The development of thermoregulation and the emergence from the pouch of the marsupial bandicoot *Isoodon macrourus*. Physiol. Zool., 58:299–302.

Gittleman, J. L., and P. H. Harvey. 1982. Carnivore home-range size needs and ecology. Behav. Ecol. Sociobiol., 10:57–63.

Gittleman, J. L., and S. D. Thompson. 1988. Energy allocation in mammalian reproduction. Am. Zool., 28:863–875.

Glazier, D. S. 1985a. Energetics of litter size in five species of *Peromyscus* with generalizations for other mammals. J. Mammal., 66:629–642.

——. 1985b. Relationship between metabolic rate and energy expenditure for lactation in *Peromyscus*. Comp. Biochem. Physiol., 80A:587–590.

Goldizen, A. W. 1987a. Facultative polyandry and the role of infant-carrying in wild saddle-back tamarins (*Saguinus fuscicollis*). Behav. Ecol. Sociobiol., 20:99–109.

——. 1987b. Tamarins and marmosets: communal care of offspring. Pp. 34–43 *in* Primate Societies (B. Smuts, D. L. Cheney, R. M. Seyfarth, R. W. Wrangham, and T. T. Struhsaker, eds.). University of Chicago Press, Chicago. 578 pp.

Goldstein, D. L. 1988. Estimates of daily energy expenditure in birds: the time-energy budget as an integrator of laboratory and field studies. Am. Zool., 28:829–845.

Goldstein, D. L., and K. Nagy. 1985. Resource utilization by desert quail: time and energy, food and water. Ecology, 66:378–387.

Gould, S. J. 1977. Ontogeny and Phylogeny. Harvard University Press, Cambridge. 501 pp.

Graham, N. M. 1964. Energy exchanges of pregnant and lactating ewes. Austral. J. Agric. Res., 15:127–141.

Grigor, M. R., J. E. Allan, J. M. Carrington, A. Carne, A. Geursen, D. Young, M. P. Thompson, E. B. Haynes, and R. A. Coleman. 1987. Effect of dietary protein and food restriction on milk production and composition, maternal tissues and enzymes in lactating rats. J. Nutr., 117:1247–1258.

Gross, J. E., Z. Weng, and B. A. Wunder. 1985. Effects of food quality and energy needs: changes in gut morphology and capacity of *Microtus ochrogaster*. J. Mammal., 66:661–667.

Hadjipieris, G., and W. Holmes. 1966. Studies on feed intake and feed utilization by sheep. I. The voluntary feed intake of dry, pregnant and lactating ewes. J. Agric. Sci., 66:217–233.

Haim, A., and F. le R. Fourie. 1980. Heat production in nocturnal (*Praomys natalensis*) and diurnal (*Rhabdomys pumilo*) South African rodents. S. Afr. J. Zool., 15:91–94.

Hammel, H. T. 1976. On the origin of endothermy in mammals. Isr. J. Med. Sci., 12:905–915.

Hanwell, A., and M. Peaker. 1977. Physiological effects of lactation on the mother. Symp. Zool. Soc. Lond., 41:297–312.

Harvey, P. H. 1986. Energetic costs of reproduction. Nature, 321:648–649.

Harvey, P. H., and T. Clutton-Brock. 1981. Primate home-range size and metabolic needs. Behav. Ecol. Sociobiol., 8:151–154.

——. 1985. Life history variation in primates. Evolution, 39:559–581.

Havera, S. 1979. Energy and nutrient cost of lactation in fox squirrels. J. Wildl. Manage., 43:957–965.

Hayssen, V. 1984. Basal metabolic rate and the intrinsic rate of increase: an empirical and theoretical reexamination. Oecologia, 64:419–421.

Hayssen, V., R. Lacy, and P. J. Parker. 1985. Metatherian reproduction: transitional or transcending? Am. Nat., 126:617–632.

Heasley, J. E. 1983. Energy allocation in response to reduced food intake in pregnant and lactating laboratory mice. Acta Theriol., 28:55–71.

Hennemann, W. W., III. 1983. Relationship among body mass, metabolic rate and the intrinsic rate of natural increase. Oecologia, 56:104–108.

——. 1984. Intrinsic rates of natural increase of altricial and precocial eutherian mammals: the potential price of altriciality. Oikos, 43:363–368.

Hennemann, W. W., III, S. D. Thompson, and M. J. Konecny. 1983. Metabolism of crab-eating foxes, *Cerdocyon thous*: ecological influences on the energetics of canids. Physiol. Zool., 56:319–324.

Hill, R. W. 1976. Ontogeny of homeothermy in neonatal *Peromyscus leucopus*. Physiol. Zool., 49:292–306.

——. 1983. Thermal physiology and energetics of *Peromyscus*: ontogeny, body temperature, metabolism, insulation, and microclimatology. J. Mammal., 64:19–37.
Hinds, D. S., and R. E. MacMillen. 1984. Energy scaling in marsupials and eutherians. Science, 225:335–337.
——. 1985. Scaling of energy metabolism and evaporative water loss in heteromyid rodents. Physiol. Zool., 58:282–298.
Hofman, M. A. 1983. Evolution of brain size in neonatal and adult placental mammals: a theoretical approach. J. Theor. Biol., 105:317–331.
Hoversland, A. S., J. Metcalfe, and J. T. Parer. 1974. Adjustments in maternal blood gases, acid-base balance, and oxygen consumption in the pregnant pygmy goat. Biol. Reprod., 10:589–595.
Huesner, A. A. 1987. What does the power function reveal about structure and function in animals of different size? Annu. Rev. Physiol., 49:121–133.
Hulbert, A. J., and T. J. Dawson. 1974. Standard metabolism and body temperature of perameloid marsupials from different environments. Comp. Biochem. Physiol., 47A: 583–590.
Illingworth, P. J., R. T. Jung, P. W. Howie, P. Leslie, and T. E. Isles. 1986. Diminution in energy expenditure during lactation. Br. Med. J., 292:437–441.
Innes, D. G., and J. Millar. 1981. Body weight, litter size, and energetics of reproduction in *Clethrionomys gapperi* and *Microtus pennsylvanicus*. Can. J. Zool., 59:785–789.
Jolicoeur, L., J. Asselin, and J. Morisset. 1980. Trophic effects of gestation and lactation on rat pancreas. Biomed. Res., 1:482–488.
Jolly, A. 1984. The puzzle of female breeding priority. Pp. 197–216 *in* Female Primates: Studies by Female Primatologists (M. Small, ed.). A. R. Liss, New York.
Kaczmarski, F. 1966. Bioenergetics of pregnancy and lactation in the bank vole. Acta Theriol., 11:409–417.
Kenagy, G. J. 1987. Energy allocation for reproduction in the golden-mantled ground squirrel. Symp. Zool. Soc. Lond., 57:259–273.
Kenagy, G. J., and G. A. Bartholomew, Jr. 1985. Seasonal reproductive patterns in five coexisting California desert rodent species. Ecol. Monogr., 55:371–397.
King, J. R. 1974. Seasonal allocation of time and energy resources in birds. Pp. 4–85 *in* Avian Energetics (R. A. Paynter, ed.). Nuttall Ornithological Club, Cambridge, Mass.
King, J. R., and M. E. Murphy. 1985. Periods of nutritional stress in the annual cycles of endotherms: fact or fiction? Am. Zool., 25:955–964.
Kirkwood, J. K. 1983. A limit to metabolisable energy intake in mammals and birds. Comp. Biochem. Physiol., 75A:1–3.
Kleiber, M. 1975. The Fire of Life: An Introduction to Animal Energetics. Krieger, Boca Raton, Fla. 451 pp.
Kleiber, M., H. H. Cole, and A. H. Smith. 1943. Metabolic rate of rat fetuses, in vitro. J. Cell. Comp. Physiol., 22:167–176.
Knight, C. H., E. Maltz, and A. H. Docherty. 1986. Milk yield and composition in mice: effects of litter size and lactation. Comp. Biochem. Physiol., 84A:127–133.
Koteja, P. 1987. On the relation between basal and maximum metabolic rate in mammals. Comp. Biochem. Physiol., 87A:205–208.
Krol, E. 1985. Reproductive energy budgets of hedgehogs during lactation. Zesz. Nauk. Fili UW 48 Biol. 10:105–117.
Kurland, J. A., and J. D. Pearson. 1986. Ecological significance of hypometabolism in nonhuman primates: allometry, adaptation, and deviant diets. Am. J. Phys. Anthropol., 71:445–457.
Kurta, A., K. A. Johnson, and T. H. Kunz. 1987. Oxygen consumption and body temperature of female little brown bats (*Myotis lucifugus*) under simulated roost conditions. Physiol. Zool., 60:386–397.

Lavigne, D. M., S. Innes, G. A. J. Worthy, K. M. Kovacs, O. J. Schmitz, and J. P. Hickie. 1986. Metabolic rates of seals and whales. Can. J. Zool., 64:279–284.

Lechner, A. J. 1978. The scaling of maximal oxygen consumption and pulmonary dimensions in small mammals. Resp. Physiol., 34:29–44.

Lee, A. K., and A. Cockburn. 1985. The Evolutionary Ecology of Marsupials. Cambridge University Press, Cambridge.

Lee, P. C. 1987. Nutrition, fertility, and maternal investment in primates. J. Zool., 213:409–422.

——. 1988. Ecological constraints and opportunities: interactions, relationships, and social organization of primates. Pp. 297–312 *in* Ecology and Behavior of Food-Enhanced Primate Groups (J. E. Fa and C. H. Southwick, eds.). A. R. Liss, New York.

Lee, P. C., and C. J. Moss. 1986. Early maternal investment in male and female African elephant calves. Behav. Ecol. Sociobiol., 18:353–361.

Leon, M., P. G. Croskerry, and G. K. Smith. 1978. Thermal control of mother-young contact in rats. Physiol. Behav., 21:793–811.

Leon, M., C. Fischette, P. Chee, and B. Woodside. 1983. Energetic limits on reproduction: interaction of thermal and dietary factors. Physiol. Behav., 30:937–943.

Leon, M., and B. Woodside. 1983. Energetic limits on reproduction: maternal energy intake. Physiol. Behav., 30:945–957.

Lillegraven, J. A., S. D. Thompson, B. K. McNab, and J. L. Patton. 1987. The origin of eutherian mammals. Biol. J. Linn. Soc., 32:281–336.

Lochmiller, R. L., E. C. Hellgren, and W. E. Grant. 1987. Influence of moderate nutritional stress during gestation on reproduction of collared peccaries (*Tayassu tajacu*). J. Zool., 211:321–328.

Lochmiller, R. L., J. B. Whelan, and R. L. Kirkpatrick. 1982. Energetic cost of lactation in *Microtus pinetorum*. J. Mammal., 63:475–481.

——. 1983. Seasonal energy requirements of adult pine voles, *Microtus pinetorum*. J. Mammal., 64:345–350.

Lodge, G. A. 1957. The utilization of dietary energy by lactating sows. J. Agric. Sci., 44:200–210.

Lodge, G. A., and D. P. Heaney. 1973. Energy cost of pregnancy in single- and twin-ewes. Can. J. Anim. Sci., 53:479–489.

Loudon, A. S. I. 1987. The reproductive energetics of lactation in a seasonal macropodid marsupial: comparison of marsupial and eutherian herbivores. Symp. Zool. Soc. Lond., 57:127–148.

Loudon, A. S. I., and P. A. Racey, eds. 1987. Reproductive Energetics in Mammals. Symposia of the Zoological Society of London 57. Clarendon Press, Oxford.

Loveridge, G. G. 1986. Bodyweight changes and energy intake of cats during gestation and lactation. Anim. Technol., 37:7–15.

Low, B. S. 1978. Environmental uncertainty and the parental strategies of marsupials and placentals. Am. Nat., 112:197–213.

Loy, J. 1988. Effects of supplementary feeding on maturation and fertility in primate groups. Pp. 153–166 *in* Ecology and Behavior of Food-Enhanced Primate Groups (J. E. Fa and C. H. Southwick, eds.). A. R. Liss, New York.

Lucas, A., G. Ewing, S. B. Roberts, and W. A. Coward. 1987. How much energy does the breast fed infant consume and expend? Br. Med. J., 295:75–77.

Markussen, K. A., A. Rognmo, and A. S. Blix. 1985. Some aspects of thermoregulation in newborn reindeer calves (*Rangifer tarandus tarandus*). Acta Physiol. Scand., 123: 215–220.

Martin, R. D. 1984a. Body size, brain size and feeding strategies. Pp. 73–103 *in* Food Acquisition and Processing in Primates (D. J. Chivers, B. A. Wood, and A. Bilsborough, eds.). Plenum Press, New York.

——. 1984b. Scaling effects and adaptive strategies. Symp. Zool. Soc. Lond., 51:87–117.

Mattingly, K. D., and P. A. McClure. 1982. Energetics of reproduction in large littered cotton rats (*Sigmodon hispidus*). Ecology, 63:183–195.

——. 1985. Energy allocation during lactation in cotton rats (*Sigmodon hispidus*) on a restricted diet. Ecology, 66:928–937.

Maust, L. E., R. E. McDowell, and N. W. Hooven. 1972. Effect of summer weather on performance of Holstein cows in three stages of lactation. J. Dairy Sci., 55:1133–1139.

McClure, P. A. 1981. Sex-biased litter reduction in food-restricted woodrats (*Neotoma floridana*). Science, 211:1058–1060.

——. 1987. The energetics of reproduction and life histories of cricetine rodents (*Neotoma floridana* and *Sigmodon hispidus*). Symp. Zool. Soc. Lond., 57:241–258.

McClure, P. A., and J. C. Randolph. 1980. Relative allocation of energy to growth and development of homeothermy in the eastern wood rat (*Neotoma floridana*) and hispid cotton rat (*Sigmodon hispidus*). Ecol. Monogr., 50:199–219.

McNab, B. K. 1974. The energetics of endotherms. Ohio J. Sci., 74:370–380.

——. 1978. The comparative energetics of neotropical marsupials. J. Comp. Physiol., 125B:115–128.

——. 1979. Climatic adaptation in the energetics of heteromyid rodents. Comp. Biochem. Physiol., 62A:813–820.

——. 1980. Food habits, energetics, and the population biology of mammals. Am. Nat., 116:106–124.

——. 1983. Energetics, body size, and the limits to endothermy. J. Zool., 199:1–29.

——. 1986a. Food habits, energetics, and the reproduction of marsupials. J. Zool., 208:595–614.

——. 1986b. The influence of food habits on the energetics of eutherian mammals. Ecol. Monogr., 56:1–19.

——. 1987. The reproduction of marsupial and eutherian mammals in relation to energy expenditure. Symp. Zool. Soc. Lond., 57:29–39.

——. 1988. Complications inherent in scaling the basal rate of metabolism in mammals. Q. Rev. Biol., 63:25–54.

McNab, B. K., and P. Morrison. 1963. Body temperature and metabolism in subspecies of *Peromyscus* from arid and mesic environments. Ecol. Monogr., 33:63–82.

Merson, M. H., and R. L. Kirkpatrick. 1981. Relative sensitivity of reproductive activity and body-fat level to food restriction in white-footed mice. Am. Midl. Nat., 102: 305–312.

Migula, P. 1969. Bioenergetics of pregnancy and lactation in the European vole. Acta Theriol., 14:167–179.

Millar, J. S. 1975. Tactics of energy partitioning in breeding *Peromyscus*. Can J. Zool., 53:967–976.

——. 1977. Adaptive features of mammalian reproduction. Evolution, 31:370–386.

——. 1978. Energetics of reproduction in *Peromyscus leucopus*: the cost of lactation. Ecology, 59:1055–1061.

——. 1979. Energetics of lactation in *Peromyscus maniculatus*. Can. J. Zool., 57:1015–1019.

——. 1987. Energy reserves in breeding small rodents. Symp. Zool. Soc. Lond., 57: 231–241.

Moehlman, P. D. 1986. Ecology of cooperation in canids. Pp. 64–86 *in* Ecological Aspects of Social Evolution (D. I. Rubenstein and R. W. Wrangham, eds.). Princeton University Press, Princeton. 551 pp.

Morrison, P. R., and J. H. Petajan. 1962. The development of temperature regulation in the opossum, *Didelphis marsupialis virginiana*. Physiol. Zool., 35:52–65.

Morrison, S. D. 1955. The total energy metabolism of non-pregnant rats. J. Physiol., 127:479–497.
——. 1956. The total energy and water metabolism during pregnancy in the rat. J. Physiol., 134:650–654.
Morton, S. R., H. F. Recher, S. D. Thompson, and R. Braithwaite. 1982. Comments on the relative advantages of marsupial and eutherian reproduction. Am. Nat., 210:128–134.
Mount, L. E. 1960. The influence of huddling and body size on the metabolic rate of the young pig. J. Agric. Sci., 55:101–105.
——. 1979. Energy Metabolism. Butterworths, London.
Mover, H., S. Hellwing., and A. Ar. 1988. Energetic cost of gestation in the white-toothed shrew *Crocidura russula monacha* (Soricidae, Insectivora). Physiol. Zool., 61:17–26.
Mulder, M. B. 1987. Resources and reproductive success in women with an example from the Kipsigis of Kenya. J. Zool., 213:489–505.
Murlin, J. R. 1910. The metabolism of development. I. Energy metabolism in the pregnant dog. Am. J. Physiol., 26:134–155.
Myers, P., and L. L. Master. 1983. Reproduction by *Peromyscus maniculatus*: size and compromise. J. Mammal., 64:1–18.
Myrcha, A. 1962. Variations in the length and weight of the alimentary tract of *Clethrionomys glareolus* (Schreber, 1780). Acta Theriol., 9:139–148.
——. 1965. Length and weight of the alimentary tract of *Apodemus flavicollis* (Melchior, 1834). Acta Theriol., 10:225–228.
Myrcha, A., L. Ryskowski, and W. Walkowa. 1969. Bioenergetics of pregnancy and lactation in the white mouse. Acta Theriol., 12:161–166.
Nagy, K. A. 1987. Field metabolic rate and food requirement scaling in mammals and birds. Ecol. Monogr., 57:111–128.
Nagy, K. A., and G. G. Montgomery. 1980. Field metabolic rate, water flux, and food consumption in three-toed sloths (*Bradypus variagatus*). J. Mammal., 61:465–472.
Nagy, K. A., R. S. Seymour, A. K. Lee, and R. Braithwaite. 1978. Energy and water budgets in free-living *Antechinus stuartii* (Marsupialia: Dasyuridae). J. Mammal., 59:60–68.
Nagy, K. A., and G. C. Suckling. 1985. Field energetics and water balance of sugar gliders, *Petaurus breviceps* (Marsupialia: Petauridae). Austral. J. Zool., 33:683–698.
Nicoll, M. E., and S. D. Thompson. 1987. The energetics of reproduction in therian mammals: didelphids and tenrecs. Symp. Zool. Soc. Lond., 57:1–27.
Oddy, V. H., J. M. Gooden, and E. F. Annison. 1984. Partitioning of nutrients in Merino ewes. I. Contribution of skeletal muscle, the pregnant uterus, and the lactating mammary gland to total energy expenditure. Aust. J. Biol. Sci., 37:375–388.
Oftedal, O. T. 1984. Milk composition, milk yield and energy output at peak lactation: a comparative review. Symp. Zool. Soc. Lond., 51:33–85.
——. 1985. Pregnancy and lactation. Pp. 215–238 *in* Bioenergetics of Wild Herbivores (R. J. Hudson and R. G. White, eds.). CRC Press, Boca Raton, Fla.
Oftedal, O. T., D. J. Boness, and R. A. Tedman. 1987. The behavior, physiology, and anatomy of lactation in the pinnipedia. Pp. 175–245 *in* Current Mammalogy, vol. 1. (H. Genoways, ed.). Plenum, New York.
Oftedal, O. T., and J. L. Gittleman. 1988. Patterns of energy output during reproduction in carnivores. Pp. 355–377 *in* Carnivore Behavior and Biology (J. L. Gittleman, ed.). Cornell University Press, Ithaca. 620 pp.
Oliveras, D., and M. Novak. 1986. A comparison of paternal behaviour in the meadow vole *Microtus pennsylvanicus*, the pine vole *M. pinetorum* and the prairie vole *M. ochrogaster*. Anim. Behav., 34:519–526.

Ortiz, C. L., B. J. LeBoeuf, and D. P. Costa. 1984. Milk intake of elephant seal pups: an index of parental investment. Am. Nat., 124:416–422.

Oswald, C., and P. A. McClure. 1987. Energy allocation during concurrent pregnancy and lactation in Norway rats with delayed and undelayed implantation. J. Exp. Zool., 241:343–357.

Ota, K., and A. Yokohama. 1967. Body weight and food consumption of lactating rats nursing various sizes of litters. J. Endocrinol., 38:263–268.

Packard, G. C., and T. J. Boardman. 1988. The misuse of ratios, indices, and percentages in ecophysiological research. Physiol. Zool., 61:1–10.

Pagel, M. D., and P. H. Harvey. 1988a. Recent developments in the analysis of comparative data. Q. Rev. Biol., 63:413–440.

Pagel, M. D., and P. H. Harvey. 1988b. How mammals produce large-brained offspring. Evolution, 42:948–957.

Parker, P. J. 1977. An ecological comparison of marsupial and placental patterns of reproduction. Pp. 273–286 *in* The Biology of Marsupials (B. Stonehouse and D. Gilmore, eds.). Macmillan, London.

Partridge, G. C., G. E. Lobley, and R. A. Fordyce. 1986. Energy and nitrogen metabolism of rabbits during pregnancy, lactation, and concurrent pregnancy and lactation. Br. J. Nutr., 56:199–207.

Payne, P. R., and E. Wheeler. 1967. Comparative nutrition in pregnancy. Nature, 215:1134–1136.

Peaker, M., R. G. Vernon, and C. H. Knight. 1984. Physiological strategies in lactation. Academic Press, London.

Perrin, M. R., and J. R. Clarke. 1987. A preliminary investigation of the bioenergetics of pregnancy and lactation of *Praomys natalensis* and *Saccostomus campestris*. S. Afr. J. Zool., 22:77–82.

Peterson, A. D., and B. R. Baumgardt. 1971. Influence of level of energy demand on the ability of rats to compensate for diet dilution. J. Nutr., 101:1069–1074.

Pianka, E. R. 1976. Natural selection of optimal reproductive tactics. Am. Zool., 16:775–784.

Pike, I. L. 1981. Comparative studies of embryo metabolism in early pregnancy. J. Reprod. Fertil. Suppl., 29:203–213.

Poczopko, P. 1979. Metabolic rate and body size relationships in adult and growing homeotherms. Acta Theriol., 24:125–136.

Pommerenke, W. T., H. F. Haney, and W. J. Meek. 1930. Energy metabolism of pregnant rabbits. Am. J. Physiol., 93:249–257.

Pond, C. M. 1983. Parental feeding as a determinant of ecological relationships in Mesozoic terrestrial ecosystems. Acta Palaeontol. Pol., 28:215–224.

——. 1984. Physiological and ecological importance of energy storage in the evolution of lactation: evidence for a common pattern of anatomical organization of adipose tissue in mammals. Symp. Zool. Soc. Lond., 51:1–32.

Pond, W. G., and J. H. Maner. 1974. Swine Production in Temperate and Tropical Environments. W. H. Freeman, San Francisco.

Pond, W. G., and H. J. Mersmann. 1988. Severe restriction of dietary protein or total feed during gestation in rats: effects on progeny during postnatal life. Nutr. Rep. Int., 37:1167–1177.

Poo, L. J., W. Lew, and T. Addis. 1939. Protein anabolism of organs and tissues during pregnancy and lactation. J. Biol. Chem., 128:69–77.

Porter, W. P., and P. A. McClure. 1984. Climate effects on growth and reproduction potential in *Sigmodon hispidus* and *Peromyscus maniculatus*. Pp. 173–181 *in* Winter Ecology of Small Mammals (J. F. Merritt, ed.). Carnegie Museum, Pittsburgh. 380 pp.

Portman, O. 1970. Nutritional requirements (NRC) of non-human primates. Pp. 87–115 *in* Feeding and Nutrition of Non-Human Primates (R. Harris, ed.). Academic Press, New York.

Prentice, A. M., and A. Prentice. 1988. Energy costs of lactation. Annu. Rev. Nutr., 8:63–79.

Prentice, A. M., and R. G. Whitehead. 1987. The energetics of human reproduction. Symp. Zool. Soc. Lond., 57:275–304.

Racey, P. A. 1973. Environmental factors affecting the length of gestation in heterothermic bats. J. Reprod. Fertil. Suppl., 19:175–189.

——. 1981. Environmental factors affecting the length of gestation in mammals. Pp. 199–213 *in* Environmental Factors in Mammal Reproduction (D. Gilmore and B. Cook, eds.). Macmillan, London.

Racey, P. A., and J. R. Speakman. 1987. The energy costs of pregnancy and lactation in heterothermic bats. Symp. Zool. Soc. Lond., 57:107–125.

Racey, P. A., and S. M. Swift. 1981. Variations in gestation length in a colony of pipistrelle bats (*Pipistrellus pipistrellus*) from year to year. J. Reprod. Fertil., 61:123–129.

Rahn, H. 1980. Comparison of embryonic development in birds and mammals: birth weight, time and cost. Pp. 124–137 *in* A Companion to Animal Physiology (C. R. Taylor, K. Johansen, and L. Bolis, eds.). Cambridge University Press, Cambridge.

Randolph, P. A., J. C. Randolph, K. Mattingly, and M. M. Foster. 1977. Energy costs of reproduction in the cotton rat, *Sigmodon hispidus*. Ecology, 58:31–45.

Rasmussen, D. T., and M. K. Izard. 1988. Scaling of growth and life history traits relative to body size, brain size, and metabolic rate in lorises and galagos (Lorisidae, Primates). Am. J. Phys. Anthropol., 75:357–367.

Rattray, R., W. N. Garrett, N. E. East, and N. Hinman. 1974. Growth, development and composition of the ovine conceptus and mammary gland during pregnancy. J. Anim. Sci., 38:613–626.

Richard, A. F., and M. E. Nicoll. 1987. Female social dominance and basal metabolism in a Malagasy primate, *Propithecus verreauxi*. Am. J. Primatol., 12:309–314.

Robbins, C. T., R. S. Podbielancik-Norman, D. L. Wilson, and E. D. Mould. 1981. Growth and nutrient consumption of elk calves compared to other ungulate species. J. Wildl. Manage., 45:172–186.

Roberts, M., E. Maliniak, and M. Deal. 1984. The reproductive biology of the rock cavy, *Kerodon rupestris*, in captivity: a study of reproductive adaptation in a trophic specialist. Mammalia, 48:253–266.

Roberts, S. B., and W. A. Coward. 1985. The effects of lactation on the relationship between metabolic rate and ambient temperature in the rat. Ann. Nutr. Metab., 29: 19–22.

Robinson, J. J., I. MacDonald, C. Fraser, and J. G. Gordon. 1980. Studies on reproduction in prolific ewes, 6. The efficiency of energy utilization for conceptus growth. J. Agric. Sci., 94:331–338.

Roby, D., and R. Ricklefs. 1986. Daily energy expenditure by adult Leach's Storm-petrels during the nesting cycle. Physiol. Zool., 59:661–678.

Rose, R. W. 1987. Reproductive energetics of two Tasmanian rat-kangaroos (Potorinae: Marsupialia). Symp. Zool. Soc. Lond., 57:149–165.

Sacher, G. A., and E. F. Staffeldt. 1974. Relation of gestation time to brain weight for placental mammals: implications for the theory of vertebrate growth. Am. Nat., 108:593–615.

Sadlier, R. M. F. S. 1982. Energy consumption and subsequent partitioning in lactating black-tailed deer. Can. J. Zool., 60:382–386.

Sailer, L. D., S. J. Gaulin, J. S. Boster, and J. A. Kurland. 1985. Measuring the relationship between dietary quality and body size in primates. Primates, 26:14–27.

Sandiford, I., and T. Wheeler. 1924. The basal metabolism before, during, and after pregnancy. J. Biol. Chem., 62:329–352.
Sandiford, I., T. Wheeler, and W. M. Boothby. 1931. Metabolic studies during pregnancy and menstruation. Am. J. Physiol., 96:191–202.
Scheck, S. H. 1982. A comparison of thermoregulation and evaporative water loss in the hispid cotton rat, *Sigmodon hispidus texianus* from Northern Kansas and south-central Texas. Ecology, 63:361–369.
Schmitz, O. J., and D. M. Lavigne. 1984. Intrinsic rate of increase, body size, and specific metabolic rate in marine mammals. Oecologia, 62:305–309.
Shield, J. 1966. Oxygen consumption during pouch development of the macropodid marsupial *Setonix brachyurus*. J. Physiol., 187:257–270.
Sibley, R. M. 1981. Strategies of digestion and defecation. Pp. 109–114 *in* Physiological Ecology: An Evolutionary Approach to Resource Use (C. R. Townsend and P. Calow, eds.). Sinaur, Sunderland, Mass. 353 pp.
Silk, J. B. 1986. Eating for two: behavioral and environmental correlates of gestation length among free-ranging baboons (*Papio cynocephalus*). Int. J. Primatol., 7:583–602.
Slonaker, J. R. 1924. The effect of copulation, pregnancy, pseudo-pregnancy and lactation on the voluntary activity and food consumption of the albino rat. Am. J. Physiol., 71:362–394.
Smith, A. P., K. A. Nagy, M. R. Fleming, and B. Green. 1982. Energy requirements and water turnover in free-living Leadbeater's possums, *Gymnobelideus leadbeateri* (Marsupialia: Petauridae). Aust. J. Zool., 30:737–779.
Smith, B. W., and J. J. McManus. 1975. The effects of litter size on the bioenergetics and water requirements of lactating *Mus musculus*. Comp. Biochem. Physiol., 51A: 111–115.
Smith, N. E., and R. L. Baldwin. 1974. Effects of breed, pregnancy, and lactation on energy use in dairy cattle. J. Dairy Sci., 57:1055–1060.
Smuts, B. B., D. L. Cheney, R. M. Seyfarth, R. W. Wrangham, and T. T. Struhsaker. 1987. Primate societies. University of Chicago Press, Chicago. 578 pp.
Speakman, J. R., and P. A. Racey. 1987. The energetics of pregnancy and lactation in the brown long-eared bat, *Plecotus auritus*. Pp. 367–393 *in* Recent Advances in the Study of Bats (M. B. Fenton, P. A. Racey, and J. M. V. Rayner, eds.). Cambridge University Press, Cambridge.
Stearns, S. C. 1976. Life history tactics: a review of the ideas. Q. Rev. Biol., 51:3–47.
Stebbins, L. 1977. Energy requirements during reproduction of *Peromyscus maniculatus*. Can. J. Zool., 55:1701–1704.
Stenseth, N. C., E. Framstad, P. Migula, P. Trojan, B. Wojciechowska-Trojan. 1980. Energy models for the common vole *Microtus arvalis*: energy as a limiting resource for reproductive output. Oikos, 34:1–22.
Stevens, E. D. 1973. The evolution of endothermy. J. Theor. Biol., 38:597–611.
Stewart, R. E. A. 1986. Energetics of age-specific reproductive effort in female Harp seals, *Phoca groenlandica*. J. Zool., 208:503–517.
Studier, E. H. 1979. Bioenergetics of growth, pregnancy and lactation in the laboratory mouse, *Mus musculus*. Comp. Biochem. Physiol., 64A:473–481.
Studier, E. H., V. L. Lysengen, and M. J. O'Farrell. 1973. Biology of *Myotis thysanodes* and *M. lucifugus* (Chiroptera: Vespertilionidae). II. Bioenergetics of pregnancy and lactation. Comp. Biochem. Physiol., 44A:467–471.
Studier, E. H., and M. J. O'Farrell. 1976. Biology of *Myotis thysanodes* and *M. lucifugus* (Chiroptera: Vespertilionidae). III. Metabolism, heart rate, breathing rate, evaporative water loss, and general energetics. Comp. Biochem. Physiol., 44A:467–471.

Taigen, T. L. 1983. Activity metabolism of anuran amphibians: implications for the origin of endothermy. Am. Nat., 121:94–109.

Taylor, C. R., G. M. O. Maloiy, E. R. Weibel, V. A. Langman, J. M. Z. Kamau, H. J. Seehermann, N. C. Heglund. 1981. Design of the mammalian respiratory system. III. Scaling maximum aerobic capacity to body mass: wild and domestic mammals. Resp. Physiol., 44:25–37.

Thompson, S. D. 1985. Subspecific differences in metabolism, thermoregulation, and torpor in the western harvest mouse, *Reithrodontomys megalotis*. Physiol. Zool., 58:430–444.

——. 1987. Body size, duration of parental care, and the intrinsic rate of natural increase in eutherian and metatherian mammals. Oecologia, 71:201–209.

Thompson, S. D., and M. E. Nicoll. 1986. Basal metabolic rate and energetics of reproduction in therian mammals. Nature, 321:690–693.

Tissier, M., M. Theriez, A. Purroy, and A. Brelurut. 1979. Energy utilization by ewes during pregnancy and lactation. Pp. 329–333 *in* Energy Metabolism (L. E. Mount, ed.). Butterworths, London.

Tracy, C. R. 1977. Minimum size of mammalian homeotherms: role of the thermal environment. Science, 198:1034–1035.

Trayhurn, P., J. B. Douglas, and M. M. McGuckin. 1982. Brown adipose tissue thermogenesis is "suppressed" during lactation in mice. Nature, 298:59–60.

Trillmich, F. 1986. Maternal investment and sex-allocation in the Galapagos fur seal, *Arctocephalus galapagoensis*. Behav. Ecol. Sociobiol., 19:157–164.

Trojan, P., and B. Wojciechowska. 1967. Resting metabolic rate during pregnancy and lactation in the European common vole *Microtus arvalis* (Pall.). Ekol. Pol. Ser. A, 15:811–817.

Tuomi, J., T. Hakala, and E. Haukioja. 1983. Alternative concepts of reproductive effort, costs of reproduction and selection in life-history evolution. Am. Zool., 23:25–34.

Tyler, N. J. C. 1987. Body composition and energy balance of pregnant and non-pregnant Svalbard reindeer during winter. Symp. Zool. Soc. Lond., 57:203–229.

Tytler, P., and P. Calow. 1985. Fish Energetics: New Perspectives. Johns Hopkins University Press, Baltimore. 349 pp.

Verhagen, J. M. F., A. A. M. Kloosterman, A. Slijkhuis, and M. W. A. Verstegen. 1987a. Effect of ambient temperature on energy metabolism in growing pigs. Anim. Prod. 44:427–433.

Verhagen, J. M. F., J. J. M. Michels, and W. G. P. Schouten. 1987b. The relation between body temperature, metabolic rate and climatic environment in young-growing pigs. J. Therm. Biol., 13:1–8.

Verhagen, J. M. F., M. W. A. Verstegen, T. P. A. Geuyen, and B. Kemp. 1986. Effect of environmental temperature and feeding level on heat production and lower critical temperature of pregnant sows. J. Anim. Physiol., 55:246–256.

Waldschmidt, A., and E. F. Müller. 1988. A comparison of postnatal thermal physiology and energetics in an altricial (*Gerbillus perpallidus*) and precocial (*Acomys cahirinus*) rodent species. Comp. Biochem. Physiol., 90A:169–181.

Wang, G. H. 1925. The changes in the amount of daily food intake of the albino rat during pregnancy and lactation. Am. J. Physiol., 71:736–741.

Watson, A. 1970. Animal Populations in Relation to Their Food Resources. Blackwell Scientific, Edinburgh.

Weiner, J. 1977. Energy metabolism of the roe deer. Acta Theriol., 22:3–24.

——. 1987. Limits to energy budget and tactics in energy investments during reproduction in the Djungarian hamster (*Phodopus sungorus sungorus* Pallas 1770). Symp. Zool. Soc. Lond., 57:167–187.

Weiner, J., and G. Heldmaier. 1987. Metabolism and thermoregulation in two races of Djungarian hamsters: *Phodopus sungorus sungorus* and *P. sungorus campbelli*. Comp. Biochem. Physiol., 86A:639–642.

White, R. G., I. D. Hume, and J. V. Nolan. 1988. Energy expenditure and protein turnover in three species of wallabies (Marsupialia: Macropodidae). J. Comp. Physiol., 158:237–246.

Whittam, R. 1964. The interdependence of metabolism and active transport. P. 139 *in* The Cellular Functions of Membrane Transport (J. F. Hoffman, ed.). Prentice-Hall, Englewood Cliffs, N.J.

Wilkinson, L. 1986. SYSTAT: The System for Statistics. Systat, Evanston, Ill.

Williams, J., and K. Nagy. 1984. Validation of the doubly-labeled water technique for measuring energy metabolism in savannah sparrows. Physiol. Zool., 57:325–328.

Woodside, B., R. Wilson, P. Chee, and M. Leon. 1981. Resource partitioning during reproduction in the Norway rat. Science, 211:76–77.

Wunder, B. 1978. Implications of a conceptual model for the allocation of energy resources by small mammals. Pp. 68–75 *in* Populations of Small Mammals under Natural Conditions (D. P. Snyder, ed.). University of Pittsburgh Press, Pittsburgh. 217 pp.

Yousef, M. K. 1985a. Stress Physiology in Livestock. I. Basic Principles. CRC Press, Boca Raton, Fla.

——. 1985b. Stress Physiology in Livestock. II. Ungulates. CRC Press, Boca Raton, Fla. 261 pp.

Summary and Conclusion

The study of animal energetics is an evolving field; what began as an exploration of the physical source of animal heat has developed into the study of the economies of animals (McNab: Chapter 1). Historically, the rate of heat loss was proposed to dictate the rate of heat production and therefore of energy consumption: A variety of physical models were developed to explain the maintenance of a particular metabolic rate and body temperature. Newer models examine phylogenetic and ecological factors as well as the physics of heat transfer (McNab: Chapter 1). Several physiological characteristics associated with mammals emphasize the use of energy: high metabolic rates, endothermy, gestation, lactation, and complex territorial and mating behaviors. Variations in locomotor pattern, metabolic rate, thermoregulatory capacity, and reproductive strategy have all been suggested to contribute to more efficient use of energy, which in turn is proposed to enhance survival and reproductive success. The role that energetics has played in shaping theories of mammalian evolution is reflected in the statements by Eisenberg (1981, chap. 16) suggesting that energy requirements can influence feeding strategies and that feeding strategies have affected the evolution of body size, demographic strategies, and mating systems. These statements reflect the fact that energetic constraints are often assumed to limit survival and reproduction. Two major questions raised by several of the chapters in this volume are, how correct is that assumption, and are there data to support it?

One behavior that is influenced by both ecology and phylogeny, and that is assumed to consume considerable amounts of energy, is locomotion. Changes in locomotor pattern are often assumed to have evolved toward forms that reduce energy consumption, resulting in greater availability of energy for other functions. It is possible to test the assumption that morphological and behavioral changes resulted in more efficient energy expenditure during locomotion by studying species that exhibit extreme patterns of locomotion. Examples of such extremes are provided by bats, which fly, and aquatic or semiaquatic mammals, which swim. Fish (Chapter 3) reviews the differences in the morphology of mammals with different degrees of adaptation to an aquatic mode of life. He concludes that the convergence of swimming mode and body form in highly aquatic mammalian species indicates the importance of these morphological changes to the optimization of energy use. However, ecological and historical constraints have limited the evolution of optimal designs. Similar models could be developed to determine whether the energetic cost of bat flight varies with body shape or wing shape or both in ways similar to that seen in swimming mammals, with consequent differences in energy consumption.

While phylogenetic and ecological comparisons may aid in explaining patterns of energy consumption among groups of animals, it may also be possible to infer the evolution of certain behaviors by comparing patterns of energy use when animals are exposed to different challenges. MacMillan and Hinds (Chapter 2) examine patterns in data on metabolic rate for several groups of heteromyid rodents. On the basis of their comparisons of metabolism of animals under standard conditions, during exercise, and while cold-stressed, MacMillen and Hinds suggest that bipedal locomotion may have arisen independently in two groups of heteromyid rodents (*Microdipodops* and *Dipodomys*). When comparing maximum metabolic rates between species, one can also ask about the "costs" of maintaining a high metabolic potential. Unfortunately, little is known about the energetic cost of maintaining the "machinery" for a high maximum metabolism.

Although researchers do not agree as to the mechanistic interpretation, it is universally accepted that body mass is the primary determinant of metabolic rate. However, the identification of other determinants (i.e., the causes of residual variation) is a fertile area for future study. While differences in taxonomic affiliation and ecology have been investigated as ultimate contributors to this variation and are discussed in Chapter 1, the proximal (physiological) causes have received very little attention. Since thyroxine (T_4) and triiodothyronine (T_3) are the primary hormones regulating basal metabolic rate, variation in their availability and metabolism by tissues may greatly influence basal metabolic rate. Tomasi and Gleit (Chapter 4) review and analyze the rates of T_4 utilization in mammals. They provide a comparative framework for investigating how environmental, ecological, and evolutionary factors may alter thyroid function and (presumably) metabolic rate. Although suggestive, the data are not unambiguous; more data on species with relatively elevated or depressed metabolic rates are needed for a more definitive interpretation. This analysis of proximal mechanisms yields several questions about the biochemical factors that limit metabolic rate. For example:

1. What are the limiting factors for maximum metabolism? Are they related to gas exchange, or are they cellular and, hence, quantitatively different among tissues?
2. How is energy allocated among tissues within an individual at any point in time?

The regulation of biochemical energetics and energy allocation is primarily endocrine, and many questions also remain as to the mechanisms for translating the energy state (Chapter 5) of an animal into hormonal signals (for example, thyroid function and reproductive function).

While endothermy and the associated high metabolic rate are considered to be characteristic of mammals, many mammals do not maintain a high basal metabolic rate at all times. Regulated reductions in metabolic rate are one mechanism by which mammals can reduce their energy requirements during periods of reduced energy supplies. The corresponding reduction in body temperature results in periods of torpor or hibernation. If reducing body temperature can result in significant energy savings, why do mammals maintain a higher temperature? How are the benefits of a high metabolic rate or body temperature or both balanced with costs? French (Chapter 6) finds that there are advantages to having an elevated body temperature (for example, increased growth rate, see below) that outweigh the energetic benefits of lowered body temperature. His cost-benefit analysis also points out that the pressures on mammals may change seasonally with changes in food availability and the thermal environment. French proposes that the length of time that a mammal spends in torpor is proportional to the energy deficit experienced by the individual. How is this energy deficit detected and translated into a propensity for torpor?

The neonatal period in altricial young is one developmental state during which the costs of maintaining a high body temperature are often thought to have been circumvented. Hill (Chapter 7) examines the thermoregulatory capacity of altricial young and concludes that many altricial mammals are capable of thermoregulating. The benefits of euthermy to increased growth may be very important. An interesting avenue for future study in this area would be to search for ecological parameters (litter size, nest site, diet of parents, predators, and so on) that determine or are correlated with a species' position on the altricial-to-precocial continuum.

A theme that is first stated by McNab (Chapter 1), but which reappears throughout the book, is the need to observe animals in laboratory *and* field settings. In laboratory studies, physiological processes can be measured under a variety of controlled conditions to demonstrate the functional capacity (i.e., physiological limits) of the animal. Field studies, or at least naturalistic studies, are necessary to determine how much of this functional range the animals actually use. One method to test whether a physiological process has been optimized would be to compare the physiological range used by animals in natural settings with their functional capacities, as determined in the laboratory (Lindstedt and Jones, 1987; Wunder, Chapter 5).

Many theories of mammalian life history patterns assume that mammals face periods of reduced energy supplies during which the energetic requirements for reproduction cannot be met, but direct assessments of energy-stress in the field have been difficult to obtain. Wunder (Chapter 5) discusses a series of measurements that can be made to determine if animals are energy-limited in the field. He presents data from laboratory experiments and field-caught animals and discusses the morphological changes associated with the enhancement of energy acquisition when animals are cold-stressed and during pregnancy and lactation. The data suggest that the energetic cost of reproduction may exceed winter thermoregulatory costs and that the requirements to meet increased thermal demands during cold-stress may preclude the assimilation of sufficient energy to support reproduction during the winter. He suggests that in the future, data from field studies can be correlated with changes in energy assimilation determined in the laboratory to assess whether (and to what extent) free-living animals are energetically stressed. We would then be in a position to quantify different types of "stress" and to determine whether decreases in energy availability or increases in energy demand (thermoregulation, migration, reproduction, and so on) create a more severe problem.

If life history strategies are based on the premise that energy supplies appear intermittently in the environment, then periods of reproductive activity must coincide with the appearance of resources sufficient to support reproduction (Bronson, 1989; Loudon and Racey, 1987). Experiments using food restriction have shown that a decreased energetic input can produce specific reductions in reproductive hormones (i.e., luteinizing hormone) and reproductive output (Bronson, 1987). Mammals are capable of using cues in the environment (i.e., photoperiod, plant compounds) to regulate reproduction, and it is assumed that the use of these cues reflects their ability to predict the availability of energy. Mechanistically, the allocation of energy between reproductive and nonreproductive systems is under the control of the endocrine system. Chapters 8 and 9 discuss the role of the endocrine system in controlling mammalian reproduction, with particular emphasis on the variability between individuals and how they respond to environmental signals. Horton and Rowsemitt (Chapter 8) emphasize that there may be many sources of variability (i.e., genetic polymorphisms, sexual dimorphisms, conflicting environmental signals, developmental plasticity) that may not result in a uniform response of a population to a given environmental cue. They propose that these sources of variation must be examined carefully in order to understand the endocrine control of reproductive responses to environmental cues and how this control may have evolved. One of the mechanisms for this evolution may be the existence of genetic polymorphisms in the neuroendocrine system. Blank (Chapter 9) presents data on the reproductive consequences of a genetic polymorphism in reproductive responses to short photoperiod by *Peromyscus maniculatus*. He demonstrates that individual mice collected from natural populations will either maintain or cease reproductive function in response to short photoperiod. The energetic advantages and disadvantages of breeding during winter months are discussed, as well as the differences in thermogenic abilities accompanying different reproductive responses. Animals in natural populations probably employ a variety of reproductive and metabolic strategies to meet the demands of their environment. Questions that remain unresolved regarding the evolution of reproductive strategies include:

1. Under what energetic circumstances is reproduction an all-or-none event? When is it "graded"?

2. How unpredictable must the environment be to maintain a genetic polymorphism for energetic parameters?

3. What limits the amount of energy that can be channeled to reproduction?

Considering the diversity of mammals and the weight placed on the assumption that energy limits reproduction we have data on reproductive costs for very few species. Unfortunately, even these measurements are often indirect assessments of energetic demands and expenditures. Thompson (Chapter 10) examines the assumptions about energy demands during pregnancy and lactation. His analysis compares the efficiencies of energy transfer during gestation and lactation to examine the effect of allocation patterns on the energetic cost of reproduction, on reproductive effort, and on the speed of reproduction. He finds that few longitudinal studies present sufficient data to compare allocation strategies during reproduction. Furthermore, he finds that there are great variations in allocation patterns among even closely related taxa. His analysis suggests that more descriptive and experimental data must be accumulated, especially direct measurements of energy use and investment, before we can develop general models. Perhaps with more data, we will someday be able to correlate ecological parameters with a species' position on the short gestation (long lactation) to long gestation (short lactation) continuum.

As stated in the preface, the authors have all addressed the diversity of ways that mammals cope with energetic constraints. Variability, both intra- and interspecific, must be viewed as a positive and instructive feature of mammalian biology: one that is worthy of study and that must be understood to answer broader questions. The significance of this variability will be revealed primarily through interdisciplinary collaboration. In concert with this, it must be recognized that most studies of energetic and reproductive traits examine one parameter at a time. However, suites of characters may evolve simultaneously (Lynch, 1986; Merritt, 1986).

Our aims in assembling this volume were to present recent advances in energetic aspects of mammalian physiology to biologists from a variety of subdisciplines and to stimulate discussion of the ways in which these physiological processes contribute to and limit the ecology, evolution, and life histories of mammals. In the book *New Directions in Ecological Physiology*, Feder suggested that ecological physiologists should think of themselves as organismal biologists first and as ecological physiologists second (Feder, 1987). He wanted researchers to focus attention on the broad question of organismal complexity and its evolution, rather than on the enumeration of physiological adaptations. While Feder's suggestion was not discussed a priori with the contributors to this book, many examples of this philosophy are included herein. We hope that this volume will continue to stimulate discussions among scientists from diverse backgrounds and that these discussions will yield a better understanding of how the study of energetics can interface with a myriad of subdisciplines within mammalian biology.

Literature Cited

Bronson, F. H. 1987. Environmental regulation of reproduction in rodents. Pp. 204–230 *in* Psychobiology of Reproductive Behavior: An Evolutionary Approach. D. Crews, (ed.). Prentice-Hall, Englewood Cliffs, N.J.

——. 1989. Mammalian Reproductive Biology. University of Chicago Press, Chicago. 325 pp.

Eisenberg, J. F. 1981. The Mammalian Radiations. University of Chicago Press, Chicago. 610 pp.

Feder, M. E. 1987. New directions in ecological physiology: conclusions. Pp. 347–351 *in* New Directions in Ecological Physiology (M. E. Feder, A. F. Bennett, W. W. Burggren, and R. B. Huey, eds.). Cambridge University Press, Camridge.

Lindstedt, S. L., and J. H. Jones. 1987. Symmorphosis: the concept of optimal design. Pp. 289–309 *in* New Directions in Ecological Physiology (M. E. Feder, A. F. Bennett, W. W. Burggren, and R. B. Huey, eds.). Cambridge University Press, Cambridge.

Loudon, A. S. I., and P. A. Racey, eds. 1987 Reproductive Energetics in Mammals. Symposia of the Zoological Society of London 57. Clarendon Press, Oxford. 371 pp.

Lynch, C. B. 1986. Genetic basis of cold adaptation in laboratory and wild mice, *Mus domesticus*. Pp. 497–504 *in* Living in the Cold (H. C. Heller, X. J. Musacchia, and L. C. H. Wang, eds.). Elsevier Science, New York.

Merritt, J. F. 1986. Winter survival adaptations of the short-tailed shrew (*Blarina brevicauda*) in an Appalachian montane forest. J. Mammal. 67:450–464.

Index